Y0-BZE-224

PHYSICAL CHEMISTRY FOR THE BIOLOGICAL SCIENCES

THE WILEY BICENTENNIAL–KNOWLEDGE FOR GENERATIONS

Each generation has its unique needs and aspirations. When Charles Wiley first opened his small printing shop in lower Manhattan in 1807, it was a generation of boundless potential searching for an identity. And we were there, helping to define a new American literary tradition. Over half a century later, in the midst of the Second Industrial Revolution, it was a generation focused on building the future. Once again, we were there, supplying the critical scientific, technical, and engineering knowledge that helped frame the world. Throughout the 20th Century, and into the new millennium, nations began to reach out beyond their own borders and a new international community was born. Wiley was there, expanding its operations around the world to enable a global exchange of ideas, opinions, and know-how.

For 200 years, Wiley has been an integral part of each generation's journey, enabling the flow of information and understanding necessary to meet their needs and fulfill their aspirations. Today, bold new technologies are changing the way we live and learn. Wiley will be there, providing you the must-have knowledge you need to imagine new worlds, new possibilities, and new opportunities.

Generations come and go, but you can always count on Wiley to provide you the knowledge you need, when and where you need it!

William J. Pesce
President and Chief Executive Officer

Peter Booth Wiley
Chairman of the Board

PHYSICAL CHEMISTRY FOR THE BIOLOGICAL SCIENCES

Gordon G. Hammes
Department of Biochemistry
Duke University

WILEY-INTERSCIENCE

A John Wiley & Sons, Inc., Publication

Published by John Wiley & Sons, Inc., Hoboken, New Jersey
Published simultaneously in Canada

Wiley Bicentennial Logo: Richard J. Pacifico

Library of Congress Cataloging-in-Publication Data:

Hammes, Gordon G., 1934-
Physical chemistry for the biological sciences / Gordon G. Hammes.
p. cm.
Includes bibliographical references (p.).
ISBN 978-0-470-12202-0
1. Physical biochemistry. 2. Thermodynamics. 3. Chemical kinetics. 4. Biomolecules – Spectra. 5. Spectrum analysis. I. Title.

QP517.P49H348 2007
572′.43–dc22 2006052998

Printed in the United States of America

10 9 8 7 6 5 4 3 2 1

PREFACE

Biology is the study of living species. The historic origin of biology is descriptive in nature, a classification and description of the various biological species. Modern biology is far different and seeks to understand living phenomena on a molecular basis. The incredible amount of information available and the databases of this information are staggering, the most obvious example being the nucleotide sequence of the human genome. In essence, biology has moved from a qualitative to a quantitative science. Inevitably, this requires a theoretical framework and associated mathematics. Physical chemistry provides this framework for molecular structure and chemical reactions, the components of all biological systems that ultimately must be understood.

Traditionally, physical chemistry has been a major training component for chemists, but not for biologists. This has been attributed to the relatively sophisticated mathematical underpinnings of rigorous physical chemistry. However, the concepts of physical chemistry can be understood and applied to biology with a minimum of mathematics.

This volume attempts to present physical chemistry in conceptual terms using mathematics only at an upper level of elementary calculus, a level required for all science students. Nevertheless, the approach is quantitative in nature, with explicit calculations and numerical problems. Examples from biology are used to illustrate the principles, and problems are appended at the end of each chapter. This book is intended to serve as a one-semester introduction to physical chemistry for undergraduate biology majors and as a refresher course for first-year graduate students. This book combines two volumes published earlier, *Thermodynamics and Kinetics for the Biological Sciences and Spectroscopy for the Biological Sciences*. These two books have been integrated with some additions and modification. The most notable addition is a chapter on the hydrodynamics of macromolecules. Hydrodynamics is the basis of several important laboratory techniques used in molecular biology, and understanding the underlying concepts will permit better use of the methods and development of new methods.

We begin with a discussion of thermodynamics, a subject that provides a convenient framework for all equilibrium phenomena. This is followed by chemical kinetics, the quantitative description of the time dependence of chemical reactions. For both subjects, multiple applications to biology are presented. The concepts associated with spectroscopy and structure determination are then considered. These topics deal with the molecular nature of matter and the techniques used to characterize molecules and their interactions. The concluding section of the book includes the important subjects of ligand binding to macromolecules, hydrodynamics, and mass spectrometry. The coverage of this book represents the minimal knowledge that every biologist should have to understand biological phenomenon in molecular terms (in my opinion!).

I am indebted to my colleagues at Duke for their encouragement and assistance. In particular, Professors Jane and David Richardson, Lorena Beese, Leonard Spicer, Terrance Oas, Michael Fitzgerald, and Harvey Sage who have provided vital expertise. A special thanks also goes to Darla Henderson who as a Wiley editor has provided both encouragement and professional assistance in the preparation of this volume. As always, my wife Judy has provided her much appreciated (and needed) support.

GORDON G. HAMMES
Duke University

CONTENTS

THERMODYNAMICS

CHAPTER 1

Heat, Work, and Energy

1.1 INTRODUCTION

Thermodynamics is deceptively simple or exceedingly complex, depending on how you approach it. In this book, we will be concerned with the principles of thermodynamics that are especially useful in thinking about biological phenomena. The emphasis will be on concepts, with a minimum of mathematics. Perhaps an accurate description might be rigor without *rigor mortis.* This may cause some squirming in the graves of thermodynamic purists, but the objective is to provide a foundation for researchers in experimental biology to use thermodynamics. This includes cell biology, microbiology, molecular biology, and pharmacology, among others. Excellent texts are available that present a more advanced and complete exposition of thermodynamics (cf. Refs. 1 and 2).

In point of fact, thermodynamics can provide a useful way of thinking about biological processes and is indispensable when considering molecular and cellular mechanisms. For example, what reactions and coupled physiological processes are possible? What are the allowed mechanisms involved in cell division, in protein synthesis? What are the thermodynamic considerations that cause proteins, nucleic acids, and membranes to assume their active structures? It is easy to postulate biological mechanisms that are inconsistent with thermodynamic principles—but just as easy to postulate those that are consistent. Consequently, no active researcher in biology should be without a rudimentary knowledge of the principles of thermodynamics. The ultimate goal of this exposition is to understand what determines equilibrium in biological systems, and how these equilibrium processes can be coupled together to produce living systems, even though we recognize that living organisms are not at equilibrium. Thermodynamics provides a unifying framework for diverse systems in biology. Both a qualitative and a quantitative understanding are important and will be developed.

The beauty of thermodynamics is that a relatively small number of postulates can be used to develop the entire subject. Perhaps the most important part of this development is to be very precise with regard to concepts and definitions, without getting bogged down with mathematics. Thermodynamics is a macroscopic theory,

Physical Chemistry for the Biological Sciences by Gordon G. Hammes

not molecular. As far as thermodynamics is concerned, molecules need not exist. However, we will not be purists in this regard: If molecular descriptions are useful for understanding or introducing concepts, they will be used. We will not hesitate to give molecular descriptions of thermodynamic results, but we should recognize that these interpretations are not inherent in thermodynamics itself. It is important to note, nevertheless, that large collections of molecules are assumed so that their behavior is governed by Boltzmann statistics; that is, the normal thermal energy distribution is assumed. This is almost always the case in practice. Furthermore, thermodynamics is concerned with time-independent systems, that is, systems at equilibrium. Thermodynamics has been extended to nonequilibrium systems, but we will not be concerned with the formal development of this subject here.

The first step is to define the *system*. A thermodynamic system is simply that part of the universe in which we are interested. The only caveat is that the system must be large relative to molecular dimensions. The system could be a room, it could be a beaker, it could be a cell, etc. An *open system* can exchange energy and matter across its boundaries, for example, a cell or a room with open doors and windows. A *closed system* can exchange energy but not matter, for example, a closed room or box. An *isolated system* can exchange neither energy nor matter, for example, the universe or, approximately, a closed Dewar. We are free to select the system as we choose, but it is very important that we specify what it is. This will be illustrated as we proceed. The *properties* of a system are any measurable quantities characterizing the system. Properties are either *extensive*, proportional to the mass of the system, or *intensive*, independent of the mass. Examples of extensive properties are mass and volume. Examples of intensive properties are temperature, pressure, and color.

1.2 TEMPERATURE

We are now ready to introduce three important concepts: temperature, heat, and work. None of these are unfamiliar, but we must define them carefully so that they can be used as we develop thermodynamics.

Temperature is an obvious concept, as it simply measures how hot or cold a system is. We will not belabor its definition and will simply assert that thermodynamics requires a unique temperature scale, namely, the Kelvin temperature scale. The Kelvin temperature scale is related to the more conventional Celsius temperature scale by the definition

$$T_{\text{Kelvin}} = T_{\text{Celsius}} + 273.16 \tag{1-1}$$

Although the temperature on the Celsius scale is referred to as "degrees Celsius," by convention degrees are not stated on the Kelvin scale. For example, a temperature of 100°C is 373 K. (Thermodynamics is entirely logical—some of the conventions used are not.) The definition of *thermal equilibrium* is very simple: When two systems are at the same temperature, they are at thermal equilibrium.

1.3 HEAT

Heat flows across the system boundary during a change in the state of the system because a temperature difference exists between the system and its surroundings. We know of many examples of heat: Some chemical reactions produce heat, such as the combustion of gas and coal. Reactions in cells can produce heat. By convention, heat flows from higher temperature to lower temperature. This fixes the sign of the heat change. It is important to note that this is a convention and is not required by any principle. For example, if the temperature of the surroundings decreases, heat flows to the system, and the sign of the heat change is positive (+). A simple example will illustrate this sign convention as well as the importance of defining the system under consideration.

Consider two beakers of the same size filled with the same amount of water. In one beaker, A, the temperature is 25°C, and in the other beaker, B, the temperature is 75°C. Let us now place the two beakers in thermal contact and allow them to reach thermal equilibrium (50°C). This situation is illustrated in Figure 1-1. If the system is defined as A, the temperature of the system increases, so the heat change is positive. If the system is defined as B, the temperature of the system decreases, so the heat change is negative. If the system is defined as A and B, no heat flow occurs across the boundary of the system, so the heat change is zero! This illustrates how important it is to define the system before asking questions about what is occurring.

The heat change that occurs is proportional to the temperature difference between the initial and final states of the system. This can be expressed mathematically as

$$q = C(T_\mathrm{f} - T_\mathrm{i}) \tag{1-2}$$

where q is the heat change, the constant C is the *heat capacity*, T_f, is the final temperature, and T_i is the initial temperature. This relationship assumes that the heat capacity is constant, independent of the temperature. In point of fact, the heat capacity often changes as the temperature changes, so that a more precise definition puts this relationship in differential form:

$$\mathrm{d}q = C\,\mathrm{d}T \tag{1-3}$$

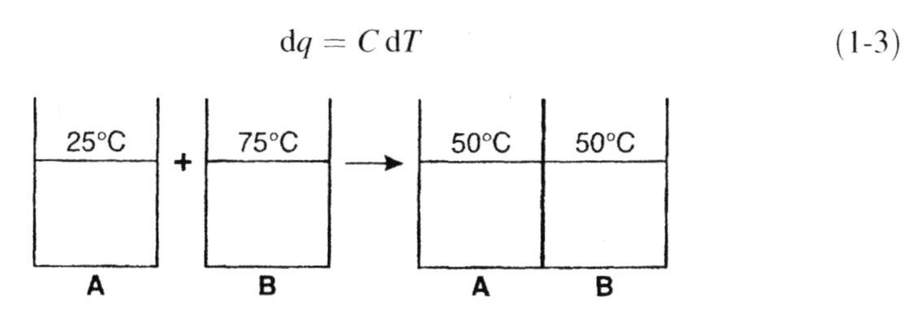

FIGURE 1-1. Illustration of the establishment of thermal equilibrium and importance of defining the *system* carefully. Two identical vessels filled with the same amount of liquid, but at different temperatures, are placed in contact and allowed to reach thermal equilibrium. A discussion of this figure is given in the text.

Note that the heat change and the heat capacity are extensive properties—the larger the system, the larger the heat capacity and the heat change. Temperature, of course, is an intensive property.

1.4 WORK

The definition of *work* is not as simple as that for heat. Many different forms of work exist, for example, mechanical work, such as muscle action, and electrical work, such as ions crossing charged membranes. We will use a rather artificial, but very general, definition of work that is easily understood. Work is a quantity that can be transferred across the system boundary and can always be converted to lifting and lowering a weight in the surroundings. By convention, work done on a system is positive: this corresponds to lowering the weight in the surroundings.

You may recall that mechanical work, w, is defined as the product of the force in the direction of movement, F_x, times the distance moved, x, or in differential form

$$\mathrm{d}w = F_x \mathrm{d}x \tag{1-4}$$

Therefore, the work to lower a weight is $-mgh$, where m is the mass, g is the gravitational constant, and h is the distance the weight is lowered. This formula is generally useful: for example, mgh is the work required for a person of mass m to walk up a hill of height h. The work required to stretch a muscle could be calculated with Eq. 1-4 if we knew the force required and the distance the muscle was stretched. Electrical work, for example, is equal to $-EIt$, where E is the electromotive force, I is the current, and t is the time. In living systems, membranes often have potentials (voltages) across them. In this case, the work required for an ion to cross the membrane is $-zF\Psi$, where z is the valence of the ion, F is the Faraday (96,489 coulombs/mole), and Ψ is the potential. A specific example is the cotransport of Na^+ and K^+, Na^+ moving out of the cell and K^+ moving into the cell. A potential of $-70\,\mathrm{mV}$ is established on the inside so that the electrical work required to move a mole of K^+ ions to the inside is $-(1)(96,489)(0.07) = -6750$ Joules. ($\Psi = \Psi_{\mathrm{outside}} - \Psi_{\mathrm{inside}} = +70\,\mathrm{mV}$.) The negative sign means that work is done by the system.

Although not very biologically relevant, we will now consider in some detail pressure–volume work, or $P - V$ work. This type of work is conceptually easy to understand, and calculations are relatively easy. The principles discussed are generally applicable to more complex systems, such as those encountered in biology. As a simple example of $P - V$ work, consider a piston filled with a gas, as pictured in Figure 1-2. In this case, the force is equal to the external pressure, P_{ex}, times the area, A, of the piston face, so the infinitesimal work can be written as

$$\mathrm{d}w = -P_{\mathrm{ex}} A\,\mathrm{d}x = -P_{\mathrm{ex}}\,\mathrm{d}V \tag{1-5}$$

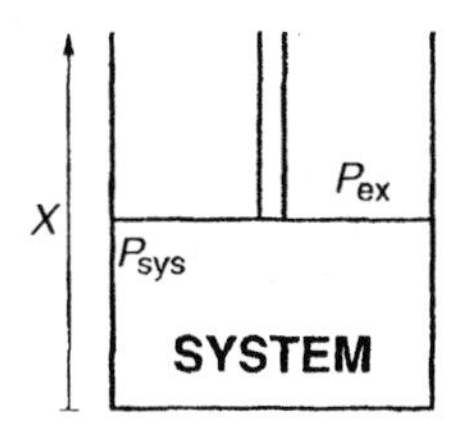

FIGURE 1-2. Schematic representation of a piston pushing on the system. P_{ex} is the external pressure and P_{sys} is the pressure of the system.

If the piston is lowered, work is done on the system and is positive, whereas if the piston is raised, work is done by the system and is negative. Note that the work done on or by the system by lowering or raising the piston depends on what the external pressure is. Therefore, the work can have any value from 0 to ∞, depending on how the process is done. This is a very important point: The work associated with a given change in state depends on *how* the change in state is carried out.

The idea that work depends on how the process is carried out can be illustrated further by considering the expansion and compression of a gas. The $P - V$ isotherm for an ideal gas is shown in Figure 1-3. An ideal gas is a gas that obeys the ideal gas law, $PV = nRT$ (n is the number of moles of gas and R is the gas constant). The behavior of most gases at moderate pressures is well described by this relationship. Let us consider the expansion of the gas from P_1, V_1 to P_2, V_2. If this expansion is done with the external pressure equal to zero, that is, into a vacuum, the work is zero. Clearly, this is the minimum amount of work that can be done for this change

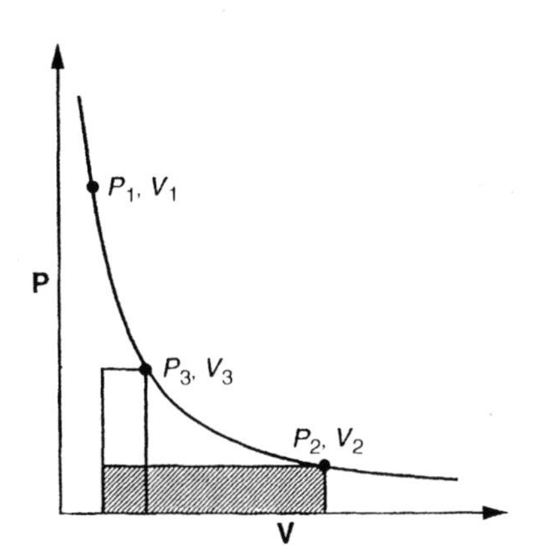

FIGURE 1-3. A $P - V$ isotherm for an ideal gas. The narrow rectangle with both hatched and open areas is the work done in going from P_1,V_1 to P_3,V_3 with an external pressure of P_3. The hatched area is the work done by the system in going from P_1,V_1 to P_2,V_2 with an external pressure of P_2. The maximum amount of work done by the system for this change in state is the area under the curve between P_1, V_1 and P_2,V_2.

in state. Let us now carry out the same expansion with the external pressure equal to P_2. In this case, the work is

$$w = -\int_{V_1}^{V_2} P_{ex} \mathrm{d}V = -P_2(V_2 - V_1) \tag{1-6}$$

which is the striped area under the $P - V$ curve. The expansion can be broken into stages; for example, first expand the gas with $P_{ex} = P_3$ followed by $P_{ex} = P_2$, as shown in Figure 1-3. The work done by the system is then the sum of the two rectangular areas under the curve. It is clear that as the number of stages is increased, the magnitude of the work done increases. The maximum work that can be attained would set the external pressure equal to the pressure of the system minus a small differential pressure, dP, throughout the expansion. This can be expressed as

$$w_{max} = -\int_{V_1}^{V_2} P \mathrm{d}V \tag{1-7}$$

By a similar reasoning process, it can be shown that for a compression the minimum work done on the system is

$$w_{min} = -\int_{V_2}^{V_1} P \mathrm{d}V \tag{1-8}$$

This exercise illustrates two important points. First, it clearly shows that the work associated with a change in state depends on how the change in state is carried out. Second, it demonstrates the concept of a *reversible path*. When a change in state is carried out such that the surroundings and the system are not at equilibrium by an infinitesimal amount, in this case dP, during the change in state, the process is called reversible. The concept of reversibility is only an ideal—it cannot be achieved in practice. Obviously, we cannot really carry out a change in state with only an infinitesimal difference between the pressures of the system and surroundings. We will find this concept very useful, nevertheless.

Now let us think about a cycle whereby an expansion is carried out followed by a compression that returns the system back to its original state. If this is done as a one-stage process in each case, the total work can be written as

$$w_{total} = w_{exp} + w_{comp} \tag{1-9}$$

or

$$w_{total} = -P_2(V_2 - V_1) - P_1(V_1 - V_2) \tag{1-10}$$

or

$$w_{total} = (P_1 - P_2)(V_2 - V_1) > 0 \tag{1-11}$$

In this case, net work has been done on the system. For a reversible process, however, the work associated with compression and expansion is

$$w_{\text{exp}} = -\int_{V_1}^{V_2} P\,\mathrm{d}V \tag{1-12}$$

and

$$w_{\text{comp}} = -\int_{V_2}^{V_1} P\,\mathrm{d}V \tag{1-13}$$

so that the total work for the cycle is equal to zero. Indeed, for reversible cycles the net work is always zero.

To summarize this discussion of the concept of work, the work done on or by the system depends on how the change in state of the system occurs. In the real world, changes in state always occur irreversibly, but we will find the concept of a reversible change in state to be very useful.

Heat changes also depend on how the process is carried out. Generally, a subscript is appended to q, for example, q_P and q_V for heat changes at constant pressure and volume, respectively. As a case in point, the heat change at constant pressure is greater than that at constant volume if the temperature of a gas is raised. This is because not only must the temperature be raised, but the gas must also be expanded.

Although this discussion of gases seems far removed from biology, the concepts and conclusions reached are quite general and can be applied to biological systems. The only difference is that exact calculations are usually more difficult. It is useful to consider why this is true. In the case of ideal gases, a simple equation of state is known, $PV = nRT$, that is obeyed quite well by real gases under normal conditions. This equation is valid because gas molecules, on average, are quite far apart and their energetic interactions can be neglected. Collisions between gas molecules can be approximated as billiard balls colliding. This situation obviously does not prevail in liquids and solids where molecules are close together and the energetics of their interactions cannot be neglected. Consequently, simple equations of state do not exist for liquids and solids.

1.5 DEFINITION OF ENERGY

The first law of thermodynamics is basically a definition of the energy change associated with a change in state. It is based on the experimental observation that heat and work can be interconverted. Probably the most elegant demonstration of this is the experimental work of James Prescott Joule in the late 1800s. He carried out experiments in which he measured the work necessary to turn a paddle wheel in water and the concomitant rise in temperature of the water. With this rather primitive experiment, he was able to calculate the conversion factor between work and heat with amazing accuracy, namely, to within 0.2%. The first law states that the energy change, ΔE, associated with a change in state is

$$\Delta E = q + w \tag{1-14}$$

Furthermore, the energy change is the same regardless of how the change in state is carried out. In this regard, energy clearly has quite different properties than heat and work. This is true for both reversible and irreversible processes. Because of this property, the energy (usually designated the internal energy in physical chemistry textbooks) is called a *state function*. State functions are extremely important in thermodynamics, both conceptually and practically.

Obviously we cannot prove the first law, as it is a basic postulate of thermodynamics. However, we can show that without this law events could occur that are contrary to our experience. Assume, for example, that the energy change in going from state 1 to state 2 is greater than the negative of that for going from state 2 to 1 because the changes in state are carried out differently. We could then cycle between these two states and produce energy as each cycle is completed, essentially making a perpetual motion machine. We know that such machines do not exist, consistent with the first law. Another way of looking at this law is as a statement of the conservation of energy.

It is important that thermodynamic variables are not just hypothetical—we must be able to relate them to laboratory experience, that is, to measure them. Thermodynamics is developed here for practical usage. Therefore, we must be able to relate the concepts to what can be done in the laboratory. How can we measure energy changes? If we only consider $P - V$ work, the first law can be written as

$$\Delta E = q - \int_{V_1}^{V_2} P_{\text{ex}} \mathrm{d}V \tag{1-15}$$

If the change in state is measured at constant volume, then

$$\Delta E = q_V \tag{1-16}$$

At first glance, it may seem paradoxical that a state function, the energy change, is equal to a quantity whose magnitude depends on how the change in state is carried out, namely, the heat change. However, in this instance we have specified how the change in state is to occur, namely, at constant volume. Therefore, if we measure the heat change at constant volume associated with a change in state, we have also measured the energy change.

Temperature is an especially important variable in biological systems. If the temperature is constant during a change in state, the process is *isothermal*. On the other hand, if the system is insulated so that no heat escapes or enters the system during the change in state ($q = 0$), the process is *adiabatic*.

1.6 ENTHALPY

Most experiments in the laboratory and in biological systems are done at constant pressure, rather than at constant volume. At constant pressure,

$$\Delta E = q_P - P(V_2 - V_1) \tag{1-17}$$

or

$$E_2 - E_1 = q_P - P(V_2 - V_1) \tag{1-18}$$

The heat change at constant pressure can be written as

$$q_P = (E_2 + PV_2) - (E_1 + PV_1) \tag{1-19}$$

This relationship can be simplified by defining a new state function, the *enthalpy*, H:

$$H = E + PV \tag{1-20}$$

The enthalpy is obviously a state function since E, P, and V are state functions. The heat change at constant pressure is then equal to the enthalpy change:

$$q_P = \Delta H = H_2 - H_1 \tag{1-21}$$

For biological reactions and processes, we will usually be interested in the enthalpy change rather than the energy change. It can be measured experimentally by determining the heat change at constant pressure.

As a simple example of how energy and enthalpy can be calculated, let us consider the conversion of liquid water to steam at 100°C and 1 atm, that is, boiling water:

$$H_2O(\ell, 1\text{ atm}, \ 100°\text{C}) \rightarrow H_2O(\text{g}, \ 1\text{ atm}, \ 100°\text{C}) \tag{1-22}$$

The heat required for this process, ΔH $(= q_P)$, is 9.71 kilocalories/mol. What is ΔE for this process? This can be calculated as follows:

$$\Delta E = \Delta H - \Delta(PV) = \Delta H - P\Delta H$$
$$\Delta V = V_g - V_\ell = 22.4\text{ liters/mol} - 18.0 \times 10^{-3}\text{ liters/mol} \approx PV_g \approx RT$$
$$\Delta E = \Delta H - RT = 9710 - 2(373) = 8970\text{ calories/mol}$$

Note that the Kelvin temperature must be used in thermodynamic calculations and that ΔH is significantly greater than ΔE.

Let us do a similar calculation for the melting of ice into liquid water

$$H_2O(s,\ 273\ K,\ 1\ atm) \rightarrow H_2O(\ell,\ 273\ K,\ 1\ atm) \tag{1-23}$$

In this case, the measured heat change, ΔH (= q_P), is 1.44 kcal/mol. The calculation of ΔE parallels the previous calculation.

$$\begin{aligned}
\Delta E &= \Delta H - P\,\Delta V \\
\Delta V &= V_\ell - V_s \approx 18.0\,\text{ml/mol} - 19.6\,\text{ml/mol} \approx -1.6\,\text{ml/mol} \\
P\,\Delta V &\approx -1.6\,\text{ml atm} = -0.04\,\text{cal} \\
\Delta E &= 1440 + 0.04 = 1440\,\text{cal/mol}
\end{aligned}$$

In this case, ΔE and ΔH are essentially the same. In general, they do not differ greatly in condensed media, but the differences can be substantial in the gas phase.

The two most common units for energy are the calorie and the joule (J). (One calorie equals 4.184 J.) The official MKS unit is the joule, but many research publications use the calorie. We will use both in this text, in order to familiarize the student with both units.

1.7 STANDARD STATES

Only changes in energy states can be measured. Therefore, it is arbitrary what we set as the zero for the energy scale. As a matter of convenience, a common zero has been set for both the energy and the enthalpy. Elements in their stablest forms at 25°C (298 K) and 1 atm are assigned an enthalpy of zero. This is called a *standard state* and is usually written as H°_{298}. The superscript means 1 atm, and the subscript is the temperature in Kelvin.

As an example of how this concept is used, consider the formation of carbon tetrachloride from its elements:

$$C\ (\text{graphite}) + 2\ Cl_2(g) \rightarrow CCl_4\ (\ell) \tag{1-24}$$

$$\Delta H = H^\circ_{298(CCl_4)} - H^\circ_{298(C)} - 2H^\circ_{298(Cl_2)}$$

$$\Delta H = H^\circ_{298(CCl_4)}$$

The quantity $H^\circ_{298(CCl_4)}$ is called the heat of formation of carbon tetrachloride. Tables of heats of formation are available for hundreds of compounds and are useful in calculating the enthalpy changes associated with chemical reactions (cf. Refs. 3,4).

In the case of substances of biological interest in solutions, the definitions of standard states and heats of formation are a bit more complex. In addition to pressure and temperature, other factors must be considered such as pH, salt concentration, metal ion concentration, etc. A universal definition has not been established. In practice, it is best to use heats of formation under a defined set of conditions, and likewise to define the standard state as these conditions. Tables of heats of formation for some compounds of biological interest are given in Appendix 4 (5). A prime is often added to the symbol for these heats of formation ($H_f^{\circ'}$). to indicate the unusual nature of the standard state. We will not make that distinction here, but it is essential that a consistent standard state is used when making thermodynamic calculations for biological systems.

A useful way of looking at chemical reactions is as algebraic equations. A characteristic enthalpy can be assigned to each product and reactant. Consider the "reaction"

$$a\mathrm{A} + b\mathrm{B} \rightleftharpoons c\mathrm{C} + d\mathrm{D} \tag{1-25}$$

For this reaction, $\Delta H = H_{\mathrm{products}} - H_{\mathrm{reactants}}$, or

$$\Delta H = dH_{\mathrm{D}} + cH_{\mathrm{C}} - aH_{\mathrm{A}} - bH_{\mathrm{B}}$$

where the H_i are molar enthalpies. At 298 K and 1 atm, the molar enthalpies of the elements are zero, whereas for compounds, the molar enthalpies are equal to the heats of formation, which are tabulated. Before we apply these considerations to biological reactions, a brief digression will be made to discuss how heats of reactions are determined experimentally.

1.8 CALORIMETRY

The area of science concerned with the measurement of heat changes associated with chemical reactions is designated as calorimetry. Only a brief introduction is given here, but it is important to relate the theoretical concepts to laboratory experiments. To begin this discussion, we will return to our earlier discussion of heat changes and the heat capacity, Eq. 1-3. Since the heat change depends on how the change in state is carried out, we must be more precise in defining the heat capacity. The two most common conditions are constant volume and constant pressure. The heat changes in these cases can be written as

$$\mathrm{d}q_V = \mathrm{d}E = C_V\,\mathrm{d}T \tag{1-26}$$

$$\mathrm{d}q_P = \mathrm{d}H = C_P\,\mathrm{d}T \tag{1-27}$$

A more exact mathematical treatment of these definitions would make use of partial derivatives, but we will avoid this complexity by using subscripts to indicate what is held constant. These equations can be integrated to give

$$\Delta E = \int_{T_1}^{T_2} C_V\,\mathrm{d}T \tag{1-28}$$

$$\Delta H = \int_{T_1}^{T_2} C_P\,\mathrm{d}T \tag{1-29}$$

Thus, heat changes can readily be measured if the heat capacity is known. The heat capacity of a substance can be determined by adding a known amount of heat to the substance and determining the resulting increase in temperature. The known amount of heat is usually added electrically since this permits very precise measurement. (Recall that the electrical heat is I^2R, where I is the current and R is the resistance of the heating element.) If heat is added repeatedly in small increments over a large temperature range, the temperature dependence of the heat capacity can be determined. Tabulations of heat capacities are available and are usually presented with the temperature dependence described as a power series:

$$C_P = a + bT + cT^2 + \cdots \tag{1-30}$$

where a, b, c,... are constants determined by experiment.

For biological systems, two types of calorimetry are commonly done—batch calorimetry and scanning calorimetry. In batch calorimetry, the reactants are mixed together and the ensuing temperature rise (or decrease) is measured. A simple experimental setup is depicted in Figure 1-4, where the calorimeter is a Dewar flask and the temperature increase is measured by a thermocouple or thermometer.

For example, if we wished to measure the heat change for the hydrolysis of adenosine 5′-triphosphate (ATP)

$$\text{ATP} + \text{H}_2\text{O} \rightleftharpoons \text{ADP} + \text{P}_\text{i} \tag{1-31}$$

a solution of known ATP concentration would be put in the Dewar at a defined pH, metal ion concentration, buffer, etc. The reaction would be initiated by adding a small amount of adenosine triphosphatase (ATPase), an enzyme that efficiently catalyzes the hydrolysis, and the subsequent temperature rise measured. The enthalpy of reaction can be calculated from the relationship

$$\Delta H = C_P \Delta T \tag{1-32}$$

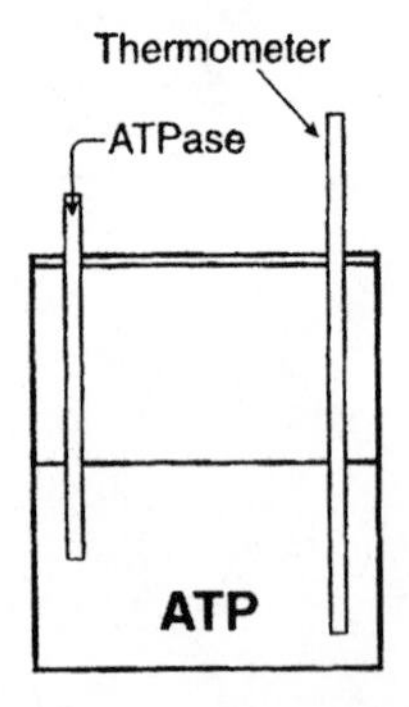

FIGURE 1-4. Schematic representation of a simple batch calorimeter. The insulated vessel is filled with a solution of ATP in a buffer containing salt and Mg^{2+}. The hydrolysis of ATP is initiated by the addition of the ATPase enzyme, and the subsequent rise in temperature is measured.

The heat capacity of the system is calculated by putting a known amount of heat into the system through an electrical heater and measuring the temperature rise of the system. The enthalpy change calculated is for the number of moles of ATP in the system. Usually, the experimental result is reported as a molar enthalpy, that is, the enthalpy change for a mole of ATP being hydrolyzed. This result can be obtained by dividing the observed enthalpy change by the moles of ATP hydrolyzed. Actual calorimeters are much more sophisticated than this primitive experimental setup. The calorimeter is well insulated, mixing is done very carefully, and very precise temperature measurements are made with a thermocouple. The enthalpy changes for many biological reactions have been measured, but unfortunately this information is not conveniently tabulated in a single source. However, many enthalpies of reaction can be derived from the heats of formation in the table in Appendix 4.

Scanning calorimetry is a quite different experiment and measures the heat capacity as a function of temperature. In these experiments, a known amount of heat is added to the system through electrical heating and the resulting temperature rise is measured. Very small amounts of heat are used, so the temperature changes are typically very small. This process is repeated automatically so that the temperature of the system slowly rises. The heat capacity of the system is calculated for each heat increment as $q_P/\Delta T$, and the data are presented as a plot of C_P versus T.

This method has been used, for example, to study protein unfolding and denaturation. Proteins unfold as the temperature is raised, and denaturation usually occurs over a very narrow temperature range. This is illustrated schematically in Figure 1-5, where the fraction of denatured protein, f_D, is plotted versus the temperature along with the corresponding plot of heat capacity, C_P, versus temperature.

As shown in Figure 1-5, the plot of heat capacity versus temperature is a smooth, slowly rising curve for the solvent. With the protein present, a peak in the curve occurs as the protein is denatured. The enthalpy change associated with

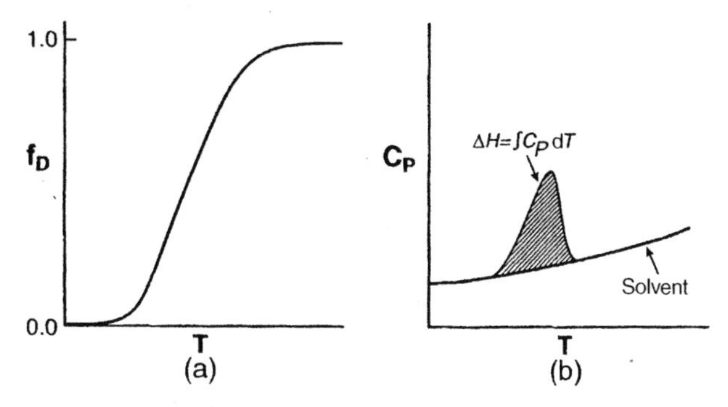

FIGURE 1-5. Schematic representation of the denaturation of a protein and the resulting change in heat capacity, C_P. In (a), the fraction of denatured protein, f_D, is shown as a function of temperature, T. In (b), the heat capacity, as measured by scanning calorimetry, is shown as a function of temperature. The lower curve is the heat capacity of the solvent. The hatched area is the excess heat capacity change due to the protein denaturing and is equal to ΔH for the unfolding.

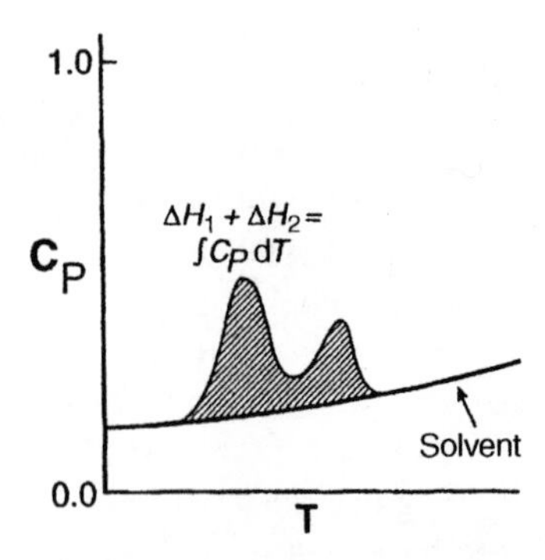

FIGURE 1-6. Schematic representation of a calorimeter scan in which the denaturation occurs in two steps. The hatched area permits the sum of the enthalpy changes to be determined, and the individual enthalpies of the unfolding reactions can be determined by a detailed analysis. As in Figure 1-5, C_P is the measured heat capacity and T is the temperature.

denaturation is the area under the peak (striped area $= \int C_P \, dT$). In some cases, the protein denaturation may occur in multiple stages, in which case more than one peak can be seen in the heat capacity plot. This is shown schematically in Figure 1-6 for a two-stage unfolding process.

The enthalpies associated with protein unfolding are often interpreted in molecular terms such as hydrogen bonds, electrostatic interactions, and hydrophobic interactions. It should be borne in mind that these interpretations are not inherent in thermodynamic quantities, which do not explicitly give information at the molecular level. Consequently, such interpretations should be scrutinized very critically.

1.9 REACTION ENTHALPIES

We now return to a consideration of reaction enthalpies. Because the enthalpy is a state function, it can be added and subtracted for a sequence of reactions—it does not matter how the reaction occurs or in what order. In this regard, chemical reactions can be considered as algebraic equations. For example, consider the reaction cycle below:

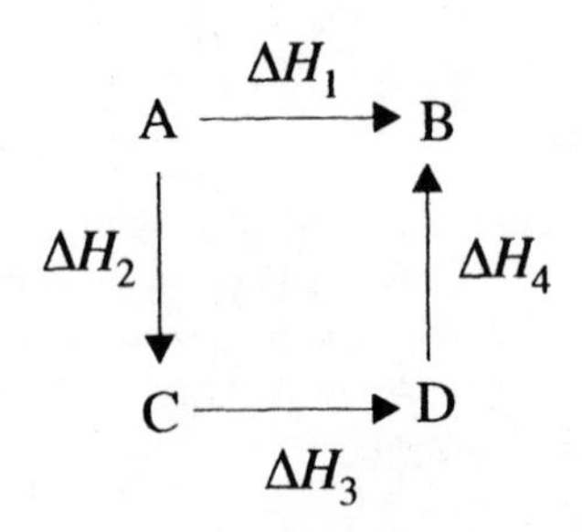

If these reactions are written sequentially, it can readily be seen how the enthalpies are related.

$$\begin{array}{ll} A \rightarrow C & \Delta H_2 \\ C \rightarrow D & \Delta H_3 \\ D \rightarrow B & \Delta H_4 \\ \hline A \rightarrow B & \Delta H_1 = \Delta H_2 + \Delta H_3 + \Delta H_4 \end{array}$$

This ability to relate enthalpies of reaction in reaction cycles in an additive fashion is often called Hess's law, although it really is derived from thermodynamic principles as discussed. We will find that this "law" is extremely useful, as it allows determination of the enthalpy of reaction without studying a reaction directly if a sequence of reactions is known that can be added to give the desired reaction.

As an illustration, we will calculate the enthalpy of reaction for the transfer of a phosphoryl group from ATP to glucose, a very important physiological reaction catalyzed by the enzyme hexokinase.

$$\text{Glucose} + \text{ATP} \rightleftharpoons \text{ADP} + \text{Glucose-6-Phosphate} \tag{1-33}$$

The standard enthalpy changes for the hydrolysis of these four compounds are given in Table 1-1. These data are for very specific conditions: $T = 298\,\text{K}$, $P = 1\,\text{atm}$, $\text{pH} = 7.0$, $\text{pMg} = 3$, and an ionic strength of 0.25 M. The ionic strength is a measure of the salt concentration that takes into account the presence of both monovalent and divalent ions $(= \frac{1}{2}\Sigma c_i z_i^2$, where c_i is the concentration of each ion, z_i is its valence, and the sum is over all of the ions present). The enthalpy change for the hexokinase reaction can easily be calculated from these data:

$$\begin{array}{ll} \text{G} + \text{P}_\text{i} \rightleftharpoons \text{G6P} + \text{H}_2\text{O} & \Delta H^\circ_{298} = 0.5\,\text{kJ/mol} \\ \text{ATP} + \text{H}_2\text{O} \rightleftharpoons \text{ADP} + \text{P}_\text{i} & \Delta H^\circ_{298} = -30.9\,\text{kJ/mol} \\ \hline \text{G} + \text{ATP} \rightleftharpoons \text{G6P} + \text{ADP} & \Delta H^\circ_{298} = -30.4\,\text{kJ/mol} \end{array}$$

The ability to calculate thermodynamic quantities for biochemical reactions that have not yet been studied is very useful. Even if data are not available to deal

TABLE 1-1

Reaction	ΔH°_{298} (kJ/mol)
$\text{ATP} + \text{H}_2\text{O}(\ell) \rightleftharpoons \text{ADP} + \text{P}_\text{i}$	−30.9
$\text{ADP} + \text{H}_2\text{O}(\ell) \rightleftharpoons \text{AMP} + \text{P}_\text{i}$	−28.9
$\text{AMP} + \text{H}_2\text{O}(\ell) \rightleftharpoons \text{A} + \text{P}_\text{i}$	−1.2
$\text{G6P} + \text{H}_2\text{O}(\ell) \rightleftharpoons \text{G} + \text{P}_\text{i}$	−0.5

with the reaction of specific interest, very often data are available for closely related reactions. Appendix 5 contains a tabulation of ΔH°_{298} for some biochemical reactions.

The enthalpy change associated with the hexokinase reaction could also be derived from the heats of formation in the table in the appendix:

$$\Delta H = H^\circ_{\mathrm{f,ADP}} + H^\circ_{\mathrm{f,G6P}} - H^\circ_{\mathrm{f,ATP}} - H^\circ_{\mathrm{f,G}}$$
$$\Delta H = -2000.2 - 2279.1 + 2981.8 + 1267.1 = -30.4\,\mathrm{kJ/mol}$$

In point of fact, the heats of formation are usually derived from measured heats of reaction as these are the primary experimental data.

A source of potential confusion is the practice of reporting enthalpies of reaction as "per mole." There is no ambiguity for the hexokinase reaction as written above. However, in many cases, the stoichiometric coefficients for reactants and products differ. For example, the reaction catalyzed by the enzyme myokinase is

$$2\ \mathrm{ADP} \rightleftharpoons \mathrm{ATP} + \mathrm{AMP} \tag{1-34}$$

Even though 2 moles of ADP are used, the reaction enthalpy is referred to as "per mole." The reaction enthalpy is always given as "per mole of reaction as it is written."

It is important, therefore, that the equation for the reaction under consideration be explicitly stated. The myokinase reaction could be written as

$$\mathrm{ADP} \rightleftharpoons \frac{1}{2}\mathrm{ATP} + \frac{1}{2}\mathrm{AMP} \tag{1-35}$$

In this case, the reaction enthalpy per mole would be one-half of that reported for Eq. 1-34.

1.10 TEMPERATURE DEPENDENCE OF THE REACTION ENTHALPY

In principle, the enthalpy changes as the pressure and temperature change. We will not worry about the dependence of the enthalpy on pressure, as it is usually very small for reactions in condensed phases. The temperature dependence of the enthalpy is given by Eq. 1-27. This can be used directly to determine the temperature dependence of reaction enthalpies. If we assume the standard state enthalpy is known for each reactant, then the temperature dependence of the enthalpy for each reactant, i, is

$$H_{T,i} = H^\circ_{298,i} + \int_{298}^{T} C_{P,i}\,\mathrm{d}T \tag{1-36}$$

If we apply this relationship to the reaction enthalpy for the generalized reaction of Eq. 1-25, we obtain the following:

$$\Delta H_T = cH_{T,\mathrm{C}} + dH_{T,\mathrm{D}} - aH_{T,\mathrm{A}} - bH_{T,\mathrm{B}}$$

$$\Delta H_T = \Delta H^\circ_{298} + \int_{298}^{T} \Delta C_P \,\mathrm{d}T$$

with

$$\Delta H^\circ_{298} = cH^\circ_{298,\mathrm{C}} + dH^\circ_{298,\mathrm{D}} - aH^\circ_{298,\mathrm{A}} - bH^\circ_{298,\mathrm{B}}$$

and

$$\Delta C_P = cC_{P,\mathrm{C}} + dC_{P,\mathrm{D}} - aC_{P,\mathrm{A}} - bC_{P,\mathrm{B}}$$

More generally,

$$\Delta H_T = \Delta H_{T_0} + \int_{T_0}^{T} \Delta C_P \,\mathrm{d}T \qquad (1\text{-}37)$$

Equation 1-37 is known as Kirchhoff's law. It can also be stated in differential form:

$$\mathrm{d}\,\Delta H/\mathrm{d}T = \Delta C_P \qquad (1\text{-}38)$$

It is important to remember that this discussion of the temperature dependence of the reaction enthalpy assumes that the pressure is constant.

The conclusion of these considerations of reaction enthalpies is that available tabulations are often sufficient to calculate the reaction enthalpy of many biological reactions. Moreover, if this is done at a standard temperature, the reaction enthalpy at other temperatures can be calculated if appropriate information about the heat capacities is known or estimated. For most chemical reactions of biological interest, the temperature dependence of the reaction enthalpy is small. On the contrary, for processes such as protein folding and unfolding, the temperature dependence is often significant and must be taken into account in data analysis and thermodynamic calculations. This will be discussed further in Chapter 3.

The first law of thermodynamics, namely, the definition of energy and its conservation, is obviously of great importance in understanding the nature of chemical reactions. As we shall see, however, the first law is not sufficient to understand what determines chemical equilibria.

REFERENCES

1. I. Tinoco, Jr., K. Sauer, & J. C. Wang, and J. D. Puglisi, *Physical Chemistry: Principles and Applications to the Biological Sciences*, 4th edition, Prentice Hall, Englewood Cliffs, NJ, 2002.

2. D. Eisenberg and D. Crothers, *Physical Chemistry with Applications to the Life Sciences*, Benjamin/Cummings, Menlo Park, CA, 1979.
3. *The NBS Tables of Thermodynamic Properties*, D. D. Wagman et al. (eds.), *J. Phys. Chem. Ref. Data, 11, Suppl.* 2, 1982.
4. D. R. Stull, E. F. Westrum, Jr., and G. C. Sinke, *The Chemical Thermodynamics of Organic Compounds*, Wiley, New York, 1969.
5. R. A. Alberty, *Arch. Biochem. Biophys.* **353**, 116 (1998).

PROBLEMS

1-1. When a gas expands rapidly through a valve, you often feel the valve get colder. This is an adiabatic expansion ($q = 0$). Calculate the decrease in temperature of 1.0 mol of ideal gas as it is expanded from 0.20 to 1.00 liters under the conditions given below. Assume a constant volume molar heat capacity, C_V, of $\frac{3}{2}R$. Note that the energy, E, of an ideal gas depends only on the temperature: It is independent of the volume of the system.

a. The expansion is irreversible with an external pressure of 1 atm and an initial temperature of 300 K.

b. The expansion is reversible with an initial temperature of 300 K.

c. Calculate ΔE for the changes in state described in parts A and B.

d. Assume that the expansion is carried out *isothermally* at 300 K, rather than adiabatically. Calculate the work done if the expansion is carried out irreversibly with an external pressure of 1.0 atm.

e. Calculate the work done if the isothermal expansion is carried out reversibly.

f. Calculate q and ΔE for the changes in state described in parts D and E.

1-2. a. Calculate the enthalpy change for the conversion of glucose [$C_6H_{12}O_6(s)$] and oxygen [$O_2(g)$] to $CO_2(aq)$ and $H_2O(\ell)$ under standard conditions. The standard enthalpies of formation of glucose(s), $CO_2(aq)$, and $H_2O(\ell)$ are −304.3, −98.7, and −68.3 kcal/mol, respectively.

b. When organisms metabolize glucose, approximately 50% of the energy available is utilized for chemical and mechanical work. Assume that 25% of the total energy from eating one mole of glucose can be utilized to climb a mountain. How high a mountain can a 70 kg person climb?

1-3. Calculate the enthalpy change for the oxidation of pyruvic acid to acetic acid under standard conditions.

$$2\ CH_3COCOOH(\ell) + O_2(g) \rightarrow 2\ CH_3COOH(\ell) + 2\ CO_2(g)$$

The heats of combustion of pyruvic acid and acetic acid under standard conditions are −227 kcal/mol and −207 kcal/mol, respectively. Heats of combustion are determined by reacting pyruvic or acetic acid with $O_2(g)$ to give

$H_2O(\ell)$ and $CO_2(g)$. *Hint*: First write balanced chemical equations for the combustion processes.

1-4. Calculate the amount of water (in liters) that would have to be vaporized at 40°C (approximately body temperature) to expend the 2.5×10^6 calories of heat generated by a person in one day (commonly called sweating). The heat of vaporization of water at this temperature is 574 cal/g. We normally do not sweat that much. What is wrong with this calculation? If 1% of the energy produced as heat could be utilized as mechanical work, how large a weight could be lifted 1 meter?

1-5. **a.** One hundred milliliters of 0.200 M ATP is mixed with an ATPase in a Dewar at 298 K, 1 atm, pH 7.0, pMg 3.0, and 0.25 M ionic strength. The temperature of the solution increases 1.48 K. What is ΔH° for the hydrolysis of ATP to adenosine 5′-diphosphate (ADP) and phosphate? Assume that the heat capacity of the system is 418 J/K.

b. The hydrolysis reaction can be written as

$$\text{ATP} + \text{H}_2\text{O} \rightleftharpoons \text{ADP} + \text{P}_\text{i}$$

Under the same conditions, the hydrolysis of ADP

$$\text{ADP} + \text{H}_2\text{O} \rightleftharpoons \text{AMP} + \text{P}_\text{i}$$

has a heat of reaction, ΔH°, of $-28.9\,\text{kJ/mol}$. Under the same conditions, calculate ΔH° for the adenylate kinase reaction:

$$2\ \text{ADP} \rightleftharpoons \text{AMP} + \text{ATP}$$

1-6. The alcohol dehydrogenase reaction

$$\text{NAD} + \text{Ethanol} \rightleftharpoons \text{NADH} + \text{Acetaldehyde}$$

removes ethanol from the blood. Use the enthalpies of formation in Appendix 4 to calculate ΔH° for this reaction. If 10.0 g of ethanol (a generous martini) is completely converted to acetaldehyde by this reaction, how much heat is produced or consumed?

CHAPTER 2

Entropy and Free Energy

2.1 INTRODUCTION

At the outset, we indicated that the primary objective of our discussion of thermodynamics is to understand chemical equilibrium in thermodynamic terms. On the basis of our discussion thus far, one possible conclusion is that chemical equilibria are governed by energy considerations and that the system will always proceed to the lowest energy state. This idea can be discarded quite quickly, as we know that some spontaneous reactions produce heat and some require heat. For example, the hydrolysis of ATP releases heat, $\Delta H^\circ_{298} = -30.9\,\mathrm{kJ/mol}$ whereas ATP and AMP are formed when ADP is mixed with myokinase, yet $\Delta H^\circ_{298} = +2.0\,\mathrm{kJ/mol}$ under identical conditions. The conversion of liquids to gases requires heat, that is, ΔH is positive, even at temperatures above the boiling point. Clearly, the lowest energy state is not necessarily the most stable state.

What factor is missing? (At this point, traditional treatments of thermodynamics launch into a discussion of heat engines, a topic we will avoid.) The missing ingredient is the consideration of the probability of a given state. As a very simple illustration, consider three balls of equal size that are numbered 1, 2, and 3. These balls can be arranged sequentially in six different ways:

123 132 213 231 312 321

The energy state of all of these arrangements is the same, yet it is obvious that the probability of the balls being in sequence (123) is 1/6, whereas the probability of the balls being out of sequence is 5/6. In other words, the probability of a disordered state is much greater than the probability of an ordered state because a larger number of arrangements of the balls exists in the disordered state.

Molecular examples of this phenomenon can readily be found. A gas expands spontaneously into a vacuum even though the energy state of the gas does not change. This occurs because the larger volume has more positions available for molecules, so a greater number of arrangements, or more technically *microstates*, of molecules are possible. Clearly, probability considerations are not sufficient by

Physical Chemistry for the Biological Sciences by Gordon G. Hammes

themselves. If this were the case, the stable state of matter would always be a gas. We know that solids and liquids are stable under appropriate conditions because they are energetically favored; that is, interactions between atoms and molecules result in a lower energy state. The real situation must involve a balance between energy and probability. This is a qualitative statement of what determines the equilibrium state of a system, but we will be able to be much more quantitative than this.

The second law states that disordered states are more probable than ordered states. This is done by defining a new state function, *entropy,* which is a measure of the disorder (or probability) of a state. Thermodynamics does not require this interpretation of the entropy, which is quasi-molecular. However, this is a much more intuitive way of understanding entropy than utilizing the traditional concept of heat engines. The more disordered a state, or the larger the number of available microstates, the higher the entropy. We already can see a glimmer of how the equilibrium state might be determined. At constant entropy, the energy should be minimized, whereas at constant energy, the entropy should be maximized. We will return to this topic a little later. First, we will define the entropy quantitatively.

2.2 STATEMENT OF THE SECOND LAW

A more formal statement of the second law is to define a new state function, the entropy, S, by the equation

$$\mathrm{d}S = \frac{\mathrm{d}q_{\mathrm{rev}}}{T} \tag{2-1}$$

or

$$\Delta S = \int \frac{\mathrm{d}q_{\mathrm{rev}}}{T} \tag{2-2}$$

The temperature scale in this definition is Kelvin. This definition is not as straightforward as that for the energy. Note that this definition requires a reversible heat change, q_{rev} or $\mathrm{d}q_{\mathrm{rev}}$, yet entropy is a state function. At first glance, this seems quite paradoxical. The meaning of this is that the entropy change must be calculated by finding a reversible path. However, all reversible paths give the same entropy change, and the calculated entropy change is correct even if the actual change in state is carried out irreversibly. Although this appears to be somewhat confusing, consideration of some examples will help in understanding this concept.

The second law also includes important considerations about entropy: For a reversible change in state, the entropy of the universe is constant, whereas for an irreversible change in state, the entropy of the universe increases.

Again, the second law cannot be proved, but we can demonstrate that without this law, events could transpire that are contrary to our everyday experience. Two examples are given below.

Without the second law, a gas could spontaneously compress! Let us illustrate this by considering the isothermal expansion of an ideal gas, V_1 to V_2 with constant T. The entropy change is

$$\Delta S = \int (\mathrm{d}q_{\mathrm{rev}}/T) = (1/T)\int \mathrm{d}q_{\mathrm{rev}} = q_{\mathrm{rev}}/T \tag{2-3}$$

For the isothermal expansion of an ideal gas, $\Delta E = 0$. (Because of the definition of an ideal gas, the energy, E, is determined by the temperature only and does not depend on the volume and pressure.) Therefore,

$$q_{\mathrm{rev}} = -w_{\mathrm{rev}} = \int_{V_1}^{V_2} P\,\mathrm{d}V = \int_{V_1}^{V_2} (nRT/V)\,\mathrm{d}V = nRT\,\ln(V_2/V_1) \tag{2-4}$$

and

$$\Delta S = nR\ln(V_2/V_1) \tag{2-5}$$

For an expansion, $V_2 > V_1$ and $\Delta S > 0$. For a compression, $V_2 < V_1$ and $\Delta S < 0$. The second law does not prohibit this situation, as we are considering the entropy change for the system, not the universe. This result is in accord with the intuitive interpretation of entropy previously discussed: A larger volume has more positions for the gas to occupy and consequently a higher entropy.

We must now consider what happens to the entropy of the surroundings. For a reversible change, $\Delta S = -q_{\mathrm{rev}}/T$, and the entropy of the universe is the sum of the entropy change for the system and that for the surroundings: $\Delta S = q_{\mathrm{rev}}/T - q_{\mathrm{rev}}/T = 0$, which is consistent with the second law. However, for an irreversible change the situation is different. Let us make this irreversible change by setting the external pressure equal to zero during the change in state. In that case, $w = 0$ and since $\Delta E = 0$, $q = 0$. Therefore, no heat is lost by the surroundings. The entropy change for the universe is

$$\Delta S = \Delta S_{\mathrm{gas}} + \Delta S_{\mathrm{surr}} = nR\ln(V_2/V_1) + 0$$

The second law says that the entropy change of the universe must be greater than zero, which requires that $V_2 > V_1$. In other words, a spontaneous compression cannot occur. This is not required by the first law.

As a second example, consider two blocks at different temperatures, T_{h} and T_{c}, where h and c designate hot and cold so $T_{\mathrm{h}} > T_{\mathrm{c}}$. We will put the blocks together so that heat is transferred from the hot block to the cold block. The entropy changes in the two blocks are given by

$$\mathrm{d}S_{\mathrm{c}} = \mathrm{d}q/T_{\mathrm{c}} \quad \text{and} \quad \mathrm{d}S_{\mathrm{h}} = -\mathrm{d}q/T_{\mathrm{h}}$$

If the two blocks are considered to be the universe, the entropy change of the universe is

$$\mathrm{d}S_c + \mathrm{d}S_h = \mathrm{d}q(1/T_c - 1/T_h) > 0$$

As predicted by the second law, the entropy of the universe increases. What if the heat flows from the cold block to the hot block? Then the sign of the heat change is reversed and

$$\mathrm{d}S_c + \mathrm{d}S_h = \mathrm{d}q(1/T_h - 1/T_c) < 0$$

This predicts that the entropy of the universe would decrease, which is impossible according to the second law. Thus, heat cannot flow from the cold bar to the hot bar. This is not prohibited by the first law.

Exceptions to the second law can be used to create perpetual motion machines, which we know are impossible. These are sometimes called perpetual motion machines of the second kind, whereas perpetual motion machines created by exceptions to the first law are called perpetual motion machines of the first kind.

2.3 CALCULATION OF THE ENTROPY

A reversible path must always be found to calculate the entropy change. At constant volume, the relationship

$$\mathrm{d}q_{\mathrm{rev}} = nC_V\mathrm{d}T \tag{2-6}$$

can be used, whereas at constant pressure

$$\mathrm{d}q_{\mathrm{rev}} = nC_P\mathrm{d}T \tag{2-7}$$

or

$$\Delta S = \int nC_P\mathrm{d}T/T \tag{2-8}$$

Entropy changes for phase changes are particularly easy to calculate since they occur at constant temperature and pressure. At constant temperature and pressure,

$$\Delta S = q_{\mathrm{rev}}/T = \Delta H/T \tag{2-9}$$

For example, for the process

$$\text{Benzene(s, 1 atm, 279K)} \rightarrow \text{Benzene}(\ell\text{, 1 atm, 279K)} \tag{2-10}$$

$\Delta H = 2380$ cal/mol and $\Delta S = 2380/279 = 8.53$ cal/(mol K) $= 8.53$ eu. Here 1 cal/(mol K) is defined as an entropy unit, eu. The entropy change for the reverse process is -8.53 eu. Note that the reverse process is not prohibited by the second law since it is the entropy change for the system, not the universe. Also note that, as expected, going from a solid to a liquid involves a positive entropy change since the liquid state is more disordered than the solid.

While it is easy to state that the entropy change can be calculated for an irreversible process by finding a reversible way of going from the initial state to the final state, this is not always easy to do. This is usually not a matter of great consequence for our considerations, but we will consider one example to illustrate the process. Let us determine the entropy change for the following process:

$$H_2O(\ell, 298\text{ K}, 1\text{ atm} \rightarrow H_2O(g, 298\text{ K}, 1\text{ atm}) \qquad (2\text{-}11)$$

This change is not a reversible change in state, as we know that the normal boiling point of water at 1 atm is 373 K. A possible reversible cycle that would go from the initial state to the final state at constant pressure is

$$\begin{array}{ccc} H_2O(\ell, 298\text{ K}, 1\text{atm} & \rightarrow & H_2O(g, 298\text{ K}, 1\text{atm}) \\ \downarrow & & \uparrow \\ H_2O(\ell, 373\text{ K}, 1\text{ atm} & \rightarrow & H_2O(g, 373\text{ K}, 1\text{ atm}) \end{array} \qquad (2\text{-}12)$$

The entropy change for the bottom process, which is reversible, is simply $\Delta H/T = 9710/373 = 26$ cal/(mol K). The entropy change for the left-hand side of the square is (Eq. 2-8)

$$\Delta S = C_P \ln(T_2/T_1) = 18 \ln(373/298) = 4 \text{ cal/(mol K)}$$

and for the right-hand side of the square is

$$\Delta S = C_P \ln(T_2/T_1) = 8.0 \ln(298/373) = -1.8 \text{ cal/(mol K)}$$

(The heat capacity of $H_2O(g)$ from 298 to 373 K is assumed to be that of $H_2O(g)$ at 373 K.) The entropy changes for these three reversible processes can be added to give the entropy change for the change in state given in Eq. 2-11: 28 cal/(mol K).

An alternative reversible path that can be constructed lowers the pressure to the equilibrium vapor pressure of water at 298 K. The corresponding constant temperature cycle is

$$\begin{array}{ccc} H_2O(\ell, 298\text{ K}, 1\text{ atm}) & \rightarrow & H_2O(g, 298\text{ K}, 1\text{ atm}) \\ \downarrow & & \uparrow \\ H_2O(\ell, 298\text{ K}, 0.0313\text{ atm}) & \rightarrow & H_2O(g, 298\text{ K}, 0.0313\text{ atm}) \end{array} \qquad (2\text{-}13)$$

In this case, we would have to calculate the change in entropy as the pressure is lowered and raised. This can easily be done but is beyond the scope of this presentation of thermodynamics. The point of this exercise is to illustrate how entropy changes can be calculated for irreversible as well as reversible processes and multiple reversible processes can be found.

In principle, the entropy can be calculated from statistical considerations. Boltzmann derived a relationship between the entropy, S, and the number of microstates, N:

$$S = k_B \ln N \tag{2-14}$$

where k_B is Boltzmann's constant, 1.38×10^{-23} J/K. It is rarely possible to determine the number of microstates although the number of microstates could be calculated from.Eq. 2-14 if the entropy is known. For a simple case, such as the three numbered balls with which we started our discussion of the second law, this calculation can readily be done. The disordered system has 3! microstates and an entropy of 1.51×10^{-23} J/K. Any ordered sequence—for example 1, 2, 3—has only one microstate, so $N = 1$, and $S = 0$. Since the disordered state has a higher entropy, this predicts that the balls will spontaneously disorder and that an ordered state is extremely unlikely.

2.4 THIRD LAW OF THERMODYNAMICS

We will not dwell on the third law as the details are of little consequence in biology. The important fact for us is that the third law establishes a zero for the entropy scale. Unlike the energy, entropy has an absolute scale. The third law can be stated as follows: The entropy of perfect crystals of all pure elements and compounds is zero at absolute zero. The tricky points of this law are the meanings of "perfect" and "pure," but we will not discuss this in detail. It is worth noting that a perfect crystal has one microstate, and therefore, an entropy of zero according to Eq. 2-14.

The absolute standard entropy can be determined from measurements of the temperature dependence of the heat capacity using the relationship

$$S^\circ_{298} = \int_0^{298} C_P \mathrm{d}T/T \tag{2-15}$$

Here the entropy at absolute zero has been assumed to be zero in accordance with the third law. A plot of C_P/T versus T gives a curve such as that shown in Figure 2-1. The area under the curve is the absolute standard entropy. Tables of S°_{298} are readily available (cf. Refs. 3 and 4 in Chapter 1). The entropy at temperatures other than 298 K can be calculated from the relationship

$$S^\circ_T = S^\circ_{298} + \int_{298}^{T} C_P\, \mathrm{d}T/T \tag{2-16}$$

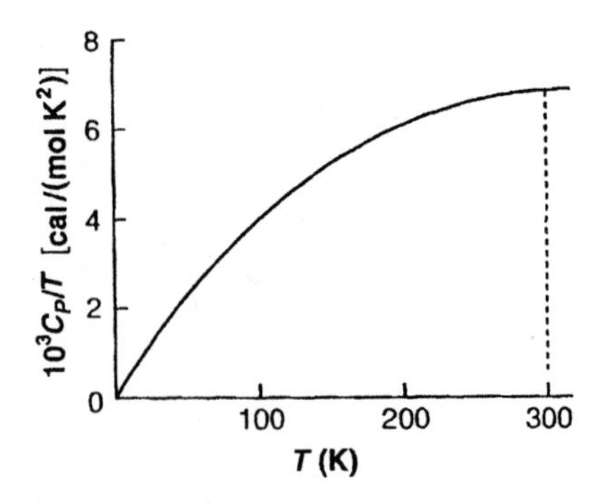

FIGURE 2-1. A plot of the constant-pressure heat capacity divided by the temperature, C_P/T, versus the temperature, T, for graphite. The absolute entropy of graphite at 300 K is the area under the curve up to the dashed line (Eq. 2-15). If phase changes occur as the temperature is changed, the entropy changes associated with the phase changes must be added to the area under the curve of the C_P/T versus T plot.

For a chemical reaction,

$$\Delta S_T = \Delta S^\circ_{298} + \int_{298}^{T} \Delta C_P \, \mathrm{d}T/T \tag{2-17}$$

These relationships are analogous to those used for the enthalpy. Indeed, entropies of reactions can be calculated in a similar fashion to enthalpy changes.

2.5 MOLECULAR INTERPRETATION OF ENTROPY

We will now consider a few examples of absolute entropies and entropy changes for chemical reactions, and how they might be interpreted in molecular terms. The absolute entropies for water as a solid, liquid, and gas at 273 K and 1 atm are 41.0, 63.2, and 188.3 J/(mol K), respectively. The molecular interpretation of these numbers is straightforward, namely, that solid is more ordered than liquid, which is more ordered than gas; therefore, the solid has the lowest entropy and the gas the highest.

Standard entropy changes for some chemical reactions are given in Table 2-1. As expected, the entropy change is negative for the first two reactions in the gas phase as the number of moles of reactants is greater than the number of moles of products. Of course, these reactions could have been written in the opposite direction. The entropy change would then have the opposite sign. At first glance, the result for the third reaction in the table is surprising, as the number of moles of reactants is greater than the number of moles of products. This results because the solvent must be included in the molecular interpretation of the observed entropy change. Ions in water interact strongly with water so that highly ordered water molecules exist around the ions.

TABLE 2-1

Reaction	ΔS°_{298}[J/(mol K)]
$H(g) + H(g) \rightleftharpoons H_2(g)$	−98.7
$2\ H_2(g) + O_2(g) \rightleftharpoons 2\ H_2O(g)$	−88.9
$H^+(aq) + OH^-(aq) \rightleftharpoons H_2O(\ell)$	+80.7
Cytidine 2′ -monophosphate + Ribonuclease A $\rightleftharpoons$ Enzyme–inhibitor complex	−54[a]

[a]H. Naghibi, A. Tamura, and J. M. Sturtevant, *Proc. Natl. Acad. Sci. USA* **92**, 5597 (1995).

When the neutral species is formed, these highly ordered water molecules become less ordered. Thus, the entropy change for water is very positive, much more positive than the expected entropy decrease for H^+ and OH^- when they form water. This simple example indicates that considerable care must be taken in interpreting entropy changes for chemical reactions in condensed media. The entropy of the entire system under consideration must be taken into account. The final entry is for the binding of a ligand to an enzyme. In this case, no reasonable interpretation of the entropy change is possible as three factors come into play: the loss in entropy as two reactants become a single entity, the changes in the structure of water, and structural changes in the protein. The fact that the entropy change is comparable to the value expected for the combination of two molecules to produce one molecule is simply fortuitous. Extreme caution should be taken in making molecular interpretations of thermodynamic changes in complex systems.

We have established all of the thermodynamic principles necessary to discuss chemical equilibrium. We will now apply these principles to develop a general framework for dealing with chemical reactions.

2.6 FREE ENERGY

In a sense, we have reached our goal. We have developed thermodynamic criteria for the occurrence of reversible and irreversible (spontaneous) processes, namely, for the universe, $\Delta S = 0$ for reversible processes and must be greater than zero for irreversible processes. Unfortunately, this is not terribly useful, as we are interested in what is happening in the system and require criteria that are easily applicable to chemical reactions. This can be achieved by defining a new thermodynamic state function, the Gibbs free energy:

$$G = H - TS \tag{2-18}$$

(J. Willard Gibbs developed the science of thermodynamics virtually single handed around the turn of the century at Yale University. To this day, his collected works are prized possessions in the libraries of people seriously interested in thermodynamics.)

At constant temperature,

$$\begin{aligned}\Delta G &= \Delta H - T\Delta S \\ &= q_P - T\Delta S = q_P - q_{\text{rev}}\end{aligned} \tag{2-19}$$

For a reversible process, $q_P = q_{\text{rev}}$, so $\Delta G = 0$. For irreversible processes, the situation is a bit more complex as the sign of the heat change must be considered. For an endothermic process, $q_P < q_{\text{rev}}$, so $\Delta G < 0$. For an exothermic process, the absolute value of q_P is greater than the absolute value of q_{rev}, so again, $\Delta G < 0$. This is the free energy change for the system and does not involve the surroundings. We now have developed criteria that tell us if a process occurs spontaneously. If $\Delta G < 0$, the change in state occurs spontaneously, whereas if $\Delta G > 0$, the reverse change in state occurs spontaneously. If $\Delta G = 0$, the system is at equilibrium (at constant pressure and temperature).

As with the energy and enthalpy, only differences in free energy can be measured. Consequently, the zero of the free energy scale is arbitrary. As for the enthalpy, the zero of the scale is taken as the elements in their stable state at 1 atm and 298 K. Again, analogous to the enthalpy, tables of the free energies of formation of compounds are available so that free energy changes for chemical reactions can be calculated (cf. Refs. 3–5 in Chapter 1). Referring back to the reaction in Eq. 1-25,

$$\Delta G^{\circ}_{298} = cG^{\circ}_{298,\text{C}} + dG^{\circ}_{298,\text{D}} - aG^{\circ}_{298,\text{A}} - bG^{\circ}_{298,\text{B}} \tag{2-20}$$

where the free energies on the right-hand side of the equation are free energies per mole. We will discuss the temperature dependence of the free energy a bit later, but it is useful to remember that at constant temperature and pressure,

$$\Delta G = \Delta H - T\Delta S \tag{2-21}$$

Standard state free energy changes are available for biochemical reactions although comprehensive tabulations do not exist, and the definition of "standard state" is more complex than in the gas phase as discussed previously. Standard state free energies of formation for some substances of biochemical interest are given in Appendix 4.

As an illustration of the concept of free energy, consider the conversion of liquid water to steam:

$$\text{H}_2\text{O}(\ell) \rightleftharpoons \text{H}_2\text{O}(\text{g}) \tag{2-22}$$

At the boiling point, $\Delta H = 9710$ cal/mol and $\Delta S = 26$ eu. Therefore,

$$\Delta G = 9710 - 26T \tag{2-23}$$

At equilibrium, $\Delta G = 0$ and Eq. 2-23 gives $T = 373$ K, the normal boiling point of water. If $T > 373$ K, $\Delta G < 0$, and the change in state is spontaneous, whereas if $T < 373$ K, $\Delta G > 0$, and the reverse process, condensation, occurs spontaneously.

Calculation of the free energy for a change in state is not always straightforward. Because the entropy is part of the definition of the free energy, a reversible process must always be found to calculate the free energy change. The free energy change, of course, is the same regardless of how the change in state is accomplished since the free energy is a state function. As a simple example, again consider the change in state in Eq. 2-11. This is not a reversible process since the two states are not in equilibrium. This also is not a spontaneous change in state so that $\Delta G > 0$. In order to calculate the value of ΔG, we must think of a reversible path for carrying out the change in state. The two cycles in Eqs. 2-12 and 2-13 again can be used. In both cases, the bottom reaction is an equilibrium process, so $\Delta G = 0$. To calculate ΔG for the top reaction, we only need to add up the ΔG values for the vertical processes. These can be calculated from the temperature (upper cycle) and pressure (lower cycle) dependence of G. Such functional dependencies will be considered shortly. It will be a useful exercise for the reader to carry out the complete calculations.

2.7 CHEMICAL EQUILIBRIA

Although we now have developed criteria for deciding whether or not a process will occur spontaneously, they are not sufficient for consideration of chemical reactions. We know that chemical reactions are generally not "all or nothing" processes; instead, an equilibrium state is reached where both reactants and products are present. We will now derive a quantitative relationship between the free energy change and the concentrations of reactants and products. We will do this in detail for the simple case of ideal gases and by analogy for reactions in liquids.

The starting point for the derivation is the definition of free energy and its total derivative:

$$G = H - TS = E + PV - TS \qquad (2\text{-}24)$$

$$\mathrm{d}G = \mathrm{d}E + P\mathrm{d}V + V\mathrm{d}P - T\mathrm{d}S - S\mathrm{d}T \qquad (2\text{-}25)$$

Since $\mathrm{d}E = \mathrm{d}q + \mathrm{d}w = T\mathrm{d}S - S\mathrm{d}V$ for a reversible process,

$$\mathrm{d}G = V\mathrm{d}P - S\mathrm{d}T \qquad (2\text{-}26)$$

(Although this relationship was derived for a reversible process, it is also valid for an irreversible process.) Let us now consider a chemical reaction of ideal gases at constant temperature. For one mole of each gas component,

$$\mathrm{d}G = V\mathrm{d}P = RT\mathrm{d}P/P \qquad (2\text{-}27)$$

We will refer all of our calculations to a pressure of 1 atm for each component. In thermodynamic terms, we have selected 1 atm as our *standard state*. If Eq. 2-27 is now integrated from $P_0 = 1$ atm to P,

$$dG = RT dP/P$$

or

$$G = G^\circ + RT \ln(P/P_0) = G^\circ + RT \ln P \tag{2-28}$$

We will now return to our prototype reaction, Eq. 1-25, and calculate the free energy change. The partial pressures of the reactants are given in parentheses:

$$a\mathrm{A}(P_\mathrm{A}) + b\mathrm{B}(P_\mathrm{B}) \rightleftharpoons c\mathrm{C}(P_\mathrm{C}) + d\mathrm{D}(P_\mathrm{D}) \tag{2-29}$$

$$\begin{aligned}\Delta G &= cG_\mathrm{C} + dG_\mathrm{D} - aG_\mathrm{A} - bG_\mathrm{B} \\ &= cG^\circ_\mathrm{C} + dG^\circ_\mathrm{D} - aG^\circ_\mathrm{A} - bG^\circ_\mathrm{B} + cRT \ln P_\mathrm{C} + dRT \ln P_\mathrm{D} - aRT \ln P_\mathrm{A} - bRT \ln P_\mathrm{B}\end{aligned}$$

or

$$\Delta G = \Delta G^\circ + RT \ln \left(\frac{P^c_\mathrm{C} P^d_\mathrm{D}}{P^a_\mathrm{A} P^b_\mathrm{B}} \right) \tag{2-30}$$

ΔG is the free energy for the reaction in Eq. 2-29 when the system is not at equilibrium. At equilibrium, at constant temperature and pressure, $\Delta G = 0$, and Eq. 2-30 becomes

$$\Delta G^\circ = -RT \ln \left(\frac{P^c_\mathrm{C} P^d_\mathrm{D}}{P^a_\mathrm{A} P^b_\mathrm{B}} \right)_\mathrm{e} = -RT \ln K \tag{2-31}$$

Here the subscript e has been used to designate equilibrium and K is the equilibrium constant.

We now have a quantitative relationship between the partial pressures of the reactants and the standard free energy change, ΔG°. The standard free energy change is a constant at a given temperature and pressure but will vary as the temperature and pressure change. If $\Delta G^\circ < 0$, then $K > 1$, whereas if $\Delta G^\circ > 0$, $K < 1$. A common mistake is to confuse the free energy change with the standard free energy change. The free energy change is always equal to zero at equilibrium and can be calculated from Eq. 2-30 when not at equilibrium. The standard free energy change is a constant representing the hypothetical reaction with all of the reactants and products at a pressure of 1 atm. It is equal to zero only if the equilibrium constant fortuitously is 1.

Biological reactions do not occur in the gas phase. What about free energy in solutions? Conceptually there is no difference. The molar free energy at constant temperature and pressure can be written as

$$G = G^\circ + RT \ln(c/c_0) \tag{2-32}$$

where c is the concentration and c_0 is the standard state concentration. A more correct treatment would define the molar free energy as

$$G = G^\circ + RT \ln a \tag{2-33}$$

where a is the thermodynamic activity and is dimensionless. However, the thermodynamic activity can be written as a product of an activity coefficient and the concentration. The activity coefficient can be included in the standard free energy, G°, which gives rise to Eq. 2-32. We need not worry about this as long as the solution conditions are clearly defined with respect to salt concentration, pH, etc. The reason it is not of great concern is that all of the aforementioned complications can be included in the standard free energy change since, in practice, the standard free energy is determined by measuring the equilibrium constant under defined conditions.

Finally, we should note that the free energy per mole at constant temperature and pressure is called the chemical potential, μ, in more sophisticated treatments of thermodynamics, but there is no need to introduce this terminology here.

If we take the standard state as 1 mol/liter, then the results parallel to Eqs. 2-28, 2-30, and 2-31 are

$$G = G^\circ + RT \ln c \tag{2-34}$$

$$\Delta G = \Delta G^\circ + RT \ln\left(\frac{c_C^c c_D^d}{c_A^a c_B^b}\right) \tag{2-35}$$

$$\Delta G^\circ = -RT \ln\left(\frac{c_C^c c_D^d}{c_A^a c_B^b}\right)_e = -RT \ln K \tag{2-36}$$

Equations 2-34–2-36 summarize the thermodynamic relationships necessary to discuss chemical equilibria.

Note that, strictly speaking, the equilibrium constant is dimensionless as all of the concentrations are ratios, the actual concentration divided by the standard state concentration. However, practically speaking, it is preferable to report equilibrium constants with the dimensions implied by the ratio of concentrations in Eq. 2-36. The equilibrium constant is determined experimentally by measuring concentrations, and attributing dimensions to this constant assures that the correct ratio of concentrations is considered and that the standard state is precisely defined.

Consideration of the free energy also allows us to assess how the energy and entropy are balanced to achieve the final equilibrium state. Since $\Delta G = \Delta H - T\Delta S$ at constant T and P, it can be seen that a change in state is spontaneous if the enthalpy change is very negative and/or the entropy change is very positive. Even if the enthalpy change is unfavorable (positive), the change in state will be spontaneous if the $T\Delta S$ term is very positive. Similarly, even if the entropy change is unfavorable ($T\Delta S$ very negative), the change in state will be spontaneous if the enthalpy change is sufficiently negative. Thus, the final equilibrium achieved is a balance between the enthalpy (ΔH) and the entropy ($T\Delta S$).

2.8 PRESSURE AND TEMPERATURE DEPENDENCE OF THE FREE ENERGY

We will now return to the pressure and temperature dependence of the free energy. At constant temperature, the pressure dependence of the free energy follows directly from Eq. 2-26, namely,

$$\mathrm{d}G = V\mathrm{d}P \tag{2-37}$$

This equation can be integrated if the pressure dependence of the volume is known, as for an ideal gas. For a chemical reaction, Eq. 2-37 can be rewritten as

$$\mathrm{d}\Delta G = \Delta V\mathrm{d}P \tag{2-38}$$

where ΔV is the difference in volume between the products and reactants. The pressure dependence of the equilibrium constant at constant temperature follows directly,

$$\mathrm{d}\Delta G^\circ = -RT\mathrm{d}\ln K = \Delta V\mathrm{d}P \tag{2-39}$$

or

$$\frac{\mathrm{d}\ln K}{\mathrm{d}P} = -\frac{\Delta V}{RT} \tag{2-40}$$

For most chemical reactions, the pressure dependence of the equilibrium constant is quite small so that it is not often considered in biological systems.

Equilibrium constants, however, frequently vary significantly with temperature. At constant pressure, Eq. 2-26 gives

$$\mathrm{d}G = -S\,\mathrm{d}T \tag{2-41}$$

or

$$\mathrm{d}\Delta G = -\Delta S\,\mathrm{d}T \tag{2-42}$$

Returning to the basic definition of free energy at constant temperature and pressure,

$$\Delta G = \Delta H - T\Delta S = \Delta H + T(\mathrm{d}\Delta G/\mathrm{d}T)$$

This equation can be divided by T^2 and rearranged as follows:

$$-\Delta G/T^2 + (\mathrm{d}\Delta G/\mathrm{d}T)/T = -\Delta H/T^2$$

$$\frac{\mathrm{d}(\Delta G/T)}{\mathrm{d}T} = -\frac{\Delta H}{T^2} \qquad (2\text{-}43)$$

Equation 2-43 is an important thermodynamic relationship describing the temperature dependence of the free energy at constant pressure and is called the Gibbs–Helmholtz equation. The temperature dependence of the equilibrium constant follows directly:

$$\frac{\mathrm{d}(\Delta G^\circ/T)}{\mathrm{d}T} = -R\frac{\mathrm{d}\ln K}{\mathrm{d}T}$$

or

$$\frac{\mathrm{d}\ln K}{\mathrm{d}T} = \frac{\Delta H^\circ}{RT^2} \qquad (2\text{-}44)$$

If ΔH° is independent of temperature, Eq. 2-44 can easily be integrated:

$$\mathrm{d}\ln K = \int_{T_1}^{T_2} \frac{\Delta H^\circ \mathrm{d}T}{RT^2}$$

$$\ln\left(\frac{K_2}{K_1}\right) = \frac{\Delta H^\circ}{R}\left(\frac{1}{T_1} - \frac{1}{T_2}\right) = \frac{(\Delta H^\circ/R)(T_2 - T_1)}{T_1 T_2} \qquad (2\text{-}45)$$

When carrying out calculations, the difference between reciprocal temperatures should never be used directly as it introduces a large error. Instead, the rearrangement in Eq. 2-45 should be used in which the difference between two temperatures occurs. With this equation and a knowledge of ΔH°, the equilibrium constant can be calculated at any temperature if it is known at one temperature.

What about the assumption that the standard enthalpy change is independent of temperature? This assumption is reasonable for many biological reactions if the temperature range is not too large. In some cases, the temperature dependence cannot be neglected and must be included explicitly in carrying out the integration of Eq. 2-44. The temperature dependence of the reaction enthalpy depends on the difference in heat capacities between the products and reactants as given by Eq. 1-38.

Examples of the temperature dependence of equilibrium constants are displayed in Figure 2-2. As predicted by Eq. 2-45, a plot of ln K versus $1/T$ is a straight line with a slope of $-\Delta H^\circ/R$. The data presented are for the binding of DNA to DNA-binding proteins (i.e., Zn fingers). The dissociation constants for binding are quite similar for both proteins at 22°C, 1.08×10^{-9} M (WT1 protein) and 3.58×10^{-9} M (EGR1 protein). However, as indicated by the data in the figure, the standard enthalpy changes are *opposite* in sign, +6.6 kcal/mol and −6.9 kcal/mol, respectively. Consequently, the standard entropy changes are also quite

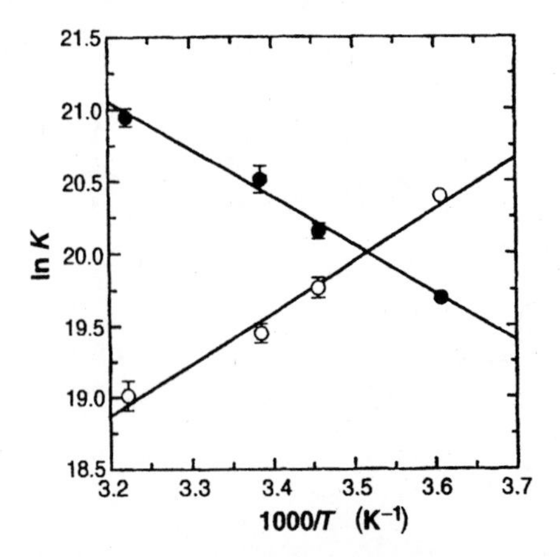

FIGURE 2-2. Temperature dependence of the equilibrium binding constant, K, for the binding of DNA to binding proteins ("zinc fingers"), WT1 (●) and EGR1 (○), to DNA. Adapted with permission from T. Hamilton, F. Borel, and P. J. Romaniuk, *Biochemistry* **37**, 2051 (1998). © 1998 American Chemical Society.

different, 63.3 eu and 15.3 eu, respectively. These data indicate that there are significant differences in the binding processes, despite the similar equilibrium constants.

2.9 PHASE CHANGES

The criterion that $\Delta G = 0$ at equilibrium at constant temperature and pressure is quite general. For chemical reactions, this means that the free energy of the products is equal to the free energy of the reactants. For phase changes of pure substances, this means that at equilibrium the free energies of the phases are equal. If we assume two phases, A and B, Eq. 2-26 gives

$$dG_A = V_A dP - S_A dT = dG_B = V_B dP - S_B dT$$

Rearrangement gives

$$\frac{dP}{dT} = \frac{\Delta S_{BA}}{\Delta V_{BA}} \tag{2-46}$$

where $\Delta S_{BA} = S_B - S_A$ and $\Delta V_{BA} = V_B - V_A$. (All of these quantities are assumed to be per mole for simplicity.) This equation is often referred to as the Clapeyron equation. Note that the entropy change can be written as

$$\Delta S_{BA} = \Delta H_{BA}/T \tag{2-47}$$

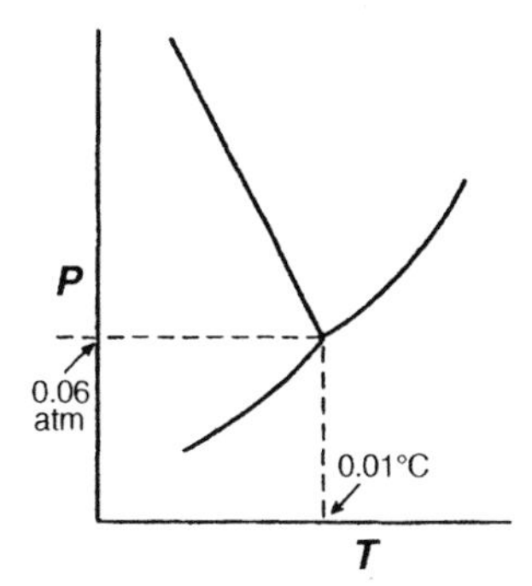

FIGURE 2-3. Schematic representation of the phase diagram of water with pressure, P, and temperature, T, as variables. The phase diagram is not to scale and is incomplete, as several different phases of solid water are known. If volume is included as a variable, a three-dimensional phase diagram can be constructed.

Equation 2-46 gives the slope of *phase diagrams*, plots of P versus T that are useful summaries of the phase behavior of a pure substance. As an example, the phase diagram of water is given in Figure 2-3.

The lines in Figure 2-3 indicate when two phases are in equilibrium and coexist. Only one phase exists in the open areas, and when the two lines meet, three phases are in equilibrium. This is called the triple point and can only occur at a pressure of 0.006 atm and 0.01°C. The slopes of the lines are given by Eq. 2-46. The number of phases that can coexist is governed by the phase rule,

$$f = c - p + 2 \tag{2-48}$$

where f is the number of degrees of freedom, c is the number of components, and p is the number of phases. This rule can be derived from the basic laws of thermodynamics. For a pure substance, the number of components is 1, so $f = 3 - p$. In the open spaces, one phase exists, so that the number of degrees of freedom is 2, which means that both T and P can be varied. Two phases coexist on the lines, which gives $f = 1$. Because only one degree of freedom exists, at a given value of P, T is fixed and vice versa. At the triple point, three phases coexist, and there are no degrees of freedom, which means that both P and T are fixed.

The study of phase diagrams is an important aspect of thermodynamics, and some interesting applications exist in biology. For example, membranes contain phospholipids. If phospholipids are mixed with water, they spontaneously form bilayer vesicles with the head groups pointing outward toward the solvent so that the interior of the bilayer is the hydrocarbon chains of the phospholipids. A schematic representation of the bilayer is given in Figure 2-4. The hydrocarbon chain can be either disordered, at high temperatures, or ordered, at low temperatures, as indicated schematically in the figure. The transformation from one form to the other behaves as a phase transition so that phase diagrams can be constructed for phospholipids. Moreover, the phase diagrams depend not only on the temperature but

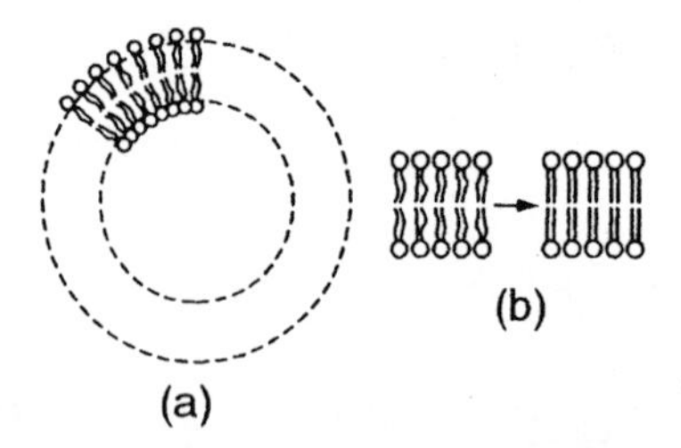

FIGURE 2-4. Schematic representation of the vesicles of phospholipids that are formed when phospholipids are suspended in water (a). Here the small circles represent the head groups and the wiggly lines the two hydrocarbon chains. A phase change can occur in the side chains from a disordered to an ordered state as sketched in (b).

also on the phospholipid composition of the bilayer. This phase transition has biological implications in that the fluidity of the hydrocarbon portion of the membrane strongly affects the transport and mechanical properties of the membranes. Phase diagrams for phospholipids and phospholipid–water mixtures can be constructed by a variety of methods, including scanning calorimetry and nuclear magnetic resonance.

2.10 ADDITIONS TO THE FREE ENERGY

As a final consideration in developing the concept of free energy, we will return to our original development of equilibrium, namely, Eq. 2-33. In arriving at this expression for the molar free energy at constant temperature and pressure, the assumption was made that only $P - V$ work occurs. This is not true in general and for many biological systems in particular. If we went all the way back to the beginning of our development of the free energy changes associated with chemical equilibria (Eq. 2-26) and derived the molar free energy at constant temperature and pressure, we would find that

$$G = G^{\circ} + RT \ln c + w_{\max} \tag{2-49}$$

where $w_{\max}$ is the maximum (reversible) non-$P - V$ work. For example, for an ion of charge z in a potential field Ψ, $w_{\max}$ is $zF\Psi$, where F is the Faraday. (We have used this relationship in Chapter 1 to calculate the work of moving an ion across a membrane with an imposed voltage. In terms of transport of an ion across a membrane, this assumes that the ion is being transported from the inside to the outside with an external membrane potential of Ψ.) This *extended chemical potential* is useful for discussing ion transport across membranes, as we shall see later. It could also be used, for example, for considering molecules in a gravitational or centrifugal field. (See Ref. 1 in Chapter 1 for a detailed discussion of this concept.)

We are now ready to consider some applications of thermodynamics to biological systems in more detail.

PROBLEMS

2-1. One mole of an ideal gas initially at 300 K is expanded from 0.2 to 1.0 liter. Calculate ΔS for this change in state if it is carried out under the following conditions. Assume a constant volume heat capacity for the gas, C_V, of $\frac{3}{2}R$.

a. The expansion is *reversible* and *isothermal.*

b. The expansion is *irreversible* and *isothermal.*

c. The expansion is *reversible* and *adiabatic* ($q = 0$).

d. The expansion is *irreversible* with an external pressure of 1 atm and *adiabatic*.

2-2. The alcohol dehydrogenase reaction, which removes ethanol from your blood, proceeds according to the following reaction:

$$\text{NAD}^+ + \text{Ethanol} \rightleftharpoons \text{NADH} + \text{Acetaldehyde}$$

Under standard conditions (298 K, 1 atm, pH 7.0, pMg 3, and an ionic strength of 0.25 M), the standard enthalpies and free energies of formation of the reactants are as follows:

	H°(kJ/mol)	G°(kJ/mol)
NAD^+	−10.3	1059.1
NADH	−41.4	1120.1
Ethanol	−290.8	63.0
Acetaldehyde	−213.6	24.1

a. Calculate ΔG°, ΔH°, and ΔS° for the alcohol dehydrogenase reaction under standard conditions.

b. Under standard conditions, what is the equilibrium constant for this reaction? Will the equilibrium constant increase or decrease as the temperature is increased?

2.3. The equilibrium constant under standard conditions (1 atm, 298 K, pH 7.0) for the reaction catalyzed by fumarase

$$\text{Fumarate} + H_2O \rightleftharpoons \text{L-malate}$$

is 4.00. At 310 K, the equilibrium constant is 8.00.

a. What is the standard free energy change, ΔG°, for this reaction?

b. What is the standard enthalpy change, ΔH°, for this reaction? Assume that the standard enthalpy change is independent of temperature.

c. What is the standard entropy change, ΔS°, at 298 K for this reaction?

d. If the concentration of both reactants is equal to 0.01 M, what is the free energy change at 298 K? As the reaction proceeds to equilibrium, will more fumarate or L-malate form?

e. What is the free energy change for this reaction when equilibrium is reached?

2-4. The following reaction is catalyzed by the enzyme creatine kinase:

$$\text{Creatine} + \text{ATP} \rightleftharpoons \text{Creatine phosphate} + \text{ADP}$$

a. Under standard conditions (1 atm, 298 K, pH 7.0, pMg 3.0, and an ionic strength of 0.25 M) with the concentrations of all reactants equal to 10 mM, the free energy change, ΔG, for this reaction is 13.3 kJ/mol. What is the standard free energy change, ΔG°, for this reaction?

b. What is the equilibrium constant for this reaction?

c. The standard enthalpies of formation for the reactants are as follows:

creatine	−540 kJ/mol
creatine phosphate	−1510 kJ/mol
ATP	−2982 kJ/mol
ADP	−2000 kJ/mol

What is the standard enthalpy change, ΔH°, for this reaction?

d. What is the standard entropy change, ΔS°, for this reaction?

2-5. a. It has been proposed that the reason ice skating works so well is that the pressure from the blades of the skates melts the ice. Consider this proposal from the viewpoint of phase equilibria. The phase change in question is

$$H_2O(s) \rightleftharpoons H_2O(\ell)$$

Assume that ΔH for this process is independent of temperature and pressure and is equal to 80 cal/g. The change in volume, ΔV, is about -9.1×10^{-5} l/g. The pressure exerted is the force per unit area. For a 180 lb person and an area for the skate blades of about 6 in.^{2} the pressure is 30 lb/in.^{2} or about 2 atm. With this information, calculate the decrease in the melting temperature of ice caused by the skate blades. (Note that 1 cal = 0.04129 l atm.) Is this a good explanation for why ice skating works?

b. A more efficient way of melting ice is to add an inert compound such as urea. (We will avoid salt to save our cars.) The extent to which the freezing point is lowered can be calculated by noting that the molar free energy of water must be the same in the solid ice and the urea solution. The molar free energy of water in the urea solution can be approximated as $G^\circ_{\text{liquid}} + RT \ln X_{\text{water}}$, where X_{water} is the mole fraction of water in the solution. The molar free energy of the solid can be written as G°_{solid}. Derive an expression for the change in the melting temperature of ice by equating the free energies in the two phases, differentiating the resulting equation with respect to temperature, integrating from a mole fraction of 1 (pure solvent) to the mole fraction of the solution, and noting that $\ln X_{\text{water}} = \ln(1 - X_{\text{urea}}) \approx -X_{\text{urea}}$. (This relationship is the series expansion of the logarithm for small values of X_{urea}. Since the concentration of water is about 55 M, this is a good approximation.) With the relationship derived, estimate the decrease in the melting temperature of ice for an 1 M urea solution. The heat of fusion of water is 1440 cal/mol.

2-6. What is the maximum amount of work that can be obtained from hydrolyzing 1 mole of ATP to ADP and phosphate at 298 K? Assume that the concentrations of all reactants are 0.01 M and ΔG° is -32.5 kJ/mol. If the conversion of free energy to mechanical work is 100% efficient, how many moles of ATP would have to be hydrolyzed to lift a 100 kg weight 1 meter high?

CHAPTER 3

Applications of Thermodynamics to Biological Systems

3.1 BIOCHEMICAL REACTIONS

From the discussion in the previous chapter, it should be clear that standard free energy and enthalpy changes can be calculated for chemical reactions from available tables of standard free energies and enthalpies of formation (e.g., Appendix 4). Even if a particular reaction has not been studied, it is frequently possible to combine the standard free energy and enthalpy changes of other known reactions to obtain the standard free energy change of the desired reaction (e.g., Appendix 5). Knowledge of the standard free energy change permits calculation of the equilibrium constant and vice versa. When considering biochemical reactions, the exact nature of the "standard state" can be confusing. Usually, the various possible ionization states of the reactants are not specified, so that the total concentration of each species is used in the expression for the equilibrium constant. For example, for ATP, the standard state would normally be 1 M, but this includes all possible states of protonation, and if magnesium ion is present, the sum of the metal complex and uncomplexed species. Furthermore, pH 7 is usually selected as the standard condition so that the activity of the hydrogen ion is set equal to 1 at pH 7 and the standard free energy of formation of the hydrogen ion is set equal to zero at pH 7. Therefore, the hydrogen ion is not usually included when writing stoichiometric equations. Finally, the activity of water is set equal to 1 for reactions in dilute solutions. This assumption is justified since the concentration of water is essentially constant. However, a word of caution is needed here as the standard free energy and enthalpy of formation for water must be explicitly included if it is a reactant. Again, the choice of standard states can readily be incorporated into the standard free energy change and standard free energy of formation. The best way to avoid ambiguity is to explicitly write the chemical reaction and the conditions (pH, salt, etc.) to which a given free energy change refers.

As a specific example, let us again consider the reaction catalyzed by hexokinase, the phosphorylation of glucose by ATP. We have previously calculated the standard enthalpy change for this reaction from the known enthalpies of hydrolysis. This can also be done for the standard free energy change of the hexokinase reaction since the standard free energies of hydrolysis are known. As before, the "standard states" are 1 M at 1 atm, 298 K, pH 7.0, pMg 3.0, and an ionic strength of 0.25 M.

	ΔG° (kJ/mol)	ΔH° (kJ/mol)
$G + P_i \rightleftharpoons G6P + H_2O$	11.6	0.5
$ATP + H_2O \rightleftharpoons ADP + P_i$	−32.5	−30.9
$G + ATP \rightleftharpoons G6P + ADP$	−20.9	−30.4

The equilibrium constant for the hexokinase reaction can be calculated from the standard free energy change:

$$\Delta G^\circ = -RT \ln K \quad \text{or} \quad K = 4630 \; \{= ([\text{G6P}][\text{ADP}])/([\text{G}][\text{ATP}])\}$$

Since the standard enthalpy change is known, −30.4 kJ/mol, the equilibrium constant can be calculated at other temperatures by use of Eq. 2-45. For example, at 310 K (37°C), the equilibrium constant is 2880. The standard entropy change for this reaction can be calculated from the relationship $\Delta G^\circ = \Delta H^\circ - T\,\Delta S^\circ$.

How are standard free energy changes for reactions measured? The obvious answer is by measurement of the equilibrium constant and use of Eq. 2-36. However, in order to determine an equilibrium constant, a measurable amount of both reactants and products must be present. For reactions that essentially go to completion, a sequence of reactions must be found with measurable equilibrium constants that can be summed to total the reaction of interest. For example, the equilibrium constant for the hydrolysis of ATP to ADP and P_i calculated from the above standard free energy change is 5×10^5 M under "standard conditions." Obviously, this constant would be very difficult to measure directly. The following reactions, however, could be used to calculate this equilibrium constant (and the standard free energy change):

$$\begin{aligned} &G + ATP \rightleftharpoons G6P + ADP \\ &G6P + H_2O \rightleftharpoons G + P_i \\ \hline &ATP + H_2O \rightleftharpoons ADP + P_i \end{aligned}$$

The equilibrium constant for the first reaction is about 4600 and for the second reaction is about 110 M, both of which can be measured experimentally. Tables of standard free energies of formation (Appendix 4) and standard free energy changes for reactions (Appendix 5) can be constructed using this methodology.

Knowledge of the equilibrium constants of biological reactions and their temperature dependencies is of great importance for understanding metabolic regulation. It can also be of practical importance in the design of laboratory experiments, for example, in the development of coupled assays.

3.2 METABOLIC CYCLES

Thermodynamics is particularly useful for understanding metabolism. Metabolism consists of many different sets of reactions, each set, or metabolic cycle, designed to utilize and produce very specific molecules. The reactions within a cycle are coupled in that the product of one reaction becomes the reactant for the next reaction in the cycle. These coupled reactions can be very conveniently characterized using thermodynamic concepts. Before considering a specific metabolic cycle, let us consider some general thermodynamic properties of coupled reactions. As a very simple illustration, consider the coupled reactions

$$\mathrm{A} \rightarrow \mathrm{B} \rightarrow \mathrm{C} \rightarrow \mathrm{D}$$

The free energy changes for the first three reactions can be written as

$$\begin{aligned} \mathrm{A} \rightarrow \mathrm{B} \quad & \Delta G_{\mathrm{AB}} = \Delta G^{\circ}_{\mathrm{AB}} + RT \ln([\mathrm{B}]/[\mathrm{A}]) \\ \mathrm{B} \rightarrow \mathrm{C} \quad & \Delta G_{\mathrm{BC}} = \Delta G^{\circ}_{\mathrm{BC}} + RT \ln([\mathrm{C}]/[\mathrm{B}]) \\ \mathrm{C} \rightarrow \mathrm{D} \quad & \Delta G_{\mathrm{CD}} = \Delta G^{\circ}_{\mathrm{CD}} + RT \ln([\mathrm{D}]/[\mathrm{C}]) \end{aligned}$$

The free energy for the overall conversion of A to D can be obtained by adding these free energies:

$$\begin{aligned} \mathrm{A} \rightarrow \mathrm{D} \quad & \Delta G_{\mathrm{AD}} = \Delta G^{\circ}_{\mathrm{AB}} + \Delta G^{\circ}_{\mathrm{BC}} + \Delta G^{\circ}_{\mathrm{CD}} + RT \ln \{([\mathrm{B}][\mathrm{C}][\mathrm{D}]/([\mathrm{A}][\mathrm{B}][\mathrm{C}])\} \\ & \Delta G_{\mathrm{AD}} = \Delta G^{\circ}_{\mathrm{AD}} + RT \ln([\mathrm{D}]/[\mathrm{A}]) \end{aligned}$$

Whether A can be converted to D depends on the standard free energies for the three individual reactions and the concentrations of A and D. The concentrations of B and C are of no consequence! Note that since the total standard free energy determines whether A will be converted to D, it is possible for one of the standard free energy changes of the intermediate steps to be very unfavorable (positive) if it is balanced by a very favorable (negative) standard free energy change.

Metabolic pathways contain hundreds of different reactions, each catalyzed by a specific enzyme. Although the thermodynamic analysis shown above indicates that only the initial and final states need be considered, it is useful to analyze a metabolic pathway to see how the individual steps are coupled to each other through the associated free energy changes. The specific metabolic pathway that we will examine is anaerobic glycolysis. Anaerobic glycolysis is the sequence of reactions that metabolizes glucose into lactate and also produces ATP, the physiological energy currency. As we have seen, the standard free energy for the hydrolysis of ATP is quite large and negative so that the hydrolysis of ATP can be coupled to reactions with an unfavorable free energy change. The sequence of reactions involved in anaerobic glycolysis is shown in Table 3-1, along with the standard free energy changes. The standard states are as usual for biochemical reactions, namely, pH 7 and 1 M for reactants, with the concentration of each reactant as the sum

TABLE 3-1. Free Energy Changes for Anaerobic Glycolysis

Reaction	ΔG° (kJ/mol)	ΔG (kJ/mol)
Part One		
Glucose + ATP ⇌ Glucose-6-P + ADP	−16.7	−33.3
Glucose-6-P ⇌ Fructose-6-P	1.7	−2.7
Fructose-6-P + ATP ⇌ Fructose-1,6-bisphosphate + ADP	−14.2	−18.6
Fructose-1-6,-bisphosphate ⇌ Dihydroxyacetone-P + Glyceraldehyde-3-P	23.9	0.7
Dihydroxyacetone-P ⇌ Glyceraldehye-3-P	7.5	2.6
Glucose + 2 ATP ⇌ 2 ADP + 2 Glyceraldehyde-3-P	2.2	−51.3
Part Two		
Glyceraldehyde-3-P + P_i + NAD^+ ⇌ 1,3-Bisphosphoglycerate + NADH	6.3	−1.0
1,3-Bisphosphoglycerate + ADP ⇌ 3-P-Glycerate + ATP	−18.9	−0.6
3-Phosphoglycerate ⇌ 2-Phosphoglycerate	4.4	1.0
2-Phosphoglycerate ⇌ Phosphoenolpyruvate + H_2O	1.8	1.1
Phosphoenolpyruvate + ADP ⇌ Pyruvate + ATP	−31.7	−23.3
Pyruvate + NADH ⇌ Lactate + NAD^+	−25.2	1.9
Glyceraldehyde-3P + P_i + 2 ADP ⇌ Lactate + 2 ATP + H_2O	−63.3	−20.9

Source: Adapted from R. H. Garrett and C. M. Grisham, *Biochemistry*, Saunders College Publishing, Philadelphia, 1995, pp. 569–597.

of all ionized species. The activity of water is assumed to be unity. (This is somewhat different from Appendixes 4 and 5, where ionic strength and, in some cases, the magnesium ion concentration are specified.)

The overall reaction for anaerobic glycolysis is

$$\text{Glucose} + 2\ P_i + 2\ \text{ADP} \rightleftharpoons 2\ \text{Lactate} + 2\ \text{ATP} + 2\ H_2O \tag{3-1}$$

In anaerobic metabolism, the pyruvate produced in the second to the last step is converted to lactate in muscle during active exercise. In aerobic metabolism, the pyruvate that is produced in the second to the last step is transported to the mitochondria, where it is oxidized to carbon dioxide and water in the citric acid cycle. The reactions in Table 3-1 can be divided into two parts. The first part produces 2 moles of glyceraldehyde-3-phosphate according to the overall reaction

$$\text{Glucose} + 2\ \text{ATP} \rightleftharpoons 2\ \text{ADP} + 2\ \text{Glyceraldehyde-3-phosphate} \tag{3-2}$$

Note that this reaction actually requires 2 moles of ATP. However, the second part of the cycle produces 4 moles of ATP with the overall reaction

$$\text{Glyceraldehyde-3-phosphate} + P_i + 2\ \text{ADP} \rightleftharpoons \text{Lactate} + 2\ \text{ATP} + H_2O \tag{3-3}$$

Since 2 moles of glyceraldehyde-3-phosphate are produced in part one of the cycle, Eq. 3-3 must be multiplied by 2 and added to Eq. 3-2 to give the overall reaction, Eq. 3-1. The standard free energies, of course, also must be multiplied by 2 for the reactions in part two of glycolysis and added to those for the reactions in part one to give the overall standard free energy change, $-124.4\,\text{kJ/mol}$. If the standard free energies of formation in Appendix 4 are used to calculate the standard free energy change for Eq. 3-3, a value of $-128.6\,\text{kJ/mol}$ is obtained. This small difference can be attributed to somewhat different standard states. Unfortunately, not all of the necessary free energies of formation are available in Appendix 4 to calculate standard free energy changes for all of the reactions in Table 3-1.

The standard enthalpy change for Eq. 3-3 is $-63.0\,\text{kJ/mol}$, so that a substantial amount of heat is produced by glycolysis. The standard entropy change can be calculated from the known standard free energy and enthalpy changes and is 220 J/(mol K). Thus, both the standard enthalpy and entropy changes are favorable for the reaction to proceed. It is interesting to compare these numbers with those for the direct oxidation of glucose:

$$\text{Glucose(s)} + 6\ O_2(g) \rightleftharpoons 6\ CO_2(g) + 6\ H_2O(\ell) \tag{3-4}$$

The standard free energy change for this reaction is $-2878.4\,\text{kJ/mol}$, the standard enthalpy change is $-2801.6\,\text{kJ/mol}$, and the standard entropy change is -259 J/(mol K). This process produces a very large amount of heat relative to that produced by the metabolic cycle. This would not be very useful for physiological systems.

As we have stressed previously, for a single reaction, it is not the standard free energy change that must be considered in determining whether products are formed, it is the free energy for the particular concentrations of reactants that are present. In order to calculate the free energy changes for the reactions in Table 3-1, the concentrations of metabolites must be known. These concentrations have been determined in erythrocytes and are summarized in Table 3-2. The free energy changes calculated with these concentrations, the standard free energy changes, and Eq. 2-35 are included in Table 3-1. The additional assumptions have been made that these concentrations are valid at 298 K, although they were determined at 310 K, and $[\text{NADH}]/[\text{NAD}^+] = 1.0 \times 10^{-3}$. (We have elected to carry out calculations at 298 K, where the standard free energies are known, rather than at the physiological temperature of 310 K. This does not alter any of the conclusions reached.)

Consideration of free energy changes, rather than standard free energy changes, produces some interesting changes. The overall standard free energy change for the first part of glycolysis is +2.2 kJ/mol, whereas the free energy change is -51.3 kJ/mol. In contrast, the standard free energy change for the second part of glycolysis is much more negative than the free energy change. Thus, when considering the coupling of chemical reactions, considerable care must be exercised in making comparisons. Of course, as we stated at the beginning, the concentrations of the intermediates are of no consequence in determining the overall free energy changes. The reader might wish to confirm that this is indeed the case.

TABLE 3-2. Steady-State Concentrations for Glycolytic Intermediates in Erythrocytes

Metabolite	Concentration (mM)
Glucose	5.0
Glucose-6-phosphate	0.083
Fructose-6-phosphate	0.014
Fructose-1,6-bisphosphate	0.031
Dihydroxyacetone phosphate	0.14
Glyceraldehyde-3-phosphate	0.019
1,3-Bisphosphoglycerate	0.001
2,3-Bisphosphoglycerate	4.0
3-Phosphoglycerate	0.12
2-Phosphoglycerate	0.030
Phosphoenolpyruvate	0.023
Pyruvate	0.051
Lactate	2.9
ATP	1.85
ADP	0.14
P_i	1.0

Source: Adapted from S. Minakami and H. Yoshikawa, *Biochem. Biophys. Res. Commun.* **18**, 345 (1965).

Before we leave our discussion of glycolysis, it is worth addressing the individual reactions in the cycle. All of the reactions are catalyzed by enzymes. If this were not the case, the reactions would occur much too slowly to be physiologically relevant.

The first step is the very favorable phosphorylation of glucose—both the standard free energy change and the free energy change are favorable. The advantage to the cell of phosphorylating glucose is that creating a charged molecule prevents it from diffusing out of the cell. Furthermore, the intracellular concentration of glucose is lowered so that if the concentration of glucose is high on the outside of the cell, more glucose will diffuse into the cell. The second step, the isomerization of glucose-6-phosphate to fructose-6-phosphate, has a somewhat unfavorable standard free energy change, but the free energy change is favorable enough for the reaction to proceed. The next step has a very favorable standard free energy change, as well as a favorable free energy change, because the phosphorylation of fructose-6-phosphate is coupled to the hydrolysis of ATP. This very irreversible step is the commitment by the cell to metabolize glucose, rather than to store it. The next two steps produce glyceraldehyde-3-phosphate, the fuel for the second half of glycolysis that produces 4 moles of ATP. Both of these reactions have very unfavorable standard free energy changes, although the free energy changes are only slightly positive. This completes the first part of glycolysis. Note that 2 moles of ATP have been utilized to produce the final product, glyceraldehyde-3-phosphate. As noted previously, the standard free energy change for this first part is actually unfavorable, but the free energy change is quite favorable.

The purpose of the second part of glycolysis is to convert a substantial portion of the metabolic energy of glucose into ATP. In comparing the standard free energy changes with the free energy changes of the individual steps, it is worth noting that the second step, the formation of 3-phosphoglycerate, has a very favorable standard free energy change, yet the free energy change is approximately zero so that this reaction is approximately at equilibrium. This is true of essentially all of the reactions in this part of glycolysis, except for the reaction that produces pyruvate, where both the standard free energy change and the free energy change are quite negative. Both the overall standard free energy change and the free energy change for the second part of glycolysis are quite favorable.

This thermodynamic analysis of glycolysis is a good example of how thermodynamics can provide a framework for understanding the many coupled reactions occurring in biology. It also illustrates how metabolism is utilized to produce molecules such as ATP that can be used to drive other physiological reactions, rather than converting most of the free energy to heat.

3.3 DIRECT SYNTHESIS OF ATP

As we have seen, the standard free energy change for the hydrolysis of ATP to ADP and P_i is -32.5 kJ/mol, so it seems unlikely that ATP would be synthesized by the reverse of this reaction. The concentrations of reactants cannot be adjusted sufficiently to make the overall free energy favorable. Yet, we know that ATP is synthesized directly from ADP and P_i in mitochondria. For many years, people in this field grappled with how this might happen: Both probable and improbable mechanisms were proposed. In 1961, Peter Mitchell proposed that the synthesis of ATP occurred due to a coupling of the chemical reaction with a proton gradient across the membrane (1). This hypothesis was quickly verified by experiments, and he received a Nobel Prize for this work.

The enzyme responsible for ATP synthesis, ATP synthase, consists of a protein "ball," which carries out the catalytic function, coupled to membrane-bound proteins through which protons can be transported. The process of ATP synthesis is shown schematically in Figure 3-1. Although this enzyme is found only in mitochondria

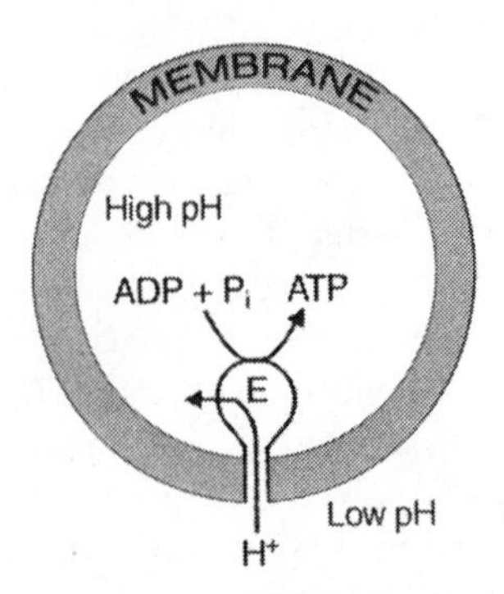

FIGURE 3-1. Schematic representation of ATP synthase, E, in mitochondria. The enzyme structure inside the mitochondria contains the catalytic sites. Protons are pumped from the outside of the membrane to the inside as ATP is synthesized.

in humans, it is quite ubiquitous in nature and is found in chloroplasts, bacteria, and yeast, among others. The chemiosmotic hypothesis states that a pH gradient is established across the membrane by a series of electron transfer reactions, and that ATP synthesis is accompanied by the simultaneous transport of protons across the membrane. The overall reaction can be written as the sum of two reactions:

$$
\begin{aligned}
&\mathrm{ADP} + \mathrm{P_i} \rightleftharpoons \mathrm{ATP} + \mathrm{H_2O} \\
&n\,\mathrm{H^+_{out}} \rightleftharpoons n\,\mathrm{H^+_{in}} \qquad (3\text{-}5) \\
\hline
&\mathrm{ADP} + \mathrm{P_i} + n\,\mathrm{H^+_{out}} \rightleftharpoons \mathrm{ATP} + n\,\mathrm{H^+_{in}} + \mathrm{H_2O}
\end{aligned}
$$

The value of n has been determined to be 3 (see Ref. 2). The free energy change for the transport of protons in Eq. 3-5 can be written as

$$\Delta G = 3RT\ln([\mathrm{H^+_{in}}]/[\mathrm{H^+_{out}}]) \qquad (3\text{-}6)$$

(The standard free energy change for this process is zero since the standard state for the hydrogen ion is the same on both sides of the membrane.)

At 298 K, a pH differential of one unit gives a free energy change of -17.1 kJ/mol. The actual physiological situation is even more favorable as a membrane potential exists whereby the membrane is more negative on the inside relative to the outside. If we utilize the extended chemical potential, Eq. 2-49, an additional term is added to Eq. 3-6 equal to 3 $F\Psi$, where 3 is the number of protons transported, F is the Faraday, and Ψ is the membrane potential. For a membrane potential of -100 mV, -29 kJ would be added to the free energy change in Eq. 3-6.

The standard free energy for the synthesis of ATP from ADP and $\mathrm{P_i}$ is $+32.5$ kJ/mol, but we need to know the free energy change under physiological conditions. Although the concentrations of the reactants are not known exactly, we can estimate that the ratio of ATP to ADP is about 100, and the concentration of phosphate is 1 – 10 mM. This makes the ratio $[\mathrm{ATP}]/([\mathrm{ADP}][\mathrm{P_i}])$ equal to 100 – 1000. The free energy change at 298 K is

$$\Delta G = 32.5 + RT\ln(100 - 1000) = 32.5 + (11.4 - 17.1) = 43.9 - 49.6\,\mathrm{kJ/mol}$$

Thus, the coupling of the synthesis of ATP to a modest proton gradient and membrane potential can readily provide the necessary free energy for the overall reaction to occur.

The principle to be learned from this example is that the coupling of free energies is very general. It can involve chemical reactions only, as in glycolysis, or it can involve other processes such as ion transport across membranes, as in this example.

3.4 ESTABLISHMENT OF MEMBRANE ION GRADIENTS BY CHEMICAL REACTIONS

In discussing ATP synthesis, we have not specified how the electrochemical gradient is established. This involves a complex sequence of coupled reactions that we shall not discuss here. Instead, we will discuss a case where the free energy associated with the hydrolysis of ATP is used to establish an ion gradient. The process of signal transduction in the nervous system involves the transport of Na^+ and K^+ across the membrane. Neuronal cells accumulate K^+ and have a deficit of Na^+ relative to the external environment. (This is also true for other mammalian cells.) When an electrical signal is transmitted, this imbalance is altered. The imbalance causes a resting membrane potential of about -70 mV. This situation is illustrated in Figure 3-2, with some typical ion concentrations.

How is the ion gradient established? This is done by a specific ATPase, the Na^+/K^+ ATPase, that simultaneously pumps ions and hydrolyzes ATP. The process can be written as the sum of two reactions:

$$\begin{array}{c} 2\ K^+_{out} + 3\ Na^+_{in} \rightleftharpoons 2\ K^+_{in} + 3\ Na^+_{out} \\ ATP + H_2O \rightleftharpoons ADP + P_i \\ \hline 2\ K^+_{out} + 3\ Na^+_{in} + ATP + H_2O \rightleftharpoons 2\ K^+_{in} + 3\ Na^+_{out} + ADP + P_i \end{array} \tag{3-7}$$

The stoichiometry has been established by experimental measurements. The free energy change for the first step is

$$\Delta G = RT \ln\{([Na^+_{out}]^3[K^+_{in}]^2)/([Na^+_{in}]^3[K^+_{out}]^2)\} + 3\ F\Psi - 2\ F\Psi$$

With $\Psi = 70$ mV, $T = 298$ K, and the concentrations of Na^+ and K^+ in Figure 3-2, $\Delta G = 41.3$ kJ/mol. Note that the membrane potential produces a favorable (negative) free energy change for the transport of K^+ and an unfavorable free energy change (positive) for the transport of Na^+. We have previously calculated that ΔG for ATP hydrolysis is $-(43.9 - 49.6)$ kJ/mol. Therefore, the hydrolysis of ATP is sufficient to establish the ion gradient and accompanying membrane potential. This process is called *active transport*.

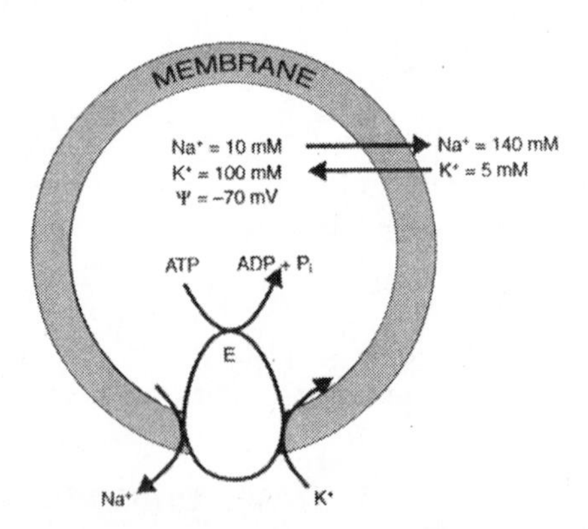

FIGURE 3-2. Schematic representation of the Na^+ and K^+ gradients in a cell. The free energy necessary for the creation of these gradients is generated by the enzymatic hydrolysis of ATP. The Na^+/K^+ ATPase is designated as E, and typical concentrations of the ions and membrane potential, Ψ, are given.

The coupling of free energies in biological systems can be used to understand many of the events occurring, and many other interesting examples exist. However, we will now turn to consideration of protein and nucleic acid structures in terms of thermodynamics.

3.5 PROTEIN STRUCTURE

An extensive discussion of protein structure is beyond the scope of this treatise, but various aspects of protein structure are considered throughout this text. Many excellent discussions of protein structure are available (cf. Refs. 3–5). A few aspects of protein structure are now discussed in terms of thermodynamic considerations. Proteins, of course, are polymers of amino acids, and the amino acids contain both polar and nonpolar groups. The structures of the 20 common amino acids are given in Appendix 2, and the covalent structure of a polypeptide chain is shown in Figure 3-3. We will first consider the role of nonpolar groups in proteins, that is, amino acid side chains containing methylene and methyl groups and aromatic residues.

As a starting point, let us consider the thermodynamics of transferring the hydrocarbons methane and ethane from an organic solvent to water. The thermodynamic parameters can be measured for this process simply by determining the solubility of the hydrocarbons in each of the solvents. Since the pure hydrocarbon is in equilibrium with each of the saturated solutions, the two saturated solutions must be in equilibrium with each other. The transfer reaction can be written as the sum of the solubility equilibria:

$$\begin{array}{c} \text{Hydrocarbon (organic solvent)} \rightleftharpoons \text{Hydrocarbon} \\ \text{Hydrocarbon} \rightleftharpoons \text{Hydrocarbon (water)} \\ \hline \text{Hydrocarbon (organic solvent)} \rightleftharpoons \text{Hydrocarbon (water)} \end{array} \tag{3-8}$$

The transfer free energy is

$$\Delta G_t = RT \ln([X_O]/[X_W])$$

where X_O and X_W are the mole fraction solubilities in the organic solvent and water, respectively. (The use of mole fractions is the appropriate concentration scale in this instance because of standard state considerations that we will not deal with here.) Measurement of the temperature dependence of the transfer free energy will permit

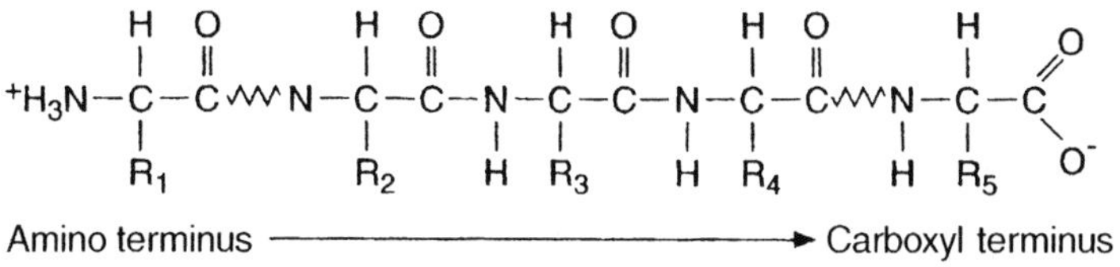

$^{+}H_3N-C(H)(R_1)-C(=O)\sim N-C(H)(R_2)-C(=O)-N(H)-C(H)(R_3)-C(=O)-N(H)-C(H)(R_4)-C(=O)\sim N(H)-C(H)(R_5)-C(=O)O^-$

Amino terminus ⟶ Carboxyl terminus

FIGURE 3-3. A polypeptide chain. The amino acid side chains are represented by R_i, and the amino and carboxyl termini are shown.

TABLE 3-3. Thermodynamic Parameters for Hydrocarbon Transfer from Organic Solvents to Water at 298 K

Transfer reaction	ΔS_t [cal/(mol K)]	ΔH_t (kcal/mol)	ΔG_t (kcal/mol)
CH_4, benzene to water	−18	−2.8	+2.6
CH_4, ether to water	−19	−2.4	+3.3
CH_4, CCl_4 to water	−18	−2.5	+2.9
C_2H_6, benzene to water	−20	−2.2	+3.8
C_2H_6, CCl_4 to water	−18	−1.7	+3.7

Source: Adapted from W. Kauzmann, *Adv. Protein Chem.* **14**, 1 (1959).

ΔH and ΔS to be determined (Eqs. 2-42 and 2-43). The thermodynamic parameters obtained at 298 K in several organic solvents are summarized in Table 3-3.

As expected, the free energy change is unfavorable—hydrocarbons do not dissolve readily in water. However, the energy change is favorable so that the negative entropy change is responsible for the unfavorable free energy. What is the reason for the negative entropy change? It is due to the fact that the normal water structure is broken by the insertion of the hydrocarbon, and the water molecules tend to form a hydrogen-bonded structure of water. Thus, it is the formation of ice-like ordered structures around the hydrocarbon that is the source of the negative entropy change.

To make this more relevant to proteins, the free energy of transfer for amino acids from a hydrocarbon-like environment (ethanol) to water has been measured. This is done by determining the solubility of the amino acids in ethanol and water. Similar to before, in both cases the amino acid in solution is in equilibrium with the solid, so that the transfer free energy can be measured:

$$\begin{array}{c}\text{Amino acid (ethanol)} \rightleftharpoons \text{Amino acid (solid)} \\ \text{Amino acid (solid} \rightleftharpoons \text{Amino acid (water)} \\ \hline \text{Amino acid (ethanol)} \rightleftharpoons \text{Amino acid (water)}\end{array} \qquad (3\text{-}9)$$

Some of the results obtained are summarized in Table 3-4. In order to interpret these results, we must remember that in ethanol the amino acid is uncharged

TABLE 3-4. Free Energy Changes for Transferring Amino Acids from Ethanol to Water at 298 K

Compound	ΔG_t (kcal/mol)	$\Delta G_{t,\text{side chain}}$ (kcal/mol)
Glycine	−4.63	—
Alanine	−3.90	+0.73
Valine	−2.94	+1.69
Leucine	−2.21	+2.42
Isoleucine	−1.69	+2.97
Phenylalanine	−1.98	+2.60
Tyrosine	−1.78	+2.85

Source: Adapted from C. Tanford, *J. Am. Chem. Soc.* **84**, 4240 (1962).

(NH_2—RCH—COOH), whereas in water it is a zwitterion (NH_3^+—RCH—COO^-). If glycine, which has no side chains, is taken as the standard, then subtracting its transfer free energy from the transfer free energies of the other amino acids will give the transfer free energy for the amino acid side chains. The overall transfer free energy for all of the amino acids is negative (favorable) because the charged amino and carboxyl groups are solvated by water, a highly favorable interaction. However, the standard free energy changes of transfer for the hydrophobic side chains are all positive, as was seen for the transfer of methane and ethane from an organic solvent into water.

What is the relevance of such data for protein structure? These data indicate that the apolar side chains of amino acid side chains would prefer to be in a nonaqueous environment. That is, they prefer to cluster together. This is, in fact, true for most proteins. The hydrophobic groups aggregate together on the interior of the protein, forming a hydrophobic core, with the more polar groups tending to be on the outside interacting with water. This important concept was enunciated by Walter Kauzmann in 1959 (6). Although the interactions associated with forming the hydrophobic core are often called hydrophobic bonds, it is important to remember that the driving force for forming a hydrophobic core is not the direct interactions between hydrophobic groups; instead, it is the release of water molecules from the ice-like structures that surround hydrophobic groups in water. The thermodynamic analysis clearly indicates that the formation of "hydrophobic bonds" is an entropy-driven process. In fact, ΔH_t for the transfer of methane and ethane from water to organic solvent is positive, that is, energetically unfavorable. This means that an increase in temperature will tend to strengthen the hydrophobic bonding—if ΔH remains positive over the temperature range under consideration.

The formation of hydrogen bonds in protein structures is prevalent, and we will now consider the thermodynamics of hydrogen bond formation. Thermodynamic studies have been made of many different types of hydrogen bonds. A few selected examples are summarized in Table 3-5. The first is the formation of a hydrogen-bonded dimer of acetic acid molecules in the gas phase:

$$2CH_3—C(=O)—O—H\,(g) \rightleftharpoons CH_3—C(=O\cdots H—O)(—O—H\cdots O=)C—CH_3\,(g) \quad (3\text{-}10)$$

TABLE 3-5. Thermodynamic Parameters for Hydrogen Bond Formation[a]

Reactant	Solvent	ΔG°_{298} (kcal/mol)	ΔH°_{298} (kcal/mol)	ΔS°_{298} [cal/(mol K)]
CH_3COOH[b]	Gas	−8.02	−15.9	−26.6
CH_3CONH_2[c]	CCl_4	−0.92	−4.2	−11
	Dioxane	0.39	−0.8	−4
	H_2O	3.1	0.0	−10

[a]Standard state is 1 M.
[b]See J. O. Halford, *J. Chem. Phys.* **9**, 859 (1941).
[c]See I. M. Klotz and J. S. Franzen, *J. Am. Chem. Soc.* **84**, 3461 (1962).

As expected, both ΔH° and ΔS° are negative. The former represents the energy produced in forming two hydrogen bonds, and the latter is due to two molecules forming a dimer (a more ordered system). As two hydrogen bonds are formed, the enthalpy change associated with a single hydrogen bond is about -7 kcal/mol.

The second example is the dimerization of *N*-methylacetamide, a good model for hydrogen bonding involving the peptide bond:

$$2CH_3{-}\overset{H}{N}{-}\overset{O}{\overset{\|}{C}}{-}CH_3 \rightleftharpoons \begin{array}{c} CH_3 \\ | \\ C{=}O\cdots\cdots HN \\ | \\ HN \\ | \\ CH_3 \end{array} \begin{array}{c} CH_3 \\ | \\ \\ | \\ C{=}O \\ | \\ CH_3 \end{array} \qquad (3\text{-}11)$$

This dimerization was studied in a variety of solvents, and the results in carbon tetrachloride, dioxane, and water are included in Table 3-5. In carbon tetrachloride, the solvent does not compete for the hydrogen bonds of *N*-methylacetamide, and the results are not very different from those for acetic acid dimerization in the gas phase, namely, ΔH° for formation of a single hydrogen bond is about -4 kcal/mol and ΔS° is about -11 cal/(mol K). However, in water, which can also form hydrogen bonds with *N*-methylacetamide and is present at a concentration of about 55 M, ΔH° is approximately zero. This indicates that there is not a significant enthalpy difference between the water–*N*-methylacetamide hydrogen bond and the *N*-methylacetamide–*N*-methylacetamide hydrogen bond. Note that the standard free energy change in water is 3 kcal/mol, so that the amount of dimer formed is very small even at 1 M *N*-methylacetamide. Dioxane has two oxygens that can accept a hydrogen bond so that ΔH° is only slightly negative.

The conclusion reached from these data (and considerably more data not presented) is that the stability of a water–protein hydrogen bond is similar to that of intramolecular protein hydrogen bonds. Therefore, a single intramolecular hydrogen bond on the surface of the protein is unlikely to be a strong stabilizing factor. On the other hand, if the hydrophobic interior of a protein excludes water, a strong hydrogen bond might exist in the interior of a protein. This statement is complicated by the fact that proteins have "breathing" motions; that is, the protein structure continually opens and closes with very small motions so that completely excluding water from the interior may be difficult.

The role of the hydrogen bond in stabilizing proteins is still a matter of some debate. The statement is often made that hydrophobic interactions are the primary source of protein stability and hydrogen bonding provides specificity (6). There is no doubt that extended hydrogen-bonded systems are extremely important structural elements in proteins. Two examples are given in Figures 3-4 and 3-5 (see colored plates), the α-helix and the β-pleated sheet. The α-helix is a spiral structure with 3.6 amino acids per turn of the helix. Every peptide carbonyl is a hydrogen bond acceptor for the peptide N–H four residues away. This structure is found in many different proteins. The β-pleated sheet is also a prevalent structure in proteins. In the β-pleated sheet, each chain is a "pleated" helix. Again, all of the peptide bonds participate in hydrogen bonding, but the bonds are all between chains, rather than intrachain.

Finally, we will say a few words about the role of electrostatic interactions in protein structures. The discussion will be confined to charge–charge interactions, even though more subtle interactions involving, for example, dipoles and/or induced dipoles are important. Many of the amino acids have side chains with ionizable groups, so that a protein contains many acids and bases. For example, a carboxyl group can ionize according to the scheme

$$\text{P-COOH} \rightleftharpoons \text{P-COO}^- + \text{H}^+ \tag{3-12}$$

Here P designates the protein. Although not shown, water plays an extremely important role in this equilibrium. Water is a strong dipole and strongly solvates ions, forming a hydration "sheath." Acetic acid is a reasonable model for this reaction: The thermodynamic parameters characterizing its ionization are $\Delta G^\circ = 6.6\,\text{kcal/mol}$, $\Delta H^\circ = 0$, and $\Delta S^\circ = -22\,\text{cal/(mol K)}$. The negative entropy change is due to the ordering of water molecules around the ions. Note that the ionization process is thermally neutral. The enthalpy changes associated with the solvation of ions are generally negative but, in this case, are balanced by the enthalpic change associated with breaking the oxygen–hydrogen bond.

The ionization constants for ionizable groups on proteins are generally not the same as those of simple model compounds. This is because the ionization process is influenced by the charge of the protein created by other ionizable groups and, in some cases, by special structural features of the protein. This factor can be included explicitly in a thermodynamic analysis of protein ionizations by writing the free energy associated with ionization as the sum of the free energy for the model compound, or the intrinsic free energy, and the free energy of interaction:

$$\Delta G_{\text{ionization}} = \Delta G^\circ_{\text{intrinsic}} + \Delta G_{\text{interaction}} \tag{3-13}$$

The ionization properties of proteins have been studied extensively, both experimentally and theoretically. For our purposes, it is important to recognize that the strong solvation of ions means that charged groups will tend to be on the outside of the protein, readily accessible to water.

A reasonable question to ask is: If there are so many charged groups on proteins, would they not influence the structure simply because charged groups of opposite sign attract, and those with the same sign repel? This is certainly the case—at very high pH, a protein becomes very negatively charged and the interactions between negative charges can eventually cause the native structure to disappear. Similarly, at very low pH, the interactions between positive charges can cause disruption of the native structure. Conversely, the formation of a salt linkage between groups with opposite charges can stabilize structures. We can make a very simple thermodynamic analysis of charge–charge interactions by recognizing that the free energy of formation of an ion pair is simply the work necessary to bring the ions to within a specified distance, a:

$$\Delta G = z_1 z_2 e^2/Da \tag{3-14}$$

Here the z_i are the ionic valences, e is the charge on an electron, and D is the dielectric constant. The assumption of a simple coulombic potential is a gross oversimplification but suffices for our purposes. A more complete potential would include the ionic environment of the medium, the structure of the ions, and the microscopic structure of the solvent. If the valences are assumed to be 1, the distance of closest approach 4 Å, and the dielectric constant of water 80, ΔG is about 1 kcal/mol. Thus, the interaction energy is not very large in water. However, the interior of a protein is more like an organic solvent, and organic solvents have dielectric constants of about 3, which would significantly increase the free energy of interaction. In fact, salt linkages are rarely found near the surface of proteins but are found in the interior of proteins, as expected.

If the free energy of ionic interactions is given by Eq. 3-14, the enthalpy and entropy can easily be calculated from Eqs. 2-42 and 2-43. Since only the dielectric constant in Eq. 3-14 is temperature dependent, this gives

$$\Delta H = -T^2[\mathrm{d}(\Delta G/T)/\mathrm{d}T] = (\Delta G/D)[\mathrm{d}(DT)/\mathrm{d}T] \tag{3-15}$$

$$\Delta S = -\mathrm{d}\Delta G/\mathrm{d}T = (\Delta G/D)(\mathrm{d}D/\mathrm{d}T) \tag{3-16}$$

In water, both of these derivatives are negative so that if ΔG is negative, both the entropy and enthalpy changes are positive. The positive entropy change can be rationalized as being due to the release of water of hydration of the ions when the ion pair is formed. The enthalpy change is a balance between the negative enthalpy change from bringing the charges closer and the positive enthalpy change associated with removing the hydration shell. Since the enthalpy change is positive, the strength of the ion pair interaction will increase as the temperature is increased—exactly as for hydrophobic interactions.

Although this is an abbreviated discussion, it is clear that we know a great deal about the thermodynamics of interactions that occur in proteins. Because of this, you might think that we could examine the amino acid sequence of any protein and predict its structure by looking at the possible interactions that occur and finding the structure that has the minimum free energy. This has been a long-standing goal of protein chemistry, but we are not yet able to predict protein structures. Why is this the case? The difficulty is that there are thousands of possible hydrogen bonding interactions, ionic interactions, hydrophobic interactions, etc. As we have seen, each of the individual interactions is associated with small free energy and enthalpy changes—in some cases, we cannot even determine the sign. The sum of the positive free energies of interactions is very large, as is the sum of the negative free energies. It is the difference between the positive and negative free energies that determines the structure. So we have two problems: accurately assessing the free energies of individual interactions and then taking the difference between two large numbers to determine which potential structure has the minimum free energy. These are formidable problems, but significant progress has been made toward achieving the ultimate goal of predicting protein structures. The possibility also exists that the structure having the lowest free energy is not the biologically relevant structure. This may be true in a few cases but is unlikely to be a problem for most proteins.

3.6 PROTEIN FOLDING

The folding of proteins into their biologically active structures has obvious physiological importance. In addition, understanding the process in molecular detail is linked directly to understanding protein structure. The study of protein folding, and the reverse process of unfolding, is a major field of research, and we will only explore a few facets of this fascinating subject. For our discussion, we will concentrate on protein unfolding as this is most easily experimentally accessible. There are many ways of unfolding proteins. When the temperature increases, proteins will eventually unfold. From a thermodynamic standpoint, this is because the $T\Delta S$ term eventually dominates in determining the free energy change, and we know that the unfolded state is more disordered than the native state at sufficiently high temperatures.

Chemical denaturants such as acid, base, urea, and guanidine chloride are also often used to unfold proteins. The role of a neutrally charged denaturant such as urea can be understood in thermodynamic terms by considering the free energy of transfer of amino acids from water to urea, again using glycine as a reference. Some representative free energies of transfer for hydrophobic side chains are given in Table 3-6. Note that the free energies are all negative, so that removing hydrophobic side chains from the interior of the protein into 8 M urea is a favorable process.

Protein unfolding can be monitored by many different methods. Probably, the most common is circular dichroism in the ultraviolet, which is quite different for native and unfolded structures. Many other spectral and physical methods work equally well, (Spectroscopic methods are discussed later in this text.) A representative plot of the fraction of denatured protein, f_D, versus temperature is shown in

TABLE 3-6. Transfer Free Energies for Selected Amino Acids from Water to 8 M Urea at 298 K

Amino acid	ΔG_t (kcal/mol)	$\Delta G_{t,\text{side chain}}$ (kcal/mol)
Glycine	+0.10	0.0
Alanine	+0.03	−0.07
Leucine	−0.28	−0.38
Phenylalanine	−0.60	−0.70
Tyrosine	−0.63	−0.73

Source: Adapted from P. L. Whitney and C. Tanford, *J. Biol. Chem.* **237**, 1735 (1962).

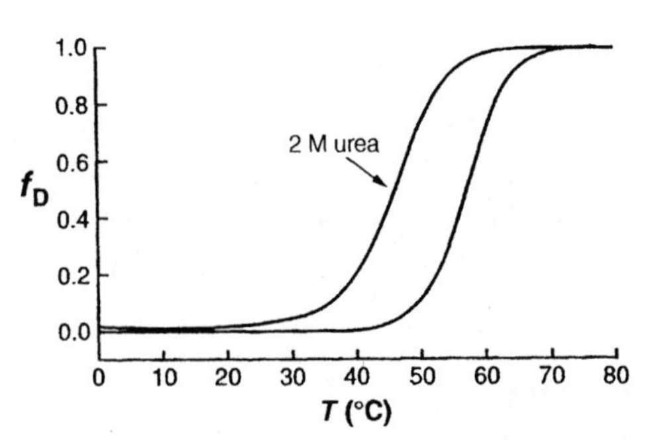

FIGURE 3-6. Fraction of denatured protein, f_D, for λ_{6-85} phage repressor protein as a function of temperature, *T*, in aqueous solution and in 2 M urea. Data from G. S. Huang and T. G. Oas, *Biochemistry* **35**, 6173 (1996).

Figure 3-6 for the N-terminal region (amino acid residues 6–85) of λ repressor, a protein from λ phage that binds to DNA and regulates transcription. Results are shown for both no urea and 2 M urea. As expected, the protein is less stable in the presence of 2 M urea. The fraction of protein denatured can be written as

$$f_D = [D]/([D] + [N]) = K/(1 + K) \tag{3-17}$$

where [D] is the concentration of denatured species, [N] is the concentration of native species, and $K(= [D]/[N])$ is the equilibrium constant for denaturation. This assumes that the unfolding process can be characterized by only two states—native and denatured. For some proteins, intermediates are formed as the protein unfolds, and a more complicated analysis must be used.

Since the equilibrium constant can be calculated at any point on the curve, and its temperature dependence can be measured, the thermodynamic parameters characterizing the unfolding can be determined. The thermodynamics of unfolding can also conveniently be studied by scanning calorimetry. It is often found that a plot of the logarithm of the equilibrium constant versus $1/T$ is not linear, as predicted, namely,

$$d \ln K/d(1/T) = -\Delta H/R \tag{3-18}$$

This means that the enthalpy change is temperature dependent, or in other words, there is a large heat capacity difference, ΔC_P, between the native and unfolded states.

$$d\Delta H/dT = \Delta C_P \tag{3-19}$$

A typical experimental result for the variation of the equilibrium constant with temperature for λ repressor is given in Figure 3-7a. The plot of ln *K* versus $1/T$ goes through a minimum: At higher temperatures, ΔH is positive, whereas at lower temperatures, it is negative. For typical chemical reactions, this plot is a straight line! In Figure 3-7b, the standard free energy change is plotted versus the temperature. The protein is most stable at the maximum in the free energy curve, about 15°C. In

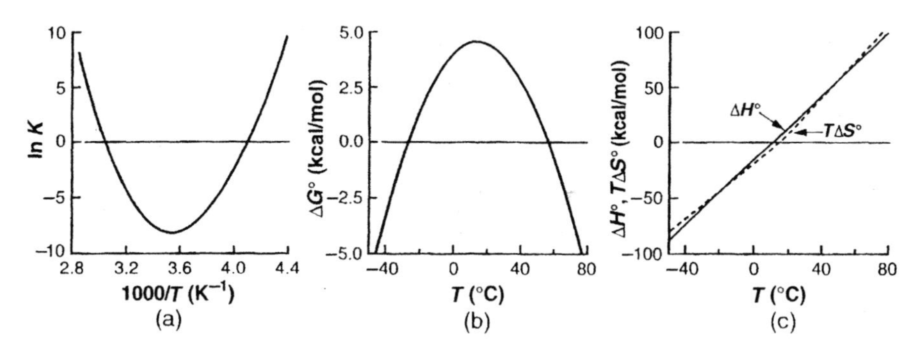

FIGURE 3-7. (a) Plot of the natural logarithm of the equilibrium constant, K, for the denaturation of λ_{6-85} phage repressor protein in aqueous solution versus the reciprocal temperature on the Kelvin scale, $1/T$. (b) Plot of ΔG° for the denaturation of λ_{6-85} phage repressor protein versus the temperature, T. (c) Plots of ΔH° and $T\,\Delta S^\circ$ versus the temperature, T, for the denaturation of λ_{6-85} phage repressor protein. Data from G. S. Huang and T. G. Oas, *Biochemistry* **35**, 6173 (1996).

Figure 3-7c, ΔH° and $T\,\Delta S^\circ$ are plotted versus T. Note that both ΔH° and ΔS° are zero at a specific temperature. The practical implication of thè positive value of ΔC_P is that the protein unfolds at both high temperatures and low temperatures. This unexpected result is not confined to λ repressor. Most proteins will unfold (denature) at low temperatures, but in many cases the temperatures where cold denaturation is predicted to occur are below 0°C.

Some thermodynamic parameters characterizing protein denaturation are given in Table 3-7 for a few proteins. In all cases, a large positive value of ΔC_P is observed. Note that at room temperature and above, ΔH° is typically large and positive, as is ΔS°, so that unfolding is an entropically driven process. Before we leave our discussion of protein folding/unfolding, one more important point should be mentioned. The transition from folded to unfolded states usually occurs over a fairly small change in temperature or denaturant concentration. This is because folding/unfolding is a highly *cooperative* process—once it starts, it proceeds with very small changes in temperature or denaturant. Cooperative processes are quite

TABLE 3-7. Representative Thermodynamic Parameters for Thermal Protein Denaturation in Aqueous Solution at 298 K

Protein	ΔG° (kJ/mol)	ΔH° (kJ/mol)	ΔS° [J/(mol K)]	ΔC_P [kJ/(mol K)]
Barnase	48.9	307	866	6.9
Chymotrypsin	45.7	268	746	14.1
Cytochrome *c*	37.1	89	174	6.8
Lysozyme	57.8	242	618	9.1
Ribonuclease A	27.0	294	896	5.2
λ Repressor$_{6-85}$[a]	17.7	90.4	244	6.0

Source: Adapted from G. I. Makhatadze and P. L. Privalov, *Adv. Protein Chem.* **47**, 307 (1995).
[a]See G. S. Huang and T. G. Oas, *Biochemistry* **35**, 6173 (1996).

prevalent in biological systems, particularly when regulation of the process is desired; cooperative processes will be discussed again in Chapter 13.

3.7 NUCLEIC ACID STRUCTURES

We will now briefly consider the structure of nucleic acids in terms of thermodynamics. Again, only a few examples will be considered. Many more complete discussions are available (3,4,7). Let us start with the well-known structure of DNA. Most DNAs consist of two chains that form a double helix. Each chain is a polymer of nucleosides linked by phosphodiester bonds as shown in Figure 3-8. Each nucleoside contains a 2′-deoxyribose sugar and a base, almost always adenine (A), thymine (T), cytosine (C), or guanine (G). The phosphodiester linkage is through the 5′ and 3′ positions on the sugars. Structures of these components are given in Appendix 3. By convention, a DNA chain is usually written so that the

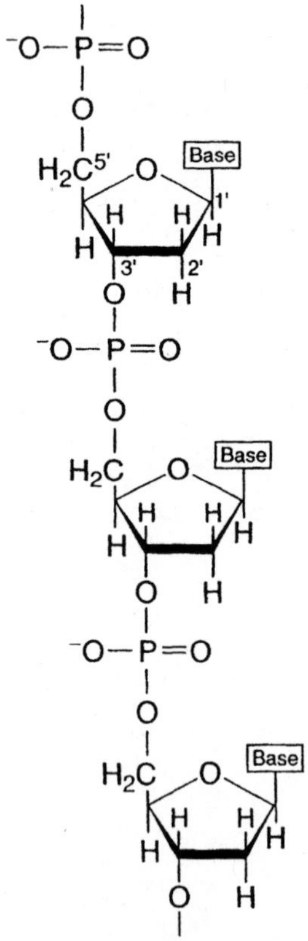

FIGURE 3-8. Structural formula of part of a DNA/RNA chain. In DNA, the bases are usually A, T, C, and G. In RNA, the T is replaced by U, and the 2′-H below the plane of the sugar is replaced by OH.

5′ end of the molecule is on the left and the 3′ on the right. The B form of the DNA double helix is shown in Figure 3-9 (see color plates). The chains are arranged in an antiparallel fashion and form a right-handed helix. The bases are paired through hydrogen bonds on the inside of the double helix, and as might be expected the negatively charged phosphodiester is on the outside. Note the similarity to protein structure as the more hydrophobic groups tend to be on the inside and the hydrated polar groups on the outside. However, DNA is not globular, unlike most proteins, and forms a rod-like structure in isolation. These rods can bend and twist to form very compact structures in cells.

Let us first examine the hydrogen-bonded pairs in DNA. Early in the history of DNA, Erwin Chargaff noted that the fraction of bases that were A was approximately equal to the fraction that were T in many different DNAs. Similarly, the fraction of bases that were G was equal to the fraction that were C. This finding was important in the postulation of the double helix structure by James Watson and Francis Crick. As shown in Figure 3-10, these bases form hydrogen-bonded pairs, with the A–T pair forming two hydrogen bonds and the G–C pair forming three hydrogen bonds. These are not the only possible hydrogen-bonded pairs that can be formed between the four bases. Table 3-8 gives the thermodynamic parameters associated with the formation of various hydrogen-bonded pairs in deuterochloroform. Instead of thymine, uracil has been used—this is the base normally found in ribonucleic acids and differs only by a methyl group in the 5 position from

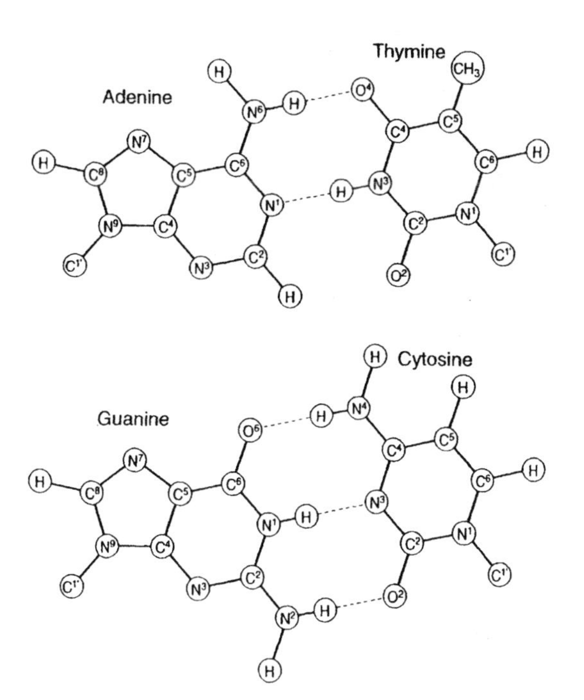

FIGURE 3-10. Formation of the hydrogen-bonded base pairs, A–T and G–C, in DNA. In RNA, U is substituted for T. All of the base rings are aromatic.

TABLE 3-8. Thermodynamic Parameters for Base Pairs in Deuterochloroform at 298 K

Base pair	$\Delta G°$ (kcal/mol)	$\Delta H°$ (kcal/mol)[a]	$\Delta S°$ [cal/(mol K)]
A–A	−0.67	−4.0	−11.4
U–U	−1.07	−4.3	−11.0
A–U	−2.72	−6.2	−11.8
C–C	−1.97	−6.3	−15
G–G	−4.1 to −5.4	(−8.5 to −10)	(−15)
G–C	−5.4 to −6.8	(−10.0 to −11.5)	(−15)

Source: Adapted from C. R. Cantor and P. R. Schimmel, *Biophysical Chemistry*, Freeman, New York, 1980, p. 325. *Data source:* Y. Kyogoku, R. C. Lord, and A. Rich, *Biochim. Biophys. Acta* **179**, 10 (1969).
[a]The $\Delta H°$ values in parentheses have been calculated assuming $\Delta S° = -15$ cal/(mol K).

thymine. It is clear that the A–U pair is favored over the A–A and U–U pairs, and that G–C is favored over G–G and C–C although the data are not very precise. There is no simple explanation for this strong preference—it must be due to more than hydrogen bonding, perhaps an electronic effect within those specific pairs. Furthermore, the preference for A–T in the former three pairs is due to a more favorable enthalpy change. On the basis of this observation, the entropy change for the latter three pairs has been assumed to be the same in order that $\Delta H°$ can be calculated.

In water solution, the hydrogen bonding between bases is considerably weaker than in organic solvents because of the competition for hydrogen bond formation with water. The assumption is that the base pair hydrogen bonds are sufficiently shielded from water so that the structure is very stable. However, we know that this is not entirely correct as some "breathing" motions occur, so some exposure to water must also occur. Factors other than hydrogen bonding must contribute to the stability of the double helix.

A consideration comes into play in nucleic acid structures that is not a major concern in proteins, namely, the interactions between the aromatic rings of the bases. The planes of the bases lie over one another so that the π electrons interact; that is, the rings are attracted to each other. This "stacking" reaction can be studied in model systems by measuring the equilibrium constant for the interaction of nucleosides in water, where the hydrogen bonding between nucleosides is negligible. Some representative equilibrium constants for dimer formation are given in Table 3-9. The interaction is quite weak, although it appears that purine–purine interactions are stronger than purine–pyrimidine interactions, which in turn are stronger than pyrimidine–pyrimidine interactions. This is the order expected for π electron interactions. The enthalpy change associated with these interactions is somewhat uncertain but is definitely negative. A value of −3.4 kcal/mol has been obtained for formation of an A–A stack (8).

Is the "stacking" interaction entirely due to π electron interactions, or is the solvent involved, as for hydrophobic interactions in proteins? The "stacking" interaction has been found to be solvent dependent, with water being the most favorable solvent. This suggests that hydrophobic interactions may be involved. However,

TABLE 3-9. Association Constants and Standard Free Energy Changes for Base Stacking in H_2O at 298 K

Base stack	K (molal^{-1})	ΔG° (kcal/mol)
A–A	12	−1.50
T–T	0.91	+0.06
C–C	0.91	+0.06
T–C	0.91	+0.06
A–T	3 to 6	−0.70 to −1.10
A–C	3 to 6	−0.70 to −1.10
G–C	4 to 8	−0.80 to −1.20

Source: Adapted from T. N. Solie and J. A. Schellman, *J. Mol. Biol.* 33, 6 (1968).

you should recall that hydrophobic interactions are endothermic and entropically driven by the release of ordered water molecules around the noninteracting hydrophobic groups. The most likely possibility is that both π electron interactions and hydrophobic effects play a role in base stacking.

3.8 DNA MELTING

One of the reasons for wanting to understand the interactions in the DNA molecule is to understand the stability of DNA since it must come apart and go together as cells reproduce. One way to assess the stability of DNA is to determine its thermal stability. This can readily be studied experimentally because the ultraviolet spectra of stacked bases differ significantly from those of unstacked bases. Thus, if the temperature is raised, the spectrum of DNA changes as it "melts." This is shown schematically in Figure 3-11. Because of the relative stability of the hydrogen bonding, the A–T regions melt at a lower temperature than the G–C regions. The melting of AT and GC polymers can be measured, and as shown in Figure 3-11, the AT structure melts at a lower temperature than a typical DNA, and the GC structure at a higher temperature. The temperature at which half of the DNA structure has disappeared is the melting temperature, T_m. If the transition from fully helical DNA to separated fragments is assumed to be a two-state system, then thermodynamic parameters can be calculated, exactly as for protein unfolding. Thus, the equilibrium constant is 1 ($\Delta G^{\circ} = 0$) at T_m. Determination of the temperature dependence of the equilibrium constant permits calculation of the standard enthalpy and free energy changes for the "melting" process.

The real situation is more complex than shown, as DNA may not melt in a well-defined single step, and mixtures of homopolynucleotides can form structures more complex than dimers, but a more detailed consideration of these points is beyond the scope of this text. However, it should be mentioned that thermal melting of single-stranded homopolymers is a useful tool for studying stacking interactions: As with DNA, the spectrum will change as the bases unstack. The thermodynamic interpretation of these results is complex (cf. Ref. 3).

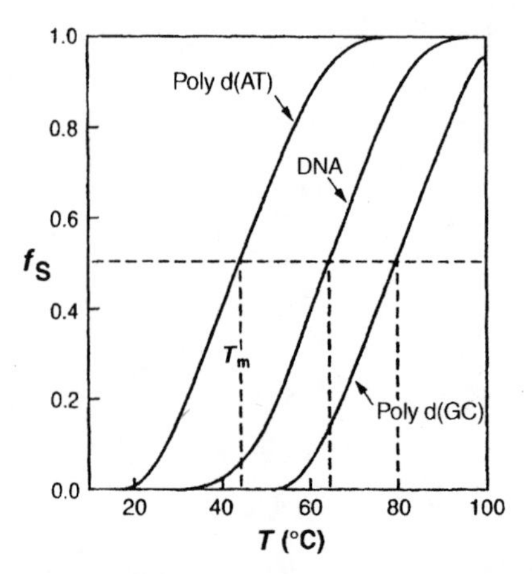

FIGURE 3-11. Hypothetical melting curves for double helix structures of poly d(AT), DNA (containing A–T and G–C base pairs), and poly d(GC). The fraction of chains not in the double helix structure, *fs*, is plotted versus the temperature. The melting temperatures, T_m, are indicated by the dashed lines. The shape of the curves and the T_m values are dependent on the length of the chains and their concentrations. This drawing assumes that the concentrations and chain lengths are comparable in all three cases.

From a practical standpoint, many of the above difficulties can be circumvented by the use of model systems to determine the thermodynamics of adding a base pair to a chain of nucleotides. Thus, for example, an oligonucleotide A_n could be added to T_n, and a complex of A_nT_n formed. The same experiment can be done with A_{n+1} and T_{n+1}. If ΔG° is determined for the formation of each of the complexes, then the difference between the two standard free energies of formation gives the standard free energy change for the formation of one A–T pair. The temperature dependence of this free energy difference can be used to obtain the standard enthalpy and entropy changes for formation of a single A–T pair. With knowledge of the thermodynamic parameters for the formation of single A–T and G–C pairs within a chain, the question can be asked: Does this permit calculation of the thermodynamic stability of a DNA of a given composition? Regrettably, this is not the case. The stability of a given DNA cannot be predicted simply by knowing the fraction of G–C pairs in the DNA.

Remarkably, if the effect of the nearest neighbors of each base pair is taken into account, a reasonable estimate of the thermodynamic stability of DNA can be obtained. This was discovered by studying many different complementary strands of short DNAs and looking for regularities (9,10). The rationale for this procedure lies in the importance of hydrogen bonding between each pair and the stacking interactions with its nearest neighbors. Ten nearest neighbor parameters suffice for determining the thermodynamic stability of a given DNA sequence, with one additional assumption, namely, the thermodynamic parameters for initiating DNA structure are different from those for adding hydrogen-bonded pairs to an existing chain. This takes into account that getting the two chains together and forming the first base pair is more difficult than adding base pairs to the double helix. Moreover, the G–C hydrogen-bonded pair nucleates the double helix formation better than an A–T pair since it is more stable. In thermodynamic terms, this can be written as

$$\Delta G^\circ = \Delta G^\circ(\text{initiation}) + \Sigma\Delta G^\circ(\text{nearest neighbors}) \tag{3-20}$$

where the sum is over all pairs of nearest neighbor interactions. The thermodynamic parameters necessary to calculate the nearest neighbor interactions are given in Table 3-10, along with those characterizing the double helix initiation. The best fit of the data on model systems utilizes two additional parameters that are usually only small corrections; this refinement will not be considered here (cf. Ref. 9).

TABLE 3-10. Thermodynamic Parameters for Determination of DNA Stability[a]

DNA pair	ΔG° (kcal/mol)	ΔH° (kcal/mol)	ΔS° [cal/(mol K)]
5′-A–A 3′-T–T	−1.4	−8.4	−23.6
5′-A–T 3′-T–A	−0.9	−6.5	−18.8
5′-T–A 3′-A–T	−0.8	−6.3	−18.5
5′-A–C 3′-T–G	−1.8	−8.6	−23.0
5′-C–A 3′-G–T	−1.6	−7.4	−19.3
5′-A–G 3′-T–C	−1.3	−6.1	−16.1
5′-G–A 3′-C–T	−1.7	−7.7	−20.3
5′-C–G 3′-G–C	−2.5	−10.1	−25.5
5′-G–C 3′-C–G	−2.5	−11.1	−28.4
5′-C–C 3′-G–G	−2.1	−6.7	−15.6
Initiation, one or more GC hydrogen bond pairs	+1.8	0.0	−5.9
Initiation, no GC hydrogen bond pairs	+2.7	0.0	−9.0

Source: Adapted from J. SantaLucia, Jr., H. T. Allawi, and P. A. Seneviratne, *Biochemistry* **35**, 3555 (1996).
[a]pH 7.0, 1 M NaCl, 298 K.

A simple example will illustrate how these data can be used. Consider the reaction

$$\begin{matrix} 5'\text{-A-G-C-T-G-3}' \\ + \\ 5'\text{-C-A-G-C-T-3}' \end{matrix} \rightleftharpoons \begin{matrix} 5'\text{-A-G-C-T-G-3}' \\ 3'\text{-T-C-G-A-C-5}' \end{matrix}$$

The sum of the nearest neighbor standard free energy changes associated with the formation of the five base pair DNA from Table 3-10 is

$$\begin{aligned} \sum \Delta G^\circ &= \begin{matrix} 5'\text{-A-G} \\ \text{T-C-5}' \end{matrix} + \begin{matrix} 5'\text{-G-C} \\ \text{C-G-5}' \end{matrix} + \begin{matrix} 5'\text{-C-T} \\ \text{G-A-5}' \end{matrix} + \begin{matrix} 5'\text{-T-G} \\ \text{A-C-5}' \end{matrix} \\ &= -1.3 - 2.5 - 1.3 - 1.6 = -6.7\,\text{kcal/mol} \end{aligned}$$

The standard free energy for initiation is 1.8 kcal/mol, so that the standard free energy change for formation of this DNA fragment is −4.9 kcal/mol. Similarly, $\Delta H^\circ = -30.7$ kcal/mol and $\Delta S^\circ = -85.8$ cal/(mol K). The temperature dependence of the standard free energy change can be calculated by assuming the standard enthalpy change is independent of temperature. This relatively simple procedure permits the thermal stability of any linear DNA to be calculated to a good approximation.

Knowledge of the thermal properties of DNA fragments is important both physiologically and practically. Knowing the stability of DNA obviously is of interest in understanding genetic replication. From a practical standpoint, knowing the stability of DNA fragments is important in planning cloning experiments. DNA probes must be used that will form stable duplexes with the target DNA. The temperature at which the duplex becomes stable can be estimated by calculating the melting temperature, T_m, that is, the temperature at which half of the strands are in the double helix conformation. The equilibrium constant for formation of the double helix duplex is

$$K = [\text{D}]/[\text{S}]^2 \tag{3-21}$$

where D is the helical duplex and S is the single strand. If the total concentration of the oligonucleotide is C_0, the concentration of single strands at T_m is $C_0/2$ and that of the duplex is $C_0/4$. If we insert these relationships into Eq. 3-21, we see that $K = 1/C_0$ at T_m. If we insert the relationship between the equilibrium constant and the standard free energy change, we obtain

$$\Delta G^\circ/(RT_m) = \ln C_0 \tag{3-22}$$

or

$$\Delta H^\circ/(RT_m) - \Delta S^\circ/R = \ln C_0$$

If Eq. 3-22 is solved for T_m, we obtain

$$T_m = \Delta H^\circ/(\Delta S^\circ + R \ln C_0) \tag{3-23}$$

For the case under discussion, if $C_0 = 0.1$ mM, $T_m = 295$ K.

In this brief discussion, we have neglected the fact that DNA is a polyelectrolyte due to the negatively charged phosphodiester backbone. The polyelectrolyte nature of DNA means that the ionic environment, particularly positively charged ions, strongly influences the structure and behavior of DNA. Normally, a negatively charged polymer such as DNA would exist as a rod, because of the charge repulsion. However, we know that DNA is packaged into a small volume in cells. This involves the twisting of the double helix, the formation of loops, etc. Interactions with metal ions and positively charged proteins are necessary for this packaging to occur. Interested readers should consult more complete descriptions of nucleic acids for information on this interesting subject (3,7).

3.9 RNA

In principle, the structure of RNA can be discussed exactly as the discussion of DNA. The principles are the same: hydrogen bonding between bases, stacking interactions, and hydrophobic interactions determine the structure. Of course, a ribose is present, rather than deoxyribose, and uracil is substituted for thymine. In addition, several modified bases are commonly found in RNAs. Unfortunately, understanding and predicting RNA structures is more complex than for DNA. Ultimately, this is because there are several quite different biological functions for RNA and thus several quite different types of RNA. Generally, RNAs do not form intermolecular double helices. Instead, double helices are formed within an RNA molecule. These can be loops, hairpins, bulges, etc. Some idea of the diversity of structures that can be formed is shown in Figure 3-12, where an RNA molecule is shown, along with the different types of structures that can be formed.

As with DNA, the RNA structures can be predicted reasonably well by considering nearest neighbor interactions. However, because of the diversity of the structures, the models are more complex. Nevertheless, quite reasonable RNA structures can be predicted using the thermodynamics derived from simple systems. We will not delve further into RNA structures here, except to say that the principles of model building have been developed sufficiently here so that the interested reader can proceed directly to current literature on this subject (11–14). As with DNA, metal ions play an important role in the biological packaging of RNA.

Finally, the interactions between RNA and DNA are of obvious physiological relevance. These have been studied quite extensively, but they are not nearly as well understood as the interactions between DNA fragments and within RNA molecules. Simple models are not yet available to calculate the properties of DNA–RNA structures. At this point, we will leave our discussion of thermody-

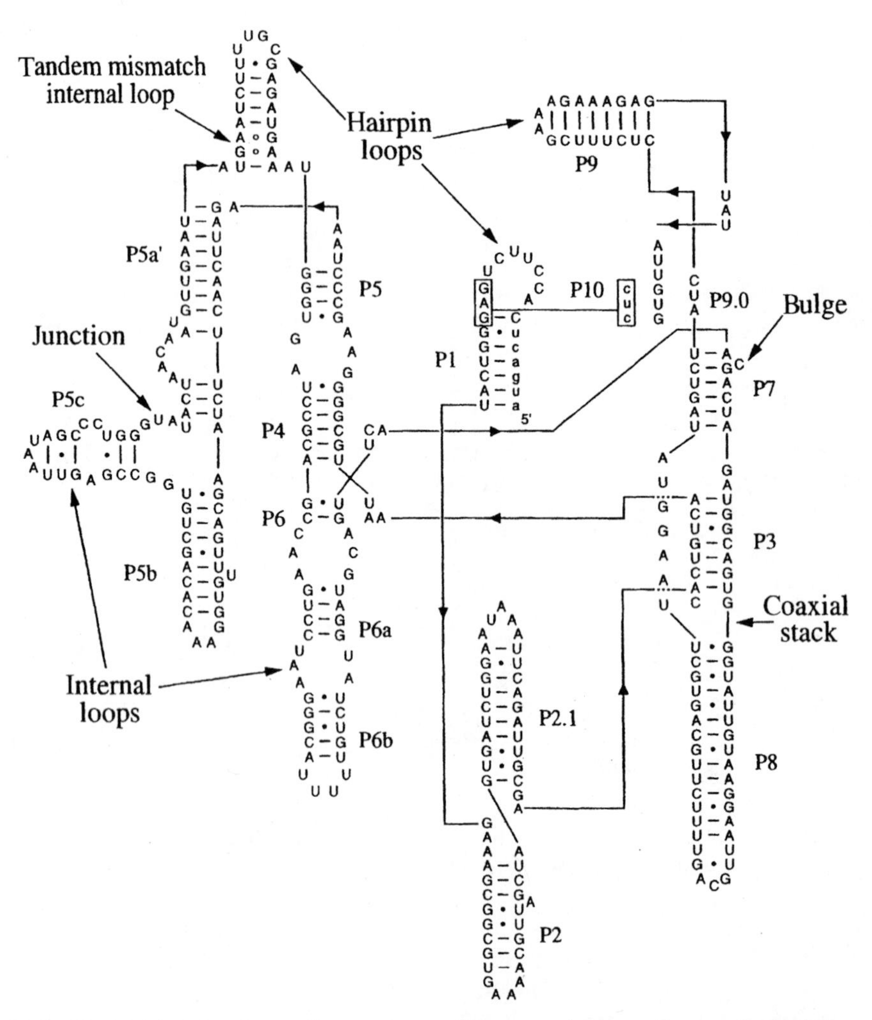

FIGURE 3-12. Proposed secondary structure of the group I intron of mouse-derived *Pneumocystis carinii.* The areas indicate secondary structures within the intron, including base-paired helices. Reproduced from J. SantaLucia, Jr. and D. H. Turner, *Biopolymers* **44**, 309 (1997). Copyright © 1997 Biopolymers. Reprinted by permission of John Wiley & Sons, Inc.

namics of biological systems, although many more interesting examples could be discussed.

REFERENCES

1. P. Mitchell, *Nature* **191**, 144 (1961).
2. T. G. Dewey and G. G. Hammes, *J. Biol. Chem.* **256**, 8941 (1981).
3. C. R. Cantor and P. R. Schimmel, *Biophysical Chemistry*, Freeman, New York, 1980.

4. J. M. Berg, J. L. Tymoczko, and L. Stryer, *Biochemistry*, 5th edition, Freeman, New York, 2002.
5. C. Branden and J. Tooze, *Introduction to Protein Structure*, 2nd edition, Garland Publishing, New York, 1999.
6. W. Kauzmann, *Adv. Protein Chem.* **14**, 1 (1959).
7. J. D. Watson, N. H. Hopkins, J. W. Roberts, J. A. Steitz, and A. M. Weiner, *Molecular Biology of the Gene*, 4th edition, Benjamin/Cummings, Menlo Park, CA, 1987.
8. K. J. Breslauer and J. M. Sturtevant, *Biophys. Chem.* **7**, 205 (1977).
9. J. SantaLucia, Jr., H. T. Allawi, and P. A. Seneviratne, *Biochemistry* **35**, 3555 (1996).
10. P. N. Borer, B. Dengler, I. Tinoco, Jr., and O. C. Uhlenbeck, *J. Mol. Biol.* **86**, 843 (1974).
11. I. Tinoco, Jr., O. C. Uhlenbeck, and M. D. Levine, *Nature* **230**, 362 (1971).
12. J. SantaLucia, Jr. and D. H. Turner, *Biopolymers* **44**, 309 (1997).
13. T. Xia, J. SantaLucia, Jr., M. E. Burkard, R. Kierzek, S. J. Schroeder, X. Jiao, C. Cox, and D. H. Turner, *Biochemistry* **37**, 14719 (1998).
14. M. Burkard, D. H. Turner, and I. Tinoco, Jr., *The RNA World*, 2nd edition, Cold Spring Harbor Laboratory Press, Cold Spring Harbor, NY, 1999, p. 233.

PROBLEMS

3-1. a. Glutamine is an important biomolecule made from glutamate. Calculate the equilibrium constant for the reaction

$$\text{Glutamate} + NH_3 \rightleftharpoons \text{Glutamine} + H_2O$$

Use the standard free energies of formation in Appendix 4 to obtain the standard free energy change for this reaction. Under physiological conditions, the concentration of ammonia is about 10 mM. Calculate the ratio [glutamine]/[glutamate] at equilibrium. (By convention, the concentration of water is set equal to 1 since its concentration does not change significantly during the course of the reaction.)

b. Physiologically, glutamine is synthesized by coupling the hydrolysis of ATP to the above reaction:

$$\text{Glutamate} + NH_3 + \text{ATP} \rightleftharpoons \text{Glutamine} + \text{ADP} + P_i$$

Calculate ΔG° and the equilibrium constant for this reaction under standard conditions (see Appendix 4). Assume that NH_3 and P_i are maintained at about 10 mM and that [ATP]/[ADP] = 1. What is the ratio of glutamate to glutamine at equilibrium? What ratio is needed to convert glutamate to glutamine spontaneously, that is, to make $\Delta G < 0$?

Do all calculations at 1 atm and 298 K. Although this is not the physiological temperature, the results are not significantly altered.

3-2. Adipose tissues contain high levels of fructose. Fructose can enter the glycolytic pathway directly through the reaction

$$\text{Fructose} + \text{ATP} \rightleftharpoons \text{Fructose-6-phosphate} + \text{ATP}$$

Assume that the standard free energy change for this reaction with the same standard state used in Table 3-1 is $-17.0\,\text{kJ/mol}$. If fructose is substituted for glucose in "glycolysis," what would the overall reaction be for the conversion of fructose to 2-glyceraldehyde, part one of glycolysis? What would the overall reaction be for the complete metabolic cycle? Calculate ΔG° and ΔG for these two reactions. Assume the concentration of fructose is 5.0 mM, and the concentrations of the other metabolites are as in Table 3-2.

3-3. Glucose is actively transported into red blood cells by coupling the transport with the hydrolysis of ATP. The overall reaction can be written as

$$\text{ATP} + \text{H}_2\text{O} + n\ \text{Glucose(outside)} \rightleftharpoons n\ \text{Glucose(inside)} + \text{ADP} + \text{P}_\text{i}$$

If the ratio of [ATP]/[ADP] $= 1$ and $[\text{P}_\text{i}] = 10\,\text{mM}$, what is the concentration gradient, [glucose(inside)]/[glucose(outside)], that is established at 298 K? Calculate this ratio for $n = 1$, $n = 2$, and $n = n$. Does this suggest a method for determining the value of n? Use the value of ΔG° for the hydrolysis of ATP in Appendix 5 for these calculations.

3-4. Derive the equation for the temperature dependence of the standard free energy change for protein denaturation when ΔC_p is not equal to zero. As the starting temperature in the derivation, use the temperature at which $\Delta G^\circ = 0$.

Hint: The easiest way to proceed is to calculate the temperature dependence of ΔH° and ΔS°. These relationships can then be combined to give the temperature dependence of ΔG° . The parameters in the final equation should be ΔC_P, the temperature, T, the temperature at which $\Delta G^\circ = 0$, T_m, and the enthalpy change at the temperature where $\Delta G^\circ = 0$, $\Delta H^\circ_{T_\text{m}}$.

3-5. Specific genes in DNA are often searched for by combining a radioactive oligonucleotide, O, with the DNA that is complementary to a sequence in the gene being sought. This reaction can be represented as

$$\text{O} + \text{DNA} \rightleftharpoons \text{Double strand}$$

For such experiments, the concentration of the oligonucleotide is much greater than that of the specific DNA sequence. Assume that the probe is 5′-GGGATCAG-3′.

a. Calculate the equilibrium constant at 298 K for the interaction of the probe with the complementary DNA sequence using the parameters in Table 3-9.

b. Calculate the fraction of the DNA present as the double strand formed with the probe if the concentration of the probe is 1.0×10^{-4} M and the temperature is 298 K.

c. Find the melting temperature for this double strand if the concentration of the complementary DNA sequence is 1.0 nM. The melting temperature in this case is when [double strand]/[DNA] = 1.

3-6. For an electrochemical cell (e.g., a battery), the reversible work is $-nF\mathcal{E}$, where n is the number of moles of electrons involved in the chemical reaction, F is the Faraday, and $\mathcal{E}$ is the reversible voltage. This relationship is useful for considering coupled oxidation–reduction reactions in biochemical systems.

a. Use this relationship and Eq. 2-49 to derive an equation relating the free energy change for the reaction and the voltage of the cell. Your final equation should contain the standard free energy change for the reaction, the concentrations of the reactants, and the electrochemical voltage, as well as constants.

b. The voltage at equilibrium is usually designated as $\mathcal{E}^{\circ}$. How is this related to the standard free energy change for the reaction? How might the equilibrium constant for a biochemical reaction be determined from voltage measurements? For the reaction

$$\text{Malate} + \text{NAD}^{+} \rightleftharpoons \text{Oxaloacetate} + \text{NADH} + \text{H}^{+}$$

the voltage at equilibrium is -0.154 V at 298 K. Calculate the equilibrium constant for this reaction. ($F = 96,485\,\text{coulomb/mol}$ and $n = 2$ for this reaction.)

CHEMICAL KINETICS

CHAPTER 4

Principles of Chemical Kinetics

4.1 INTRODUCTION

Thermodynamics tells us what changes in state can occur, that is, the relative stability of states. For chemical reactions, it tells us what reactions can occur spontaneously. However, thermodynamics does not tell us the time scale for changes in state or how the changes in state occur; it is concerned only with the differences in the initial and final states. In terms of chemical reactions, it does not tell us *how* the reaction occurs, in other words, the molecular interactions that take place as a reaction occurs. For biological reactions, the rates are critical for the survival of the organism, and a primary interest of modern biology is the molecular events that lead to reaction. The study of the rates and mechanisms of chemical reactions is the domain of *chemical kinetics*. Thermodynamics provides no intrinsic information about mechanisms.

Many examples exist with regard to the importance of the time scale for chemical reactions. For those of you having diamond jewelry, you may be unhappy to know that the most stable state of carbon under standard conditions is graphite. So as you read this, your diamond is turning to graphite, but fortunately the time scale for this conversion is many hundreds of years. If graphite is more stable, why were diamonds formed? The answer is that the formation of diamonds did not occur under standard conditions: It is well known that at very high temperatures and pressures, graphite can be converted to diamond. In the biological realm, one of the most critical reactions is the hydrolysis of ATP to ADP and P_i. Thermodynamics tells us that the equilibrium lies far toward ADP and P_i. Yet solutions of ATP are quite stable under physiological conditions in the absence of the enzyme ATPase. A small amount of this enzyme will cause rapid and almost complete hydrolysis. Virtually all metabolic reactions occur much too slowly to sustain life in the absence of enzymes. Enzymes serve as catalysts and cause reactions to occur many orders of magnitude faster. Studies of the rates of hydrolysis under varying conditions allow us to say by what mechanism the reaction may occur, for example, how the substrates and products interact with the enzyme. Understanding the mechanism of biological processes is a research area of great current interest.

In this chapter, we will be interested in understanding some of the basic principles of chemical kinetics, with a few examples to illustrate the power of chemical kinetics. We will not delve deeply into the complex mathematics that is sometimes necessary nor the many specialized methods that are sometimes used to analyze chemical kinetics. Many treatises are available that provide a more complete discussion of chemical kinetics (1–4). Because the examples used will be relatively simple, they will not necessarily involve biological processes. The principles discussed, however, are generally applicable to all systems. The background presented here should be sufficient to get you started on utilizing chemical kinetics and reading the literature with some comprehension. The key three concepts that will be discussed first are *rates of chemical reactions, elementary reactions*, and *mechanisms of chemical reactions*.

As a simple illustration, consider the reaction of hydrogen and iodine to give hydrogen iodide in the gas phase:

$$H_2 + I_2 \rightleftharpoons 2\,HI \tag{4-1}$$

A possible mechanism for this reaction is for hydrogen and iodine to collide to produce hydrogen iodide directly. Another possible mechanism is for molecular iodine to first dissociate into iodine atoms and for the iodine atoms to react with hydrogen to produce hydrogen iodide. These possibilities can be depicted as

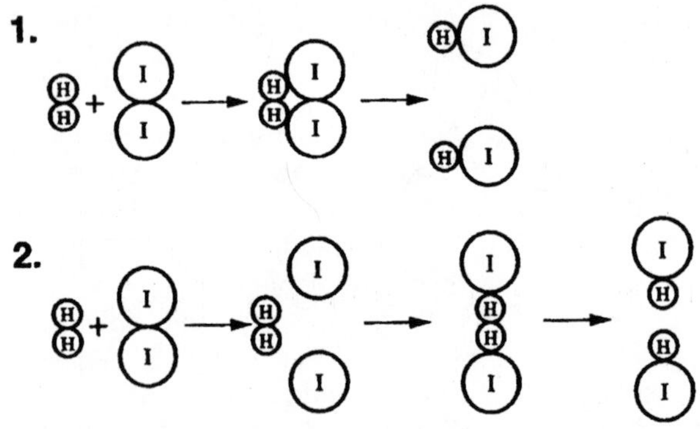

1.

2.

These two possible modes of reaction are quite distinct in terms of the chemistry occurring and are examples of two different mechanisms. We can also write these mechanisms in a more conventional manner:

1. $H_2 + I_2 \rightleftharpoons 2\,HI$ Elementary step and balanced chemical reaction (4-2)

2. $I_2 \rightleftharpoons 2I$ Elementary step

$2I + H_2 \rightleftharpoons 2HI$ Elementary step (4-3)

$H_2 + I_2 \rightleftharpoons 2\,HI$ Balanced chemical reaction

As indicated, the individual steps in the mechanism are called *elementary steps* and they must always add up to give the balanced chemical reaction. In the first mechanism, the elementary reaction and balanced chemical reaction happen to be the same, but they are conceptually quite distinct, as we shall amplify later.

A very important point to remember is that a balanced chemical reaction gives no information about the reaction mechanism. This is not so obvious in the above example, but consider the reaction

$$3\,Fe^{2+} + HCrO_4^- + 7\,H^+ \rightleftharpoons 3\,Fe^{3+} + Cr^{3+} + 4\,H_2O$$

This clearly cannot be the reaction mechanism, as it would require the 11 reactants to encounter each other simultaneously, a very unlikely event.

4.2 REACTION RATES

The rate of a chemical reaction is a measure of how fast the concentration changes. If the concentration changes an amount Δc in time interval Δt, the rate of the reaction is $\Delta c/\Delta t$. If the limit of smaller and smaller time intervals is taken, this becomes dc/dt. If we apply this definition to Eq. 4-1, starting with H_2 and I_2 as reactants, the rate could be written as $-d[H_2]/dt$, $-d[I_2]/dt$, or $+d[HI]/dt$. For every mole of hydrogen and iodine consumed, 2 mol of hydrogen iodide are formed, so that this definition of the rate does not provide a unique definition: The value of the rate depends on which reaction component is under consideration. The rate of appearance of HI is twice as great as the rate of disappearance of H_2 or I_2. This ambiguity is not convenient, so a convention is used to define the reaction rate, namely, the rate of change of the concentration divided by its coefficient in the balanced chemical reaction. This results in a unique reaction rate, R, for a given chemical reaction. By convention, R is always positive. For the case under consideration,

$$R = -\frac{d[H_2]}{dt} = -\frac{d[I_2]}{dt} = \frac{1}{2}\frac{d[HI]}{dt}$$

Consider a more complex chemical equation:

$$2\,N_2O_5 \rightarrow 4\,NO_2 + O_2$$

In this case, the reaction rate is

$$R = -\frac{1}{2}\frac{d[N_2O_5]}{dt} = -\frac{1}{4}\frac{d[NO_2]}{dt} = \frac{d[O_2]}{dt}$$

Measuring the concentration as it changes with time can be very difficult, and some of our greatest advances in understanding chemical reactions have resulted from the development of new techniques for measuring the rates of chemical reactions, particularly the rates of very fast reactions. The most convenient method of measuring reaction rates is to mix the reactants together and to measure the subsequent change in concentrations continuously using a spectroscopic technique such as light absorption. It is sometimes not possible to find a physical property to monitor continuously, so it may be necessary to stop the reaction and measure the concentration chemically or through radioactive tracers. We will not dwell on this point, except to stress that measuring the rate of a chemical reaction may not be trivial.

Once a method has been established for determining the reaction rate, the next step is to measure the dependence of the reaction rate on the concentrations of the reactants. The dependence of the rate on the concentrations is called the *rate law. The rate law cannot be predicted from the balanced chemical equation. It must be determined experimentally*. A few examples will serve to illustrate this point. Consider the hydrolysis of ATP catalyzed by an ATPase. The overall reaction is

$$\mathrm{ATP} \rightarrow \mathrm{ADP} + \mathrm{P_i} \tag{4-4}$$

and the observed dependence of the rate on concentration of ATP and enzyme in some cases is

$$R = \frac{k[\mathrm{E_0}]}{1 + K_\mathrm{m}/[\mathrm{ATP}]} \tag{4-5}$$

where $[\mathrm{E_0}]$ is the total enzyme concentration and k and K_m are constants. The rate law certainly cannot be deduced from the overall chemical reaction. In fact, the rate law is not the same for all ATPases. Many different ATPases are found in biological systems and they do not all hydrolyze ATP by the same mechanism. As another example, consider the simple reaction

$$\mathrm{H_2} + \mathrm{Br_2} \rightarrow 2\,\mathrm{HBr} \tag{4-6}$$

The experimentally determined rate law is

$$R = \frac{k[\mathrm{H_2}][\mathrm{Br_2}]^{1/2}}{1 + K[\mathrm{HBr}]/[\mathrm{Br_2}]} \tag{4-7}$$

As a final example, consider the redox reaction

$$5\,\mathrm{Br^-} + \mathrm{BrO_3^-} + 6\,\mathrm{H^+} \rightarrow 3\,\mathrm{Br_2} + 3\,\mathrm{H_2O} \tag{4-8}$$

The observed rate law is

$$R = k[\mathrm{Br^-}][\mathrm{BrO_3^-}][\mathrm{H^+}]^2 \tag{4-9}$$

These examples should emphasize the futility of attempting to predict the rate law from the balanced chemical equation.

The exponent of the concentration in the rate law is the *reaction order* with respect to that component (i.e., first order, second order, etc.). In some cases, such as Eq. 4-9, the concept of reaction order has a simple meaning. The rate law is first order with respect to $\mathrm{Br^-}$ and $\mathrm{BrO_3^-}$, and second order with respect to

H^+. For more complex rate laws, such as Eqs. 4-5 and 4-7, the concept of reaction order cannot be used for all of the concentrations, except in limiting conditions. For Eq. 4-7, the rate law is 1/2 order with respect to Br_2 at high concentrations of Br_2 and/or low concentrations of HBr. The lowercase k's in the rate laws are called *rate constants*. The dimensions of rate constants can be deduced from the rate law by remembering that the rate is usually measured as M/s. In Eq. 4-9, k, therefore, has the dimensions of $M^{-3} s^{-1}$. The capital K's in the above equations also are constants, but they are combinations of rate constants rather than individual rate constants.

4.3 DETERMINATION OF RATE LAWS

Many different methods exist for determining rate laws. Only a few methods are considered here. With the routine use of computers, numerical integration of the differential equations and simultaneous fitting of the data are possible. However, as with any experimental approach, it is best to first fit the data to simple models before embarking on complex computer fitting. Computer programs will always fit the data, but it is important to be sure that the fit is a good one and the proposed rate law makes sense.

Probably, the simplest and still often used method is the determination of initial rates. With this method, the rate of the reaction is measured at the very beginning of the reaction under conditions where the decrease in the concentrations of the reactants is so small that their concentrations in the rate law can be assumed to be the starting concentrations. In practice, this means that the concentrations should not change by more than a few percent. To illustrate the method, let us assume the rate law is

$$R = k(c_1)^a(c_2)^b \tag{4-10}$$

If c_2 is held constant and the initial rate is measured for different concentrations of c_1, the coefficient a can be determined. For example, if the concentration of c_1 is doubled and the rate increases by a factor of 4, a must be equal to 2. The same type of experiment can be done to determine b, namely, c_1 is held constant and c_2 is varied. The rate constant can readily be calculated, once the coefficients a and b are known, from the relationship $k = R/[(c_1)^a(c_2)^b]$. The determination of initial rates is especially useful for studying enzymatic reactions as we shall see later.

One of the most useful methods for determining the rate law is integration of the rate equation to obtain an analytical expression for the time dependence of the concentrations. The analytical equation is then compared with experimental data to see if it accurately describes the time dependence of the concentrations. As mentioned previously, computers can perform this integration numerically. We will consider two examples of integrated rate equations to illustrate the method.

The decomposition of nitrogen pentoxide can be written as

$$N_2O_5 \rightarrow 2\,NO_2 + \frac{1}{2}O_2 \tag{4-11}$$

Assume that the decomposition is a first-order reaction:

$$R = -\frac{d[N_2O_5]}{dt} = k[N_2O_5] \tag{4-12}$$

This equation can easily be integrated

$$-\frac{d[N_2O_5]}{[N_2O_5]} = kdt$$

or

$$-\int_{[N_2O_5]_0}^{[N_2O_5]} \frac{d[N_2O_5]}{[N_2O_5]} = \int_0^t kdt$$

and

$$\ln([N_2O_5]/[N_2O_5]_0) = -kt \tag{4-13}$$

In Eq. 4-13, $[N_2O_5]_0$ is the concentration when $t = 0$. As illustrated in Figure 4-1, this equation predicts that a plot of $\ln[N_2O_5]$ versus t should be a straight line with a slope of $-k$. The data, in fact, conform to this rate equation under most conditions.

This analysis can also be carried out for higher order reactions. Hydrogen iodide will react to give hydrogen and iodine under certain conditions:

$$2\,HI \rightarrow H_2 + I_2 \tag{4-14}$$

The rate law is

$$R = -\frac{1}{2}\frac{d[HI]}{dt} = k[HI]^2 \tag{4-15}$$

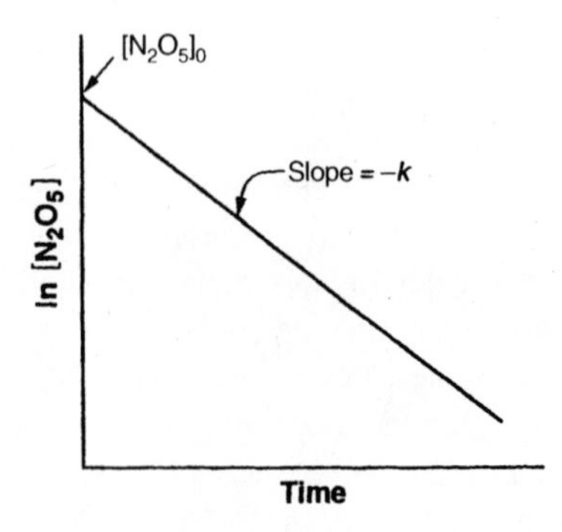

FIGURE 4-1. Demonstration of first-order kinetics. Plot of $\ln[N_2O_5]$ versus time according to Eq. 4-13. The straight line has a slope of $-k$.

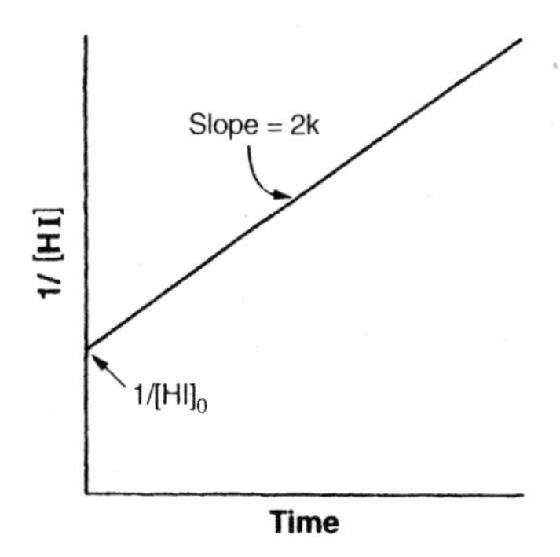

FIGURE 4-2. Demonstration of second-order kinetics. Plot of 1/[HI] versus time according to Eq. 4-16. The straight line has a slope of $2k$.

Rearrangement of this equation gives

$$-\frac{d[HI]}{[HI]^2} = 2k\,dt$$

which can be integrated to give

$$1/[HI] - 1/[HI]_0 = 2kt \qquad (4\text{-}16)$$

Here $[HI]_0$ is the concentration when $t = 0$. As illustrated in Figure 4-2, this equation predicts that a plot of 1/[HI] versus t should be a straight line with a slope of $2k$. Again, the experimental data conform to this prediction.

It is frequently, but not always, possible to integrate rate equations analytically. Good experimental design can help to make the rate law relatively simple and therefore easy to integrate. In fact, experienced researchers will try to make the reaction first order whenever possible. This might seem like a major restriction but it is not. It is often possible to convert complex rate laws to conform to *pseudo-first-order kinetics*. For example, assume the rate law is

$$R = k(c_1)f(c_2)$$

where $f(c_2)$ is a function of the concentration of c_2. The function could be very complex or a simple power of the concentration—it does not matter. If the concentration of c_2 is made much larger than the concentration of c_1, then it can be assumed to remain constant throughout the reaction, and the rate law becomes

$$R = k'(c_1)$$

where $k' = kf(c_2)$. The "constant" $kf(c_2)$ is called a "pseudo" first-order rate constant since it is constant under the experimental conditions used, but actually depends on the concentration c_2. If the concentration of c_2 is varied, the dependence of $f(c_2)$ on c_2 can be determined. It might seem restrictive to have a high concentration of c_2, but in practice it is often sufficient to have the concentration of c_2 only about a factor of 10 higher than c_1. Careful experimental design can make the job of determining the rate law much easier!

In determining the rate law, it is often necessary to use a broad range of experimental conditions before trying to interpret the rate law in terms of a chemical mechanism. As an illustration, consider the reaction

$$I^- + OCl^- \rightarrow OI^- + Cl^- \tag{4-17}$$

At constant pH, the rate law determined experimentally is

$$R = k[I^-][OCl^-] \tag{4-18}$$

If the pH is varied, it is found that k also varies. It was determined that $k = k'/[OH^-]$, where k' is a constant. Thus, a more complete rate law is

$$R = k'[I^-][OCl^-]/[OH^-] \tag{4-19}$$

The more information that can be determined about the concentration dependence of the rate the better the mechanism that can be postulated.

4.4 RADIOACTIVE DECAY

A good example of a first-order rate process is radioactive decay. Radioactive isotopes are frequently used in biological research. For example, the radioactive isotope of phosphorus, ^{32}P, gives off radiation according to the reaction

$$^{32}P \rightarrow {}^{32}S + \beta^- \tag{4-20}$$

where β^- is a high-energy electron. The rate law for radioactive decay is

$$-\frac{d[^{32}P]}{dt} = k[^{32}P] \tag{4-21}$$

Integration of this rate law as done previously (Eq. 4-13) gives

$$[^{32}P] = [^{32}P]_0 e^{-kt} \tag{4-22}$$

or

$$\ln([^{32}P]/[^{32}P]_0) = -kt$$

The rate of radioactive decay is usually given in terms of the half-life of the radioactive decay; in this case, the half-life is 14.3 days. In terms of the integrated rate law, Eq. 4-22, when half of the original radioactivity has decayed, $\ln(1/2) = -kt$, or the half-life for decay is $t_{1/2} = (\ln 2)/k$.

Thus, for radioactive decay, the half-life is constant. It does not matter when we start counting or how much radioactivity we start with, the radioactivity will decay to half of its original value in 14.3 days. This is a special property of first-order rate processes and is not true for other reaction orders where the half-life depends on the concentrations of the reactants. This is very convenient because regardless of when we start observing the rate of reaction, at $t = 0$, or at $t =$ any value, the integrated rate law is a simple single exponential.

4.5 REACTION MECHANISMS

In general, many mechanisms are possible for a given reaction. Mechanisms are proposals for how the reaction occurs. A proposed mechanism must be consistent with the experimentally observed rate law, but this is usually true for many mechanisms. *Kinetic studies can disprove a mechanism but cannot prove a mechanism.* As a practical matter, the simplest mechanism consistent with all of the data is most appropriate, but at the end of the day, all that can be said is a specific mechanism is consistent with known data. It is not possible to say that this *must* be "the" mechanism.

A mechanism consists of a combination of *elementary steps*, which must sum up to give the overall reaction. For an elementary step, the order and molecularity, the number of molecules involved in the reaction, are the same. Therefore, for elementary steps, the rate law can be written as the product of the concentrations of all reactants, each raised to the power of their stoichiometric coefficient, multiplied by a rate constant. Some examples of elementary steps and associated rate laws are given below.

$$\begin{aligned} &H_2 + I_2 \rightarrow 2\,HI, \quad && R = k[H_2][I_2] \\ &2\,I + H_2 \rightarrow 2\,HI, \quad && R = k[I]^2[H_2] \\ &O_3 \rightarrow O_2 + O, \quad && R = k[O_3] \end{aligned}$$

Remember, rate laws can be derived from the chemical equation *only* for elementry steps, and never for the balanced chemical equation of the overall reaction. If an elementary reaction is reversible, then the rate law is the difference between the rates of the forward and reverse reactions. Therefore, for the elementary step

$$2\,A + B \underset{k_r}{\overset{k_f}{\rightleftharpoons}} C + D$$

$$R = k_f[A]^2[B] - k_r[C][D]$$

We now have two criteria for a possible mechanism: (1) it must be consistent with the observed rate law and (2) the elementary steps must add up to give the overall balanced chemical reaction. Let us return to the two proposed mechanisms for the reaction of H_2 and I_2, Eqs. 4-2 and 4-3. The first mechanism contains only a single elementary step so that the rate law is

$$R = k_f[H_2][I_2] - k_r[HI]^2 \tag{4-23}$$

The second mechanism contains two elementary steps so that some assumptions must be made to derive the rate law. We will assume that iodine atoms are in rapid equilibrium with molecular iodine, or to be more specific that this equilibrium is adjusted much more rapidly than the reaction of iodine atoms with molecular hydrogen. Furthermore, the concentration of iodine atoms is assumed to be much less than that of molecular iodine. These assumptions are, in fact, known to be correct. If we now consider the second elementary step in the mechanism,

$$2\,I + H_2 \underset{k_2}{\overset{k_1}{\rightleftharpoons}} 2\,HI$$

the rate law can be written as

$$R = k_1[I]^2[H_2] - k_2[HI]^2 \tag{4-24}$$

but since the first step is always at equilibrium throughout the course of the reaction, $[I]^2 = K[I_2]$, where K is the equilibrium constant for the dissociation reaction. Substituting this relationship into Eq. 4-24 gives

$$R = k_1K[I_2][H_2] - k_2[HI]^2 \tag{4-25}$$

Equations 4-23 and 4-25 are identical in form as only the definitions of the constants are different, $k_f = k_1K$ and $k_r = k_2$. Thus, we have shown that both mechanisms are consistent with the rate law, and therefore both are possible mechanisms. Even for this simple reaction, there remains a debate as to which is the more likely mechanism.

It is easy to postulate a third possible mechanism with the following elementary steps:

$$H_2 \rightleftharpoons 2\,H$$

$$2\,H + I_2 \rightleftharpoons 2\,HI$$

By analogy with the mechanism involving iodine atoms, it can be seen that this mechanism would give the rate law of Eq. 4-25 with K now being the dissociation constant for molecular hydrogen in equilibrium with hydrogen atoms. However, K is a known constant and if this constant is combined with the value of k_1K determined experimentally, k_1 can be calculated. It is found that the value of k_1 is much larger than the rate constant characterizing the maximum rate at which two hydrogen atoms and molecular iodine encounter each other in the gas phase. Therefore, this mechanism can be ruled out as inconsistent with well-established theory. Thus, we have a third criterion for disproving a mechanism.

As a second example of how mechanisms can be deduced, let us return to the reaction of I^- and OCl^- to produce OI^- and Cl^-, Eq. 4-17. The experimentally determined rate law is given by Eq. 4-19. What is a possible mechanism? One possibility is the following scheme:

$$\begin{array}{ll} OCl^- + H_2O \rightleftharpoons HOCl + OH^- & \text{Fast, at equilibrium} \\ I^- + HOCl \xrightarrow{k_2} HOI + Cl^- & \text{Slow} \\ OH^- + HOI \rightleftharpoons H_2O + OI^- & \text{Fast} \\ \hline OCl^- + I^- \rightarrow OI^- + Cl^- & \text{Overall reaction} \end{array} \quad (4\text{-}26)$$

This mechanism is consistent with the balanced chemical equation. Now we must show that it is consistent with the observed rate law. The rate of a reaction is determined by the slowest step in the mechanism. Therefore, the rate of the overall reaction is given by the rate of the second elementary step:

$$R = k_2[I^-][HOCl] \quad (4\text{-}27)$$

If the first step is assumed to be fast and the concentration of HOCl small relative that of I^- and Cl^-, then the concentration of HOCl can be calculated from the equilibrium relationship

$$K = [HOCl][OH^-]/[OCl^-]$$
$$[HOCl] = K[OCl^-]/[OH^-]$$

Insertion of this relationship into Eq. 4-27 gives

$$R = k_2K[I^-][OCl^-]/[OH^-] \quad (4\text{-}28)$$

which is the observed rate law. Since this mechanism is consistent with the observed rate law and the balanced chemical equation, it is a possible mechanism.

You may very well be wondering if the creation of a mechanism is black magic. It is true that imagination and knowledge are important factors, but logic can be used. A species in the denominator results from a fast equilibrium prior to a rate-determining step. Therefore, what needs to be found is a first step involving one of the reactants that produces the desired species, in this case OH^-. The other product of the first step must then react with the second reactant. The remainder of the steps are fast reactions that are necessary to produce a balanced chemical reaction. Note that none of the steps after the rate-determining step play a role in determining the rate law. This is one of the simplest types of mechanisms, fast equilibria prior to and after a rate-determining step. Nature is not always so obliging, and more complex mechanisms in which several steps occur at comparable rates are often necessary to account for experimental findings.

Here is another mechanism, quite similar in concept:

$$\begin{aligned}
&OCl^- + H_2O \rightleftharpoons HOCl + OH^- && \text{fast, at equilibrium}\\
&I^- + HOCl \xrightarrow{k_2} ICl + OH^- && \text{slow}\\
&ICl + 2\,OH^- \rightleftharpoons OI^- + Cl^- + H_2O && \text{fast}\\
&\overline{\quad OCl^- + I^- \rightarrow OI^- + Cl^- \quad} && \text{overall reaction} \qquad (4\text{-}29)
\end{aligned}$$

Obviously, this gives the same rate law as the mechanism in Eq. 4-26, as all events prior to the rate-determining step are the same. However, the chemistry is quite different. In the first case, iodide attacks the oxygen; in the second case, it attacks the chlorine. These two mechanisms cannot be distinguished by kinetics. Both are equally consistent with the data.

Thus far, we have considered mechanisms in which OCl^- is the initial reactant. Now let us look at mechanisms in which I^- is the initial reactant such as

$$\begin{aligned}
&I^- + H_2O \rightleftharpoons HI + OH^- && \text{fast, at equilibrium}\\
&HI + OCl^- \rightarrow ICl + OH^- && \text{slow}
\end{aligned}$$

By analogy, it should be clear that this mechanism would give the correct rate law, and rapid reactions after the rate-determining step can be added to give the correct balanced chemical equation. However, the equilibrium constant for the first step is known, and if this is combined with the results of the kinetic experiments, the rate constant for the rate-determining step would exceed the theoretically possible value. This is because the concentration of HI is much, much smaller than that of HOCl. Thus, mechanisms of this type can be excluded on the basis of theoretical concepts.

These two examples illustrate how mechanisms can be related to the results of kinetic experiments. It is a great challenge to devise mechanisms. Once a mechanism is postulated, it is the job of the kineticist to devise experiments that will test the mechanism, often disproving it and requiring postulation of a new mechanism. However, it is important to remember that no matter how convincing the arguments, a mechanism cannot be proved. Thus far, we have considered very simple reactions that are not biological, as biological reactions are typically very complex. The purpose of this discussion has been to illustrate the principles and concepts of chemical kinetics. We will later consider the kinetic analysis of enzymatic reactions, a very relevant and timely subject.

4.6 TEMPERATURE DEPENDENCE OF RATE CONSTANTS

Before discussing the kinetic analyses of biological reactions, a few additional concepts will be described. Reaction rates are often dependent on the temperature, and typically reactions go faster as the temperature increases. The first quantitative treatment of the temperature dependence of reaction rates was developed by Arrhenius in the late 1800s. He proposed that the temperature dependence of the rate constant could be described by the equation

$$k = A\exp(-E_a/RT) \tag{4-30}$$

where A is a constant, E_a is the *activation energy*, and T is the Kelvin temperature. This equation predicts that a plot of ln k versus $1/T$ should be a straight line with a slope of $-E_a/R$. This behavior is, in fact, followed in most cases. Equation 4-30 can be differentiated with respect to temperature to give

$$\mathrm{d}(\ln k)/\mathrm{d}T = E_a/RT^2 \tag{4-31}$$

This equation can be integrated to give

$$\ln(k_{T_2}/k_{T_1}) = \frac{(E_a/R)(T_2 - T_1)}{T_1T_2} \tag{4-32}$$

Equation 4-32 permits the rate constant to be calculated at any temperature if it is known at one temperature and the activation energy has been determined. Note that these equations are similar in form to those describing the temperature dependence of the equilibrium constant except that the activation energy has replaced the standard enthalpy of reaction. In some cases, the activation energy may change with temperature, thereby making the analysis more complex.

The physical model behind the Arrhenius equation is that an energy barrier, E_a, must be overcome in order for the reaction to occur. This is shown conceptually in Figure 4-3. The reaction "path" or "coordinate" can be envisaged as the approach

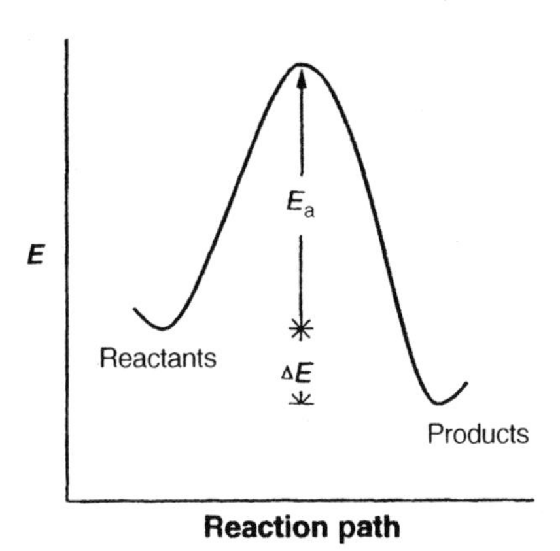

FIGURE 4-3. Schematic representation of the energy, E, versus the "reaction path" during the course of a reaction as discussed in the text. The definition of the activation energy, E_a, is indicated.

of the reactants to each other that results in the lowest activation energy. The energy difference between products, P, and reactants, R, is ΔE. For simple reactions in the gas phase, the energy can be calculated as a function of the distance between reactants. These calculations define an energy surface, and the dynamic course of the reaction on this energy surface can be determined. Even in the gas phase, this can only be done for very simple reactions. For reactions in solution, this can only be considered a conceptual model.

You might guess from your knowledge of thermodynamics that the energy is probably not the best parameter to use to characterize the dynamics of a reaction. A theory has been developed that instead discusses the reaction path in terms of the free energy. This theory is called the *transition state theory*. The basic postulate is that a *transition state* is formed in the course of the reaction, which is in equilibrium with the reactants. This model is shown schematically in Figure 4-4. In the transition state theory, the reactants go through a transition state that is at a higher free energy than reactants. They must go over this free energy barrier to produce products. In Figure 4-4, the difference in free energy between products and reactants is ΔG°, consistent with thermodynamic principles. The free energy difference between the transition state and reactants, $\Delta G^\ddagger$, is called the standard free energy of activation. In terms of transition state theory, the rate constant can be written as

$$k = (k_B T/h)\exp(-\Delta G^{\circ\ddagger}/RT) \tag{4-33}$$

where k_B is Boltzmann's constant and h is Planck's constant. (A more exact derivation includes an additional parameter, the transmission coefficient, which is usually equal to 1.) Since $\Delta G^{\circ\ddagger} = \Delta H^{\circ\ddagger} - T\Delta S^{\circ\ddagger}$, the rate constant can be restated as

$$k = (k_B T/h)\exp(\Delta S^{\circ\ddagger}/R)\exp(-\Delta H^{\circ\ddagger}/RT) \tag{4-34}$$

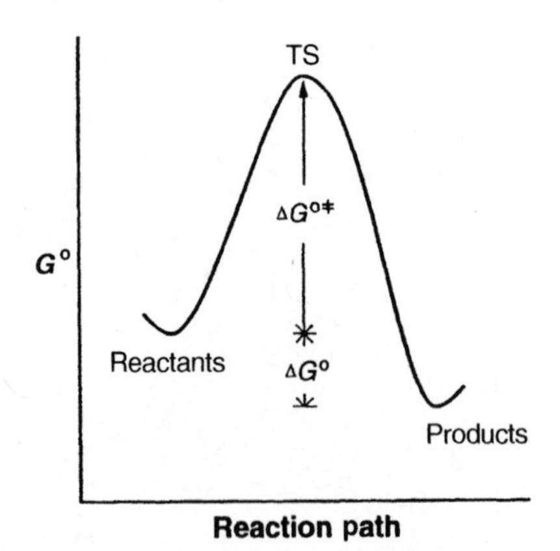

FIGURE 4-4. Schematic representation of the standard free energy, G°, versus the "reaction path" during the course of a reaction as discussed in the text. The free energy of activation, $\Delta G^{\circ\ddagger}$, is defined in this diagram. The transition state at the maximum of the free energy is indicated by TS. The reactants and products are at minima in the free energy curve.

Here $\Delta S^{\circ\ddagger}$ is the standard entropy of activation and $\Delta H^{\circ\ddagger}$ is the standard enthalpy of activation.

If the standard entropy and enthalpy of activation are assumed to be temperature independent, Eq. 4-34 can be differentiated to give

$$d(\ln k)/dT = (\Delta H^{\circ\ddagger} + RT)/(RT^2) \qquad (4\text{-}35)$$

Thus, the transition state theory predicts a temperature dependence of the rate constant very similar to the Arrhenius theory with $E_a = \Delta H^{\circ\ddagger} + RT$. Since RT is usually small relative to the standard enthalpy of activation, the activation energy and standard enthalpy of activation are usually quite similar. The Arrhenius and transition state formulations cannot be differentiated by this small difference in the temperature dependence since both the activation energy and the standard enthalpy of activation can be temperature dependent. From our knowledge of thermodynamics, we know that the enthalpy of activation is temperature dependent if a heat capacity difference exists between the transition state and the reactants. If the temperature dependence of the rate constant is to be analyzed in terms of transition state theory, it is more convenient to plot $\ln(k/T)$ versus $1/T$ as $d\ln(k/T)/dT = \Delta H^{\ddagger}/(RT^2)$ or $d\ln(k/T)/d(1/T) = -\Delta H^{\circ\ddagger}/R$.

A simple derivation of Eq. 4-33 is possible. The concentration of the transition state, TS, can be calculated from the relationship

$$[\text{TS}]/[\text{Reactants}] = \exp(-\Delta G^{\circ\ddagger}/RT)$$

where [Reactants] represents the concentrations of the reactants raised to the appropriate powers for the stoichiometric equation of the elementary step. The rate of the reaction is the concentration of the transition state multiplied by the frequency with which it passes over the free energy barrier. This frequency of passage over the barrier can be derived from an analysis of the reaction coordinate with statistical mechanics and is given by $k_B T/h$. Therefore, the rate of reaction is

$$R = (k_B T/h)[\text{TS}] = [\text{reactants}](k_B T/h)\exp(-\Delta G^{\circ\ddagger}/RT)$$

The rate constant in Eq. 4-33 follows directly from the above equation. This nonrigorous derivation provides a conceptual framework for the theory.

If a reaction goes through an intermediate, the intermediate would correspond to a minimum in the free energy, and each elementary step would have its own transition state. This is shown schematically in Figure 4-5 for a sequence of two elementary steps. The step with the highest free energy barrier is the rate-determining step.

Transition state theory is a very useful method for correlating and understanding kinetic studies. Because the framework of the theory is similar to thermodynamics, this produces a consistent way of discussing chemical reactions. The entropy and enthalpy of activation are often discussed in molecular terms. It should be remembered that, for kinetics as with thermodynamics, such interpretations should be approached with extreme caution.

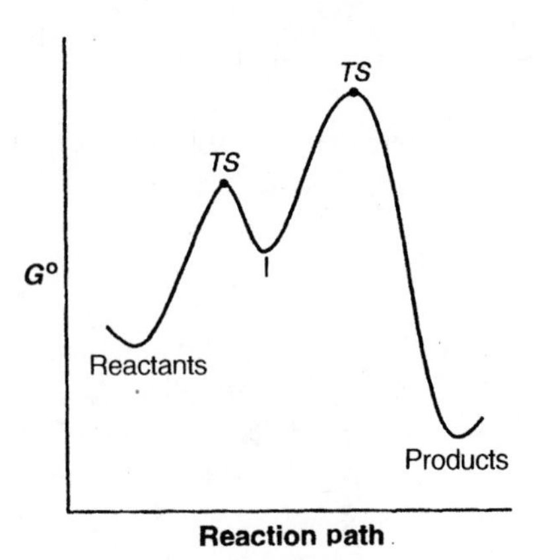

FIGURE 4-5. Schematic representation of the standard free energy, G°, versus the "reaction path" during the course of a two-step reaction. The intermediate, I, is at a minimum in the free energy whereas the transition state, TS, for each step is at a maximum. The reactants and products are also at minima in the free energy curve.

4.7 RELATIONSHIP BETWEEN THERMODYNAMICS AND KINETICS

Obviously, the principles of thermodynamics and kinetics must be self-consistent. In fact, this places some useful restrictions on the relationships between rate constants. In order to see how rate constants are related to equilibrium constants, consider the elementary step

$$\mathrm{A} + \mathrm{B} \underset{k_2}{\overset{k_1}{\rightleftharpoons}} \mathrm{AB} \tag{4-36}$$

The rate of this reaction is

$$R = -\frac{\mathrm{d[A]}}{\mathrm{d}t} = k_1[\mathrm{A}][\mathrm{B}] - k_2[\mathrm{AB}] \tag{4-37}$$

At equilibrium, the net rate of reaction must be zero. If $R = 0$, $k_1[\mathrm{A}]_\mathrm{e}[\mathrm{B}]_\mathrm{e} = k_2[\mathrm{AB}]_\mathrm{e}$, where the subscript e designates the equilibrium concentration. Thus, we see that

$$K = k_1/k_2 = [\mathrm{AB}]_\mathrm{e}/([\mathrm{A}]_\mathrm{e}[\mathrm{B}]_\mathrm{e}) \tag{4-38}$$

In this case, the equilibrium constant, K, is equal to the ratio of rate constants. Similar relationships between the rate constants and equilibrium constants can be found for more complex situations by setting the net rate equal to zero at equilibrium.

If Eq. 4-38 is cast into the framework of transition state theory, we obtain

$$K = \exp[-(\Delta G_1^{\circ\ddagger} - \Delta G_2^{\circ\ddagger})/RT] = \exp(-\Delta G^\circ/RT) \tag{4-39}$$

This result indicates the relationship between the standard free energy changes of activation and the standard free energy change for the reaction. This relationship can also be seen in the diagram of the free energy versus reaction path.

A more subtle relationship can be found if reaction cycles occur because of the *principle of detailed balance* or *microscopic reversibility*. This principle states that a mechanism for the reaction in the forward direction must also be a mechanism for the reaction in the reverse direction. Furthermore, at equilibrium, the forward and reverse rates are equal along each reaction pathway. This means that once we have found a possible mechanism for the reaction in one direction, we have also found a possible mechanism for the reaction in the other direction.

To illustrate this principle, consider the following triangular reaction mechanism:

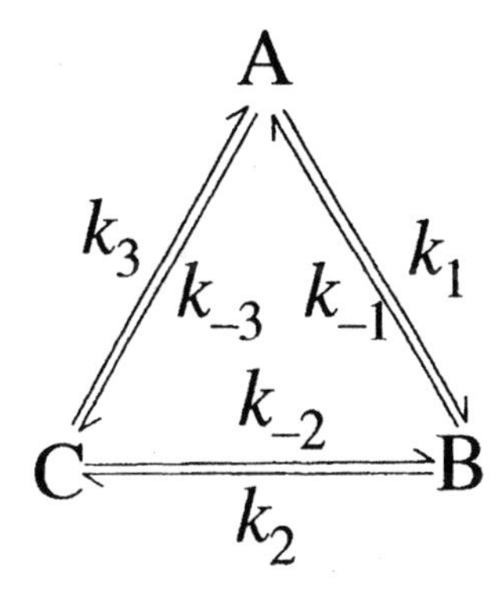

According to the principle of detailed balance, each of the individual reactions must be at equilibrium when equilibrium is attained, or

$$k_1[A]_e = k_{-1}[B]_e$$
$$k_2[B]_e = k_{-2}[C]_e$$
$$k_3[C]_e = k_{-3}[A]_e$$

If the right-hand sides of these equations are multiplied together and set equal to the left-hand sides of the equations multiplied together, we obtain

$$k_1 k_2 k_3 = k_{-1} k_{-2} k_{-3} \tag{4-40}$$

Thus, we find that the six rate constants are not independent, nor are the three equilibrium constants! This result may seem obvious, but many people have violated the principle of detailed balance in the literature. It is important to confirm that this principle is obeyed whenever reaction cycles are present.

4.8 REACTION RATES NEAR EQUILIBRIUM

Before we consider the application of chemical kinetics to biological systems, we will discuss the special case of reaction rates near equilibrium. As we shall see, the rate laws become quite simple near equilibrium, and methods exist that permit very fast reactions to be studied near equilibrium. We will begin with an elementary step that is a reversible first-order reaction, such as protein denaturation in the middle of the transition from the native to the denatured state:

$$\mathrm{N} \underset{k_r}{\overset{k_f}{\rightleftharpoons}} \mathrm{D} \tag{4-41}$$

The rate equation characterizing this system is

$$-\frac{\mathrm{d[N]}}{\mathrm{d}t} = k_\mathrm{f}[\mathrm{N}] - k_\mathrm{f}[\mathrm{D}] \tag{4-42}$$

We will now introduce new concentration variables; that is, we will set each concentration equal to its equilibrium value plus the deviation from equilibrium:

$$[\mathrm{N}] = [\mathrm{N}]_\mathrm{e} + \Delta[\mathrm{N}]$$
$$[\mathrm{D}] = [\mathrm{D}]_\mathrm{d} + \Delta[\mathrm{D}]$$

Note that the equilibrium concentrations are constants, independent of time, and mass conservation requires that $\Delta[\mathrm{N}] = -\Delta[\mathrm{D}]$. Inserting these relationships into Eq. 4-42 gives

$$\begin{aligned} -\frac{\mathrm{d}[\Delta \mathrm{N}]}{\mathrm{d}t} &= k_\mathrm{f}([\mathrm{N}]_\mathrm{e} + \Delta[\mathrm{N}]) - k_\mathrm{r}([\mathrm{D}]_\mathrm{e} - \Delta[\mathrm{N}]) \\ &= k_\mathrm{f}[\mathrm{N}]_\mathrm{e} - k_\mathrm{r}[\mathrm{D}]_\mathrm{e} + (k_\mathrm{f} + k_\mathrm{r})\Delta[\mathrm{N}] \\ &= (k_\mathrm{f} + k_\mathrm{r})\Delta[\mathrm{N}] = \Delta[\mathrm{N}]/\tau \end{aligned}$$

In deriving the above relationship, use has been made of the relationship $k_\mathrm{f}[\mathrm{N}]_\mathrm{e} = k_\mathrm{r}[\mathrm{D}]_\mathrm{e}$ and $1/\tau = k_\mathrm{f} + k_\mathrm{r}$; τ is called the relaxation time. This first-order differential equation can be integrated as before to give

$$\Delta[\mathrm{N}] = \Delta[\mathrm{N}]_0 \mathrm{e}^{-t/\tau} \tag{4-43}$$

where $\Delta[\mathrm{N}]_0$ is the deviation of N from its equilibrium value at $t = 0$.

Special note should be made of two points. First, the relaxation time can readily be obtained from experimental data simply by plotting the $\ln(\Delta[\mathrm{N}])$ versus t, and second the first-order rate constant characterizing this reaction, that is, the reciprocal relaxation time, is the sum of the two rate constants. If the equilibrium constant is known, both rate constants can be obtained from a single experiment.

How might such an experiment be carried out? In the case of protein denaturation, a small amount of denaturant such as urea might be added quickly to the solution. The ratio of the native and denatured proteins would then move to a new equilibrium value characteristic of the higher urea concentration. Alternatively, if thermal denaturation is being studied, the temperature might rapidly be raised, establishing a new equilibrium ratio. The rate of conversion of the native state to the denatured state can be measured after the experimental conditions are changed. For a reversible first-order reaction such as Eq. 4-41, the time course of the entire reaction follows a single exponential and the effective rate constant is the sum of the two rate constants. A wide range of methods exist for changing the equilibrium conditions: Concentration jumps and temperature jumps are particularly useful, but pressure jumps and electrical field jumps have also been used.

Studying rates near equilibrium is especially advantageous for higher order reactions and complex mechanisms. As a final example, we will consider the elementary step of an enzyme, E, combining with a substrate, S.

$$\mathrm{E} + \mathrm{S} \rightleftharpoons \mathrm{ES} \tag{4-44}$$

The rate equation characterizing this system is

$$-\frac{\mathrm{d}[\mathrm{E}]}{\mathrm{d}t} = k_\mathrm{f}[\mathrm{E}][\mathrm{S}] - k_\mathrm{r}[\mathrm{ES}] \tag{4-45}$$

Again, we will write the concentrations as the sum of the equilibrium concentration plus the deviation from equilibrium.

$$\begin{aligned} [\mathrm{E}] &= [\mathrm{E}]_\mathrm{e} + \Delta[\mathrm{E}] \\ [\mathrm{S}] &= [\mathrm{S}]_\mathrm{e} + \Delta[\mathrm{S}] \\ [\mathrm{ES}] &= [\mathrm{ES}]_\mathrm{e} + \Delta[\mathrm{ES}] \end{aligned}$$

Furthermore, mass conservation requires that $\Delta[\mathrm{E}] = \Delta[\mathrm{S}] = -\Delta[\mathrm{ES}]$. Inserting these relationships into Eq. 4-45 gives

$$-\frac{\mathrm{d}(\Delta[\mathrm{E}])}{\mathrm{d}t} = \{k_\mathrm{f}([\mathrm{E}]_\mathrm{e} + [\mathrm{S}]_\mathrm{e}) + k_\mathrm{r}\}\Delta[\mathrm{E}] + k_\mathrm{f}[\mathrm{E}]_\mathrm{e}[\mathrm{S}]_\mathrm{e} - k_\mathrm{r}[\mathrm{ES}]_\mathrm{e} + k_\mathrm{f}(\Delta[\mathrm{E}])^2$$

Near equilibrium, the deviation of concentrations from their equilibrium values is so small that the term $(\Delta[\mathrm{E}])^2$ can be neglected—this simplification in the rate equation results from being near equilibrium. For example, if the deviation is 10% (0.1), the square of the deviation is 1% (0.01). Furthermore, $k_\mathrm{f}[\mathrm{E}]_\mathrm{e}[\mathrm{S}]_\mathrm{e} = k_\mathrm{r}[\mathrm{ES}]_\mathrm{e}$, so that the final rate equation is

$$-\frac{\mathrm{d}[\Delta\mathrm{E}]}{\mathrm{d}t} = \frac{\Delta[\mathrm{E}]}{\tau} \tag{4-46}$$

with

$$1/\tau = k_\mathrm{f}([\mathrm{E}]_\mathrm{e} + [\mathrm{S}]_\mathrm{e}) + k_\mathrm{r} \tag{4-47}$$

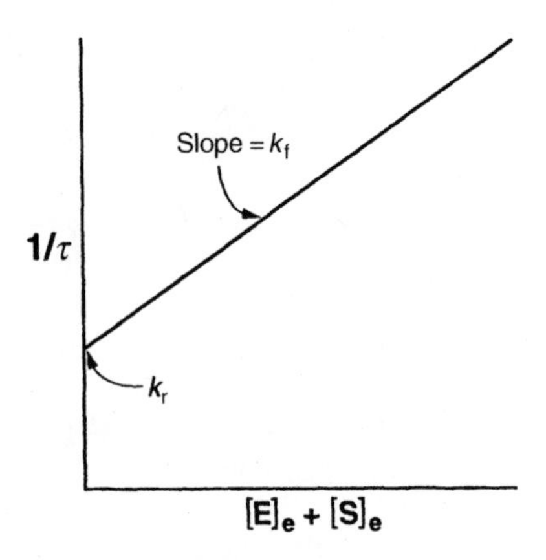

FIGURE 4-6. Schematic representation of a plot of the reciprocal relaxation time, $1/\tau$, versus the sum of the equilibrium concentrations, $[E]_e + [S]_e$, according to Eq. 4-47. As indicated, both of the rate constants can be obtained from the data.

Although the reaction of enzyme with substrate is a second-order reaction, the rate equation near equilibrium, Eq. 4-46, is the same as for a first-order reaction. In fact, all rate equations become first order near equilibrium! This is because only terms containing Δc are retained in the rate law: $(\Delta c)^2$ and higher powers are neglected. As before, Eq. 4-46 can easily be integrated, and the relaxation time can be obtained from the experimental data. If the relaxation time is determined at various equilibrium concentrations, a plot of $1/\tau$ versus $([E]_e + [S]_e)$ can be made, as shown schematically in Figure 4-6. The intercept is equal to k_r and the slope is equal to k_f. The study of reactions near equilibrium has been especially important for biological systems, as it has permitted the study of important elementary steps such as hydrogen bonding and protolytic reactions, as well as more complex processes such as enzyme catalysis and ligand binding to macromolecules.

REFERENCES

1. I. Tinoco, Jr., K. Sauer, & J. C. Wang, and J.D. Puglisi, *Physical Chemistry: Principles and Applications in Biological Sciences*, 4th edition, Prentice Hall, Englewood Cliffs, NJ, 2002.
2. C. R. Cantor and P. R. Schimmel, *Biophysical Chemistry*, Freeman, San Francisco, CA, 1980.
3. G. G. Hammes, *Enzyme Catalysis and Regulation*, Academic Press, New York, 1982.
4. A. Fersht, *Structure and Mechanism in Protein Science: A Guide to Enzyme Catalysis and Protein Folding*, Freeman, San Francisco, CA, 1999.

PROBLEMS

4-1. The activity of the antibiotic penicillin decomposes slowly when stored in a buffer at pH 7.0, 298 K. The time dependence of the penicillin antibiotic activity is given in the table below.

Time (weeks)	Penicillin activity (arbitrary units)
0	10,100
1.00	8180
2.00	6900
3.00	5380
5.00	3870
8.00	2000
10.00	1330
12.00	898
15.00	403
20.00	167

What is the rate law for this reaction, that is, what is the order of the reaction with respect to the penicillin concentration? Calculate the rate constant from the data if possible. (Data adapted from Ref. 1.)

4-2. The kinetics of the reaction

$$2Fe^{3+} + Sn^{2+} \rightarrow 2Fe^{2+} + Sn^{4+}$$

has been studied extensively in acidic aqueous solutions. When Fe^{2+} is added initially at relatively high concentrations, the rate law is

$$R = k[Fe^{3+}]^2[Sn^{2+}]/[Fe^{2+}]$$

Postulate a mechanism that is consistent with this rate law. Show that it is consistent by deriving the rate law from the proposed mechanism.

4-3. The conversion of L-malate to fumarate is catalyzed by the enzyme fumarase:

$$\begin{array}{c} COO^- \\ | \\ HO{-}C{-}H \\ | \\ H{-}C{-}H \\ | \\ COO^- \end{array} \rightleftharpoons \begin{array}{c} COO^- \\ | \\ C{-}H \\ \| \\ C{-}H \\ | \\ COO^- \end{array} + H_2O$$

The nonenzymatic conversion is very slow in neutral and alkaline media and has the rate law

$$R = k\,[\text{L-malate}]/[H^+]$$

Postulate two mechanisms for the nonenzymatic conversion and show that they are consistent with the rate law.

4-4. The radioactive decay rates of naturally occurring radioactive elements can be used to determine the age of very old materials. For example, ${}^{14}_{6}C$ is radioactive and emits a low-energy electron with a half-life of about 5730 years. Through a balance of natural processes, the ratio of ${}^{14}C/{}^{12}C$ is constant in living organisms. However, in dead organisms or material, this ratio decreases as the ^{14}C decays. Since the radioactive decay is known to be a first-order reaction, the age of the material can be estimated by measuring the decrease in the $^{14}C/^{12}C$ ratio. Suppose a piece of ancient wool is found in which the ratio has been found to decrease by 20%. What is the age of the wool?

4-5. The nonenzymatic hydration of CO_2 can be written as

$$CO_2 + H_2O \rightleftharpoons H_2CO_3$$

The reaction is found to be first order in both directions. Because the water concentration is constant, it does not appear in the expression for the equilibrium or rate equation. The first-order rate constant in the forward direction has a value of $0.0375\,s^{-1}$ at 298 K and $0.0021\,s^{-1}$ at 273 K. The thermodynamic parameters for the equilibrium constant at 298 K are $\Delta H^\circ = 1.13\,kcal/mol$ and $\Delta S^\circ = -8.00\,cal/(mol\,K)$.

a. Calculate the Arrhenius activation energy for the rate constant of the forward reaction. Also calculate the enthalpy and entropy of activation according to transition state theory at 298 K.

b. Calculate the rate constant for the reverse reaction at 273 and 298 K. Assume that ΔH° is independent of the temperature over this temperature range.

c. Calculate the Arrhenius activation energy for the rate constant of the reverse reaction. Again, also calculate the enthalpy and entropy of activation at 298 K.

4-6. A hydrogen bonded dimer is formed between 2-pyridone according to the reaction

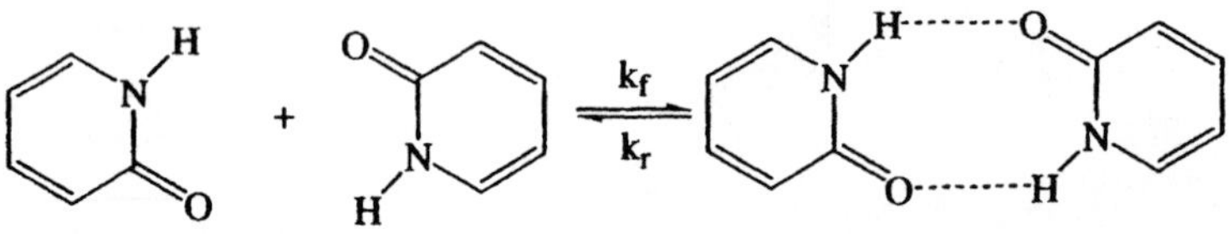

The relaxation time for this reaction, which occurs in nanoseconds, has been determined in chloroform at 298 K at various concentrations of 2-pyridone. The data obtained are [G. G. Hammes and A. C. Park, *J. Am. Chem. Soc.* **91**, 956 (1969)]

2-Pyridone (M)	$10^9\tau$ (s)
0.500	2.3
0.352	2.7
0.251	3.3
0.151	4.0
0.101	5.3

From these data calculate the equilibrium and rate constants characterizing this reaction. *Hint*: If the expression for the relaxation time is squared, the concentration dependence can be expressed as a simple function of the total concentration of 2-pyridone.

CHAPTER 5

Applications of Kinetics to Biological Systems

5.1 INTRODUCTION

We will now consider some applications of kinetic studies to biological systems. This discussion will center on enzymes, as kinetic analyses of enzymes represent a major research field and have provided considerable insight into how enzymes work. Enzymes are proteins that are incredibly efficient catalysts: They typically increase the rate of a chemical reaction by six orders of magnitude or more. Understanding how this catalytic efficiency is achieved and the exquisite specificity of enzymes has intrigued biologists for many decades and still provides a challenge for modern research. Since enzyme deficiencies are the source of many diseases, enzymes are also a target for medical research and modern therapeutics. The field of enzyme kinetics and mechanisms is so vast that we will present only an abbreviated discussion. More complete discussions are available (cf. Refs. 1-3).

We will also discuss catalysis by RNA (ribozymes), which is important in the processing of RNA in biological systems, as well as some kinetic studies of DNA denaturation and renaturation.

5.2 ENZYME CATALYSIS: THE MICHAELIS–MENTEN MECHANISM

We now consider the analysis of a simple enzymatic reaction. This discussion will introduce some new concepts that are particularly useful for analyzing complex systems. The conversion of a single substrate to product will be taken as a prototype reaction:

$$S \rightarrow P \qquad (5\text{-}1)$$

A typical example is the hydration of fumarate to give L-malate (and the reverse dehydration reaction) catalyzed by the enzyme fumarase. If a very low

---

Physical Chemistry for the Biological Sciences by Gordon G. Hammes

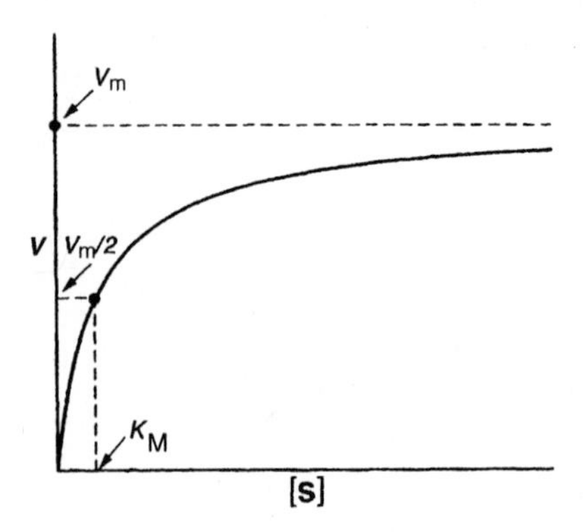

FIGURE 5-1. A plot of the initial velocity, v, versus the substrate concentration, [S], for an enzymatic reaction that can be described by the Michaelis–Menten mechanism.

concentration of enzyme is used relative to the concentration of substrate, a plot of the initial rate of the reaction, or initial velocity, v, is hyperbolic, as shown in Figure 5-1. The limiting initial velocity at high concentrations of substrate is called the *maximum velocity*, V_m, and the concentration of substrate at which the initial velocity is equal to $V_m/2$ is called the *Michaelis constant*, K_M.

A mechanism that quantitatively accounts for the dependence of the initial velocity on substrate concentration was postulated by Michaelis and Menten. It can be written in terms of elementary steps as

$$\mathrm{E} + \mathrm{S} \underset{k_2}{\overset{k_1}{\rightleftharpoons}} \mathrm{ES} \overset{k_3}{\rightarrow} \mathrm{E} + \mathrm{P} \tag{5-2}$$

where E is the enzyme and ES is a complex consisting of the enzyme and substrate. The total concentration of enzyme, $[E_0]$, is assumed to be much less than the total concentration of substrate, $[S_0]$. This mechanism is an example of *catalysis* in that the enzyme is not consumed in the overall reaction, which is greatly accelerated by the enzyme. If we tried to do an exact mathematical analysis of the mechanism in Eq. 5-2, coupled differential equations would need to be solved. Fortunately, this is not necessary. We will consider two approximations that can adequately account for the data.

The first is the *equilibrium approximation*. With this approximation, the first step is assumed to be always at equilibrium during the course of the reaction. This means that the equilibration of the first step is much more rapid than the breakdown of ES to P, or $k_3 \ll k_2$. We then have

$$\frac{\mathrm{d[P]}}{\mathrm{d}t} = v = k_3[\mathrm{ES}] \tag{5-3}$$

and

$$k_1/k_2 = [\mathrm{ES}]/([\mathrm{E}][\mathrm{S}]) \tag{5-4}$$

Furthermore, conservation of mass requires that

$$\begin{aligned}[E_0] &= [E] + [ES] \\ &= [ES](1 + [E]/[ES]) = [ES]\{1 + k_2/(k_1[S])\}\end{aligned} \tag{5-5}$$

or

$$[ES] = [E_0]/\{1 + k_2/(k_1[S])\} \tag{5-6}$$

Substitution of Eq. 5-6 into 5-3 gives

$$\frac{d[P]}{dt} = v = \frac{k_3[E_0]}{1 + k_2/(k_1[S])} \tag{5-7}$$

which can be rewritten as

$$v = \frac{V_m}{1 + K_M/[S]} \tag{5-8}$$

with

$$V_m = k_3[E_0] \tag{5-9}$$

and

$$K_M = k_2/k_1 \tag{5-10}$$

Equation 5-8 is found to account quantitatively for the data. The maximum velocity is proportional to the total enzyme concentration, and when $K_M = [S]$, $v = V_m/2$, as required.

In some cases, it is possible to obtain independent estimates or measurements of the equilibrium constant for the first step. Sometimes this independent measurement is in good agreement with the constant obtained from kinetic studies, sometimes it is not. Obviously, when the Michaelis constant and the equilibrium constant are not in agreement, the mechanism must be reexamined.

A more general analysis of the Michaelis–Menten mechanism makes use of the *steady-state approximation.* This approximation does not make any assumptions about the relative values of the rate constants but assumes that the concentrations of E and ES are very small relative to S, consistent with the experimental conditions. Under these conditions, it is assumed that the rate of change of the concentrations of E and ES is very small relative to the rate of change of the concentration of S, or

$$d[E]/dt = d[ES]/dt \approx 0$$

This is the steady-state approximation, an approximation that is very important for analyzing complex mechanisms. A careful mathematical analysis shows that this is a very good approximation under the experimental conditions used, namely, when the total substrate concentration is much greater than the total enzyme concentration.

What does the steady-state approximation mean? If we go back to the mechanism (Eq. 5-2), we find that

$$-\frac{d[ES]}{dt} = (k_2 + k_3)[ES] - k_1[E][S] \approx 0$$

or

$$k_1/(k_2 + k_3) = [ES]/([E][S]) \tag{5-11}$$

Note the similarity of this equation to the equilibrium constant (Eq. 5-4). This means that the ratio of concentrations remains constant. However, the ratio is not the equilibrium concentrations; it is the steady-state concentrations. If $k_3 \ll k_2$, the steady state is the same as the equilibrium state, so that the equilibrium condition is a special case of the steady state. For all other steady-state situations, the ratio of concentrations is always less than the equilibrium ratio.

We can now calculate the rate law for the steady-state approximation exactly as for the equilibrium approximation.

$$[E_0] = [ES](1 + [E]/[ES]) = [ES]\{1 + (k_2 + k_3)/(k_1[S])\} \tag{5-12}$$

$$v = \frac{d[P]}{dt} = k_3[ES] = \frac{k_3[E_0]}{1 + (k_2 + k_3)/(k_1[S])}$$

$$v = \frac{V_m}{1 + K_M/[S]} \tag{5-13}$$

with

$$V_m = k_3[E_0] \tag{5-14}$$

and

$$K_M = (k_2 + k_3)/k_1 \tag{5-15}$$

Thus, it is clear that the equilibrium approximation and steady-state approximation give rate laws that are indistinguishable experimentally. Most generally, the Michaelis constant is a steady-state constant, but in a limiting case it can be an equilibrium constant. As previously stated, both situations have been observed. The steady-state approximation is more general than the equilibrium approximation and is typically employed in the analysis of enzyme mechanisms.

With modern computers, data analysis is very easy, and experimental data can be fit directly to Eq. 5-8 (or Eq. 5-13) by a nonlinear least squares fitting procedure.

However, it is always a good idea to be sure that the data indeed do conform to the equation of choice by a preliminary analysis. Equation 5-8 can be linearized by taking its reciprocal:

$$1/v = 1/V_m + (K_M/V_m)/[S] \tag{5-16}$$

This is called the Lineweaver-Burke equation. A plot of $1/v$ versus $1/[S]$ is linear, and the slope and intercept can be used to calculate the maximum velocity and Michaelis constant. This equation can be used for the final data analysis providing proper statistical weighting is used. An alternative linearization of Eq. 5-8 is to multiply Eq. 5-16 by [S] to give

$$[S]/v = [S]/V_m + K_M/V_m \tag{5-17}$$

This equation predicts that a plot of $[S]/v$ versus [S] should be linear and gives a better weighting of the data than the double reciprocal plot (Eq. 5-16). Obviously, the final results should be independent of how the data are plotted and analyzed! Examples of both plots are given in Figure 5-2.

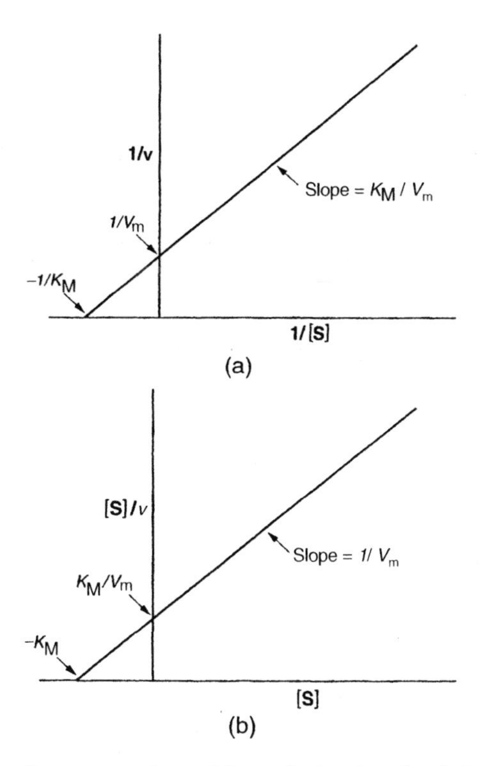

FIGURE 5-2. Linear plots commonly used for analyzing data that follow Michaelis-Menten kinetics. (a) Plot of the reciprocal initial velocity, v, versus the reciprocal substrate concentration, $1/[S]$. (b) Plot of $[S]/v$ versus [S].

The steady-state analysis can be extended to much more complex enzyme mechanisms, as well as to other biological processes, but this is beyond the scope of this presentation. How can enzyme catalysis be understood in terms of transition state theory? The simple explanation is that enzyme catalysis lowers the standard free energy of activation. This can be either an entropic or enthalpic effect. Much more detailed interpretations of enzyme catalysis in terms of transition state theory have been developed. The basic idea is that formation of the enzyme-substrate complex alters the free energy profile of the reaction such that the free energy of activation for the reaction is greatly lowered. Essentially, the free energy change associated with binding of the substrate to the enzyme is used to promote catalysis.

5.3 α-CHYMOTRYPSIN

As an example of how mechanisms can be developed for the action of enzymes, early studies of the enzyme a-chymotrypsin will be discussed. Proteolytic enzymes were among the first to be studied, not because of their intrinsic interest, but because they could easily be isolated in a reasonably pure form. In fact, for many years the availability of large quantities of pure enzyme severely restricted the range of enzymes that could be studied in mechanistic detail. With the ease of cloning and modern expression systems, this is no longer a limitation.

The enzyme a-chymotrypsin has a molecular weight of about 25,000 and consists of three polypeptide chains covalently linked by disulfides. It is an *endopeptidase*; that is, it cleaves peptide bonds in the middle of a protein. Enzymes that cleave peptide bonds at the end of a protein are called *exopeptidases.* The overall reaction can be written as

$$\mathrm{R{-}\overset{\displaystyle H}{\underset{\displaystyle NH\wr}{C}}{-}\overset{\displaystyle O}{\overset{\|}{C}}{-}\overset{\displaystyle H}{N}{\sim\sim}} \xrightarrow{H_2O} \mathrm{R{-}\overset{\displaystyle H}{\underset{\displaystyle NH\wr}{C}}{-}\overset{\displaystyle O}{\overset{\|}{C}}{-}O^{-}\ {}^{+}H_3N{\sim\sim}} \tag{5-18}$$

Experimentally, it is found that the enzyme has a strong preference for R being an aromatic group, amino acids phenylalanine, tyrosine, and tryptophan, but hydrophobic amino acids such as isoleucine, leucine, and valine are also good substrates. Studying this reaction with a protein as substrate is very difficult because the molecular structure of the substrate is changing continuously as peptide bonds are cleaved. This makes quantitative interpretation of the kinetics virtually impossible. Consequently, the first step in elucidating the molecular details of the enzyme was to develop "model" substrates, molecules that have the important features of the protein substrate but are much simpler. Chymotrypsin is also a good esterase; that is, it hydrolyzes esters. This proved very useful in elucidating the mechanism of action of the enzyme. The most successful model substrates have the general structure

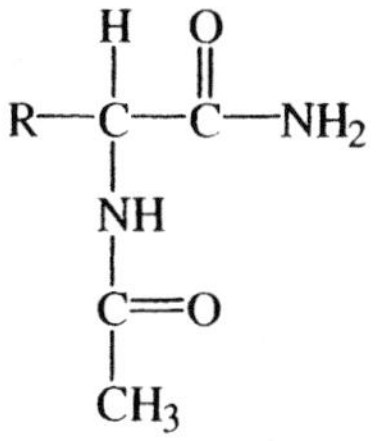

```
    H  O                    H  O
    |  ||                   |  ||
R—C—C—NH2              R—C—C—O—R′
    |                       |
    NH                      NH
    |                       |
    C=O                     C=O
    |                       |
    CH3                     CH3
   Amide                   Ester
```

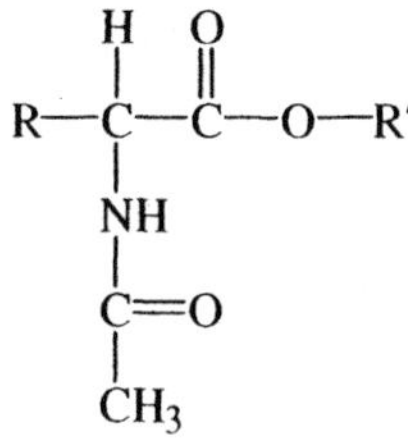

where R is the aromatic residue associated with phenylalanine, tyrosine, and tryptophan. Note that the amino group of the aromatic amino acid is acetylated. The enzyme will not work on a free amine, as might be expected for an endopeptidase. Furthermore, only L-amino acids are substrates.

We will consider some of the results obtained with tryptophan as the amino acid.

R = (indol-3-yl)methyl, NH

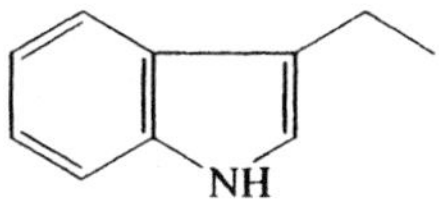

The hydrolysis of tryptophan model substrates follows Michaelis-Menten kinetics, and some of the results obtained for various esters and the amide are shown in Table 5-1. In this table, $k_{cat} = V_m/[E_0]$, and R′ is the group covalently linked to the tryptophan carboxyl. The fact that k_{cat} is essentially the same for all esters suggests that a common intermediate is formed and that the breakdown of the intermediate is the rate determining step in the mechanism. A different mechanism must occur for the amide, or there is a different slow step in the mechanism.

An inherent deficiency of steady-state kinetic studies of enzymes is that the enzyme concentration is very low. Consequently, the intermediates in the reaction sequence cannot be detected directly. Limited information about the intermediates can be obtained through steady-state kinetic studies by methods not discussed here,

TABLE 5-1. Steady-State Constants for Chymotrypsin.

R′	$k_{cat}(s^{-1})$	K_M (mM)
Ethyl	27	~5
Methyl	28	~5
p-Nitrophenyl	31	~5
Amide	0.03	~0.09

Sources: Adapted from R. J. Foster and C. Niemann, *J. Am. Chem. Soc.* **77**, 1886 (1955) and L. W. Cunningham and C. S. Brown, *J. Biol. Chem.* **221**, 287 (1956).

for example, the pH dependence of the kinetic parameters and the use of isotopically labeled substrates that alter the kinetic parameters. However, in order to study the intermediates directly, it is necessary to use sufficiently high enzyme concentrations so that the intermediates can be observed directly. This is the realm of *transient kinetics.* The difficulty for enzymatic reactions is that the reactions become very fast, typically occurring in the millisecond and microsecond range. This requires special experimental techniques. Transient kinetic studies have proved invaluable in elucidating how enzymes work.

In the case of chymotrypsin, a substrate was sought for which a color change occurred upon reaction in order that the changes in concentration could easily and rapidly be observed. The first such substrate studied was *p*-nitrophenyl acetate, which is hydrolyzed by chymotrypsin:

$$NO_2-C_6H_4-O-\overset{O}{\overset{\|}{C}}-CH_3 \xrightarrow{H_2O} NO_2-C_6H_4-O^- + CH_3-\overset{O}{\overset{\|}{C}}-O^- \tag{5-19}$$

The advantage of this substrate is that the phenolate ion product is yellow so that the time course of the reaction can easily be followed spectrophotometrically. With this substrate, it was possible to look at the establishment of the steady state, as well as the steady-state reaction. A very important finding was that the reaction becomes extremely slow at low pH. An intermediate, in fact, can be isolated by using a radioactive acetyl group in the substrate. The intermediate is an acylated enzyme. Later studies showed that the acetyl group is attached to a serine residue on chymotrypsin (4).

$$\text{Enzyme}-O-\overset{O}{\overset{\|}{C}}-CH_3$$

On the basis of the kinetic studies and isolation of the acyl enzyme, a possible mechanism is the binding of substrate, acylation of enzyme, and deacylation of the enzyme. The elementary steps can be written as

$$\begin{aligned} E + S \underset{k_{-1}}{\overset{k_1}{\rightleftharpoons}} ES \overset{k_2}{\rightarrow} \text{E-acyl} &\overset{k_3}{\rightarrow} E + \text{Acetate} \\ &+ p\text{-Nitrophenolate} \end{aligned} \tag{5-20}$$

Does this mechanism give the correct steady-state rate law? The rate law can be derived as for the simple Michaelis-Menten mechanism using the steady-state approximation for all of the enzyme species and the mass conservation relationship for enzyme:

$$[E_0] = [E] + [ES] + [\text{E-acyl}] \tag{5-21}$$

$$-\frac{d[ES]}{dt} = (k_{-1} + k_2)[ES] - k_1[E][S] \approx 0$$

$$-\frac{d[E\text{-acyl}]}{dt} = k_3[E\text{-acyl}] - k_2[ES] \approx 0$$

$$v = k_3[E\text{-acyl}] \tag{5-22}$$

When these relationship are combined, it is found that the Michaelis-Menten rate law, Eq. 5-8, is obtained, with

$$V_m = \frac{k_2 k_3[E_0]}{k_2 + k_3} \tag{5-23}$$

$$K_M = \left(\frac{k_{-1} + k_2}{k_1}\right)\left(\frac{k_3}{k_2 + k_3}\right) \tag{5-24}$$

For ester substrates, the slow step is deacylation of the enzyme, or $k_3 \ll k_2$. In this case, $k_{cat} = k_3$ and $K_M = [(k_{-1} + k_2)/k_1](k_3/k_2)$.

How did the transient kinetic studies contribute to the postulation of this mechanism? It is worth spending some time to analyze the mechanism in Eq. 5-20 in terms of transient kinetics. This analysis can serve as a prototype for understanding how transient kinetics can be used to probe enzyme mechanisms, not only for chymotrypsin but for other enzyme reactions as well. The proposed mechanism predicts that the acyl enzyme should accumulate and be directly observable. If we consider only a single turnover of the enzyme at very early times and high substrate concentrations, all of the enzyme should be converted to the acyl enzyme, followed by a very slow conversion of the acyl enzyme to the free enzyme through hydrolysis ($k_2 \gg k_3$). The slow conversion to enzyme will follow Michaelis-Menten kinetics (Eq. 5-13) initially as long as the total substrate concentration is much greater than the total enzyme concentration. Thus, the complete solution to the rate equations for the mechanism in Eq. 5-20 should contain two terms: (1) the rate of conversion of the enzyme to the acyl enzyme, and (2) the rate of the overall reaction that is limited by the rate of hydrolysis of the acyl enzyme.

To simplify the analysis, let us assume the following: k_3 is approximately zero at early times ($k_2 \gg k_3$); the formation of the initial enzyme–substrate complex, ES, is very rapid relative to the rates of the acylation and deacylation steps so that ES is in a steady state ($d[ES]/dt = 0$); and the total substrate concentration is much greater than the total enzyme concentration. The last assumption is similar to that for the steady-state approximation used earlier for the Michaelis–Menten mechanism. However, in this case the enzyme concentration is sufficiently high so that the concentration of the intermediate can be detected. The rate of formation of the intermediate is given by

$$\frac{d[E\text{-acyl}]}{dt} = k_2[ES] \tag{5-25}$$

Because the phenolate ion is formed when the acyl enzyme is formed, the rate of formation of phenolate ion is the same as the rate of formation of the acyl enzyme. This is actually a measurement of the rate of establishment of the steady state. In order to integrate this equation, we make use of the identity

$$[E] + [ES] = [ES](1 + [E]/[ES]) = [ES]\{1 + (k_{-1} + k_2)/(k_1[S])\}$$

and mass conservation

$$[E_0] = [E] + [ES] + [E\text{-acyl}]$$

or

$$[E] + [ES] + [E_0] - [E\text{-acyl}]$$

Insertion of these relationships into Eq. 5-25 gives

$$\frac{d[E\text{-acyl}]}{dt} = \frac{k_2([E_0] - [E\text{-acyl}])}{1 + (k_{-1} + k_2)/(k_1[S])} \tag{5-26}$$

Integration of this equation gives

$$[E\text{-acyl}] = [E_0](1 - e^{-k_2' t}) \tag{5-27}$$

with $k_2' = k_2/\{1 + (k_{-1} + k_2)/(k_1[S])\}$.

Note that when $t = 0$, $[E\text{-acyl}] = 0$, and when $t = \infty$, $[E\text{-acyl}] = [E_0]$, as expected. The total rate of phenolate, P_1, formation at early times is

$$[P_1] = vt + [E_0](1 - e^{-k_2' t}) \tag{5-28}$$

The time course of the reaction as described by Eq. 5-28 is shown schematically in Figure 5-3. The second term in the equation dominates at early times, and the exponential rise can be analyzed to give k_2'. The linear portion of the curve corresponds to the Michaelis–Menten initial velocity and can be used to obtain V_m and K_M. The maximum velocity for this limiting case is $k_3[E_0]$, so that both k_2 and k_3 can be determined. A more exact analysis that does not assume $k_3 = 0$ is a bit more complex, but the end result is similar. The rate equation contains the same two terms as in Eq. 5-28. However, the second term is now more complex: the exponent is $(k_2' + k_3)t$ instead of $k_2't$; and the amplitude is $[E_0][k_2'/(k_2' + k_3)]^2$ rather than $[E]_0$. The first term is the same, vt, with the maximum velocity and Michaelis constant defined by Eqs. 5-23 and 5-24.

Finally, we consider the situation when the rate of hydrolysis of the acyl enzyme is very fast relative to its formation ($k_3 \gg k_2$). We have seen previously that fast steps occurring after the rate-determining step do not enter into the rate law.

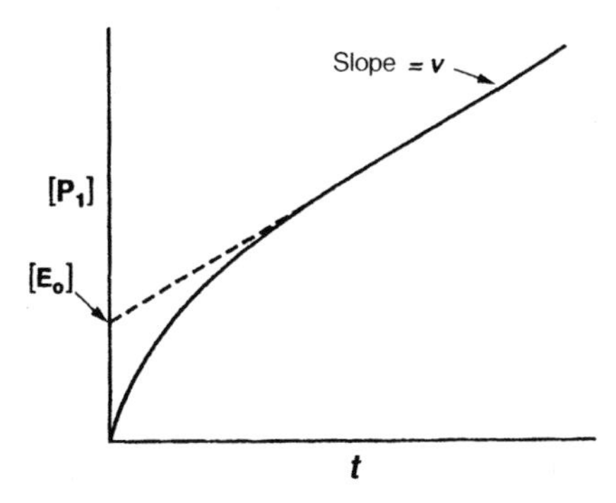

FIGURE 5-3. Schematic representation for the kinetics of an enzymatic reaction displaying "burst" kinetics. The product of the reaction, P_1, is plotted versus the time. As described by Eq. 5-28 for the limiting case of $k_2 \gg k_3$, the slope of the linear portion of the curve is the steady-state initial velocity, v, and the initial exponential time dependence is characterized by the rate constant k_2' and a "burst" amplitude of $[E_0]$.

Therefore, the mechanism is equivalent to the Michaelis–Menten mechanism with only a single intermediate (Eq. 5-2), with the rate-determining step being the acylation of the enzyme ($V_m = k_2[E_0]$). If the transient kinetics are analyzed, the rate is still given by Eq. 5-28, but the amplitude of the second term is essentially zero, rather than $[E_0]$. Thus, the transient kinetics and steady-state kinetics provide the same information. The presence of an initial "burst" of product, as shown in Figure 5-3, is commonly used as a diagnostic for the presence of an intermediate. If the burst is not present, that is, the straight line portion of the curve extrapolates through zero, this means either that an intermediate is not formed, or that its rate of disappearance is much faster than its rate of appearance. This exhaustive, perhaps exhausting, discussion of the mechanism in Eq. 5-20 is a good example of what must be done in order to arrive at a mechanism for an enzyme reaction by kinetic studies. Both steady-state kinetics and transient kinetics are necessary for a complete picture.

The proposed mechanism accounts for all of the facts presented: the observed rate law, the same k_{cat} for all ester substrates, and a common intermediate. What about amides? For amides, it turns out that the slow step is acylation of the enzyme, that is, $k_2 \ll k_3$, so that $k_{cat} = k_2$ and $K_M = (k_{-1} + k_2)/k_1$. As discussed above, if the acylation of the enzyme is slow relative to deacylation, $k_3 \gg k_2$, the intermediate does not accumulate and therefore cannot be observed directly. Literally, hundreds of experiments are consistent with this mechanism—a very remarkable achievement. Additional experiments that permitted the very fast reactions prior to the acylation of the enzyme to be studied show that this mechanism is too simple. At least one additional elementary step is required. The sequential elementary steps are (1) binding of substrate to the enzyme, (2) a conformational change of the enzyme–substrate complex, (3) acylation of the enzyme, and (4) deacylation of the enzyme.

The ultimate goal of kinetic studies is to understand the mechanism in terms of molecular structure. In the case of chymotrypsin, the three-dimensional structure of the enzyme is known and is shown in Figure 5-4a (see color plates). A well-defined binding pocket is observed for the aromatic side chain of the specific substrates of the enzyme. The pocket is lined with nonpolar side chains of amino acids and is very hydrophobic. Up to three amino acids coupled to the N-terminus of the aromatic amino acid interact with a short range of antiparallel β-sheet in the enzyme. A hydrophobic site is also observed for the amino acid attached to the C-terminus of the substrate. This explains why a free carboxyl group cannot bind and therefore why chymotrypsin is an endopentidase. As shown in Figure 5-4b (see color plates), a triad of amino acids, serine-195, histidine-57, and aspartate-102, is observed in the active site region and is found in other serine proteases. The imidazole acts as a general base during the nucleophilic attack of the serine hydroxyl on the substrate. A tetrahedral intermediate probably is formed prior to acyl enzyme formation. Water serves as the nucleophile for the hydrolysis of the acyl enzyme, with imidazole again participating as the general base. An abbreviated version of the catalytic mechanism is shown in Figure 5-5.

This greatly truncated story of the elucidation of the mechanism of action of chymotrypsin illustrates several important concepts: the steady-state approximation, the use of steady-state kinetics in determining chemical mechanisms, the importance of transient kinetics in elucidating intermediates in a mechanism, and the interpretation of the mechanism in terms of molecular structure. More complete discussions of chymotrypsin are available (1,2,5).

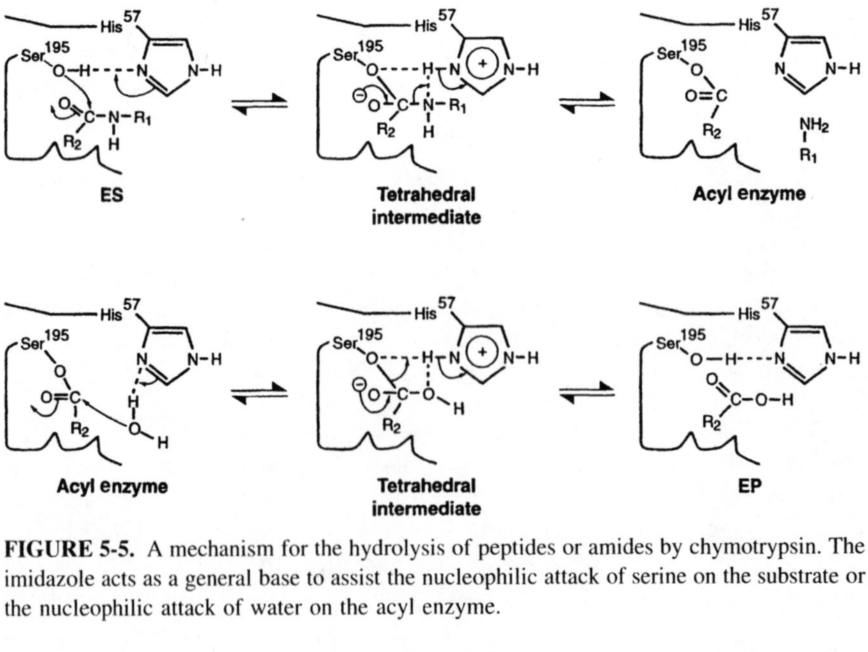

FIGURE 5-5. A mechanism for the hydrolysis of peptides or amides by chymotrypsin. The imidazole acts as a general base to assist the nucleophilic attack of serine on the substrate or the nucleophilic attack of water on the acyl enzyme.

5.4 PROTEIN TYROSINE PHOSPHATASE

In our earlier discussion of metabolism, we saw that the energy obtained in glycolysis is stored as a phosphate ester in ATP and serves as a source of free energy for biosynthesis. The importance of phosphate esters is not confined to this function. Phosphorylation and dephosphorylation of proteins play a key role in signal transduction and the regulation of many cell functions. For example, the binding of hormones to cell surfaces can trigger a cascade of such reactions that regulate metabolism within the cell. The cell processes modulated by this mechanism include T-cell activation, the cell cycle, DNA replication, transcription and translation, and programmed cell death, among many others. Both the kinases, responsible for protein phosphorylation, and the phosphatases, responsible for protein dephosphorylation, have been studied extensively. We shall restrict this discussion to the phosphatases, and primarily to one particular type of phosphatase.

The phosphorylation of proteins primarily occurs at three sites, namely, the side chains of threonine, serine, and tyrosine. Protein phosphatases can be divided into two structural classes: the serine/threonine-specific phosphatases that require metal ions, and tyrosine phosphatases that do not require metals but use a nucleophilic cysteine to cleave the phosphate. The latter class includes dual function phosphatases that can hydrolyze phosphate esters on serine and threonine, as well as tyrosine. Many reviews of these enzymes are available (cf. Refs. 6 and 7). We will discuss the mechanism of a particular example of tyrosine-specific phosphatase, namely, an enzyme found in both human and rat that has been extensively studied structurally and mechanistically (6).

The overall three-dimensional structure of the catalytic domain of the enzyme is shown in Figure 5-6a (see color plates), and the active site region is shown in Figure 5-6b (see color plates) with vanadate bound to the catalytic site. Vanadate is frequently used as a model for phosphate esters because the oxygen bonding to vanadium is similar to the oxygen bonding to phosphorus, and vanadate binds tightly to the enzyme. A cysteine rests at the base of the active-site cleft. An aspartic acid and a threonine (or serine in similar enzymes) are also near the active site and are postulated to play a role in catalysis, as discussed below.

The kinetic mechanism that has been postulated is the binding of the protein substrate followed by transfer of the phosphoryl group from tyrosine to the cysteine at the active site, and finally hydrolysis of the phosphorylated cysteine:

$$\mathrm{E + S} \underset{k_{-1}}{\overset{k_1}{\rightleftharpoons}} \mathrm{ES} \xrightarrow[k_2]{\;\text{R}-\text{C}_6\text{H}_4-\text{OH}\;\uparrow} \mathrm{E\text{-}PO_3} \xrightarrow{k_3} \mathrm{E + HPO_4} \tag{5-29}$$

As with chymotrypsin, a protein is not a convenient substrate to use because of the difficulty in phosphorylating a single specific tyrosine and the lack of a convenient method for following the reaction progress. In this case, the model substrate used was *p*-nitrophenyl phosphate. Similar to chymotrypsin, when the enzyme is phosphorylated the yellow phenolate ion is released so that the reaction progress can easily be monitored. Extensive steady-state and transient kinetic studies have been carried out, and we will only present a few selected results that summarize the findings (8,9).

The role of the cysteine was firmly established by site-specific mutagenesis, as converting the cysteine to a serine results in an inactive enzyme. The cysteine is postulated to function as a nucleophile with the formation of a cysteine phosphate. However, when transient kinetic studies were carried out with the native enzyme, a burst phase was not observed. This is shown in Figure 5-7a. As we noted previously, this indicates that either a phosphoenzyme intermediate is not formed or its hydrolysis is much faster than its formation. The aspartic acid and serine or threonine shown at the active site of the enzyme in Figure 5-6 (see color plates) are conserved in many different enzymes of this class of tyrosine phosphatases. Consequently, it was decided to mutate these residues. If the serine is mutated to alanine, the enzyme is still active, but now a burst is observed, as shown in Figure 5-7b. This burst was postulated to represent the formation of the phosphoenzyme, and the rate constants were obtained for the phosphorylation of cysteine and its hydrolysis. The results of

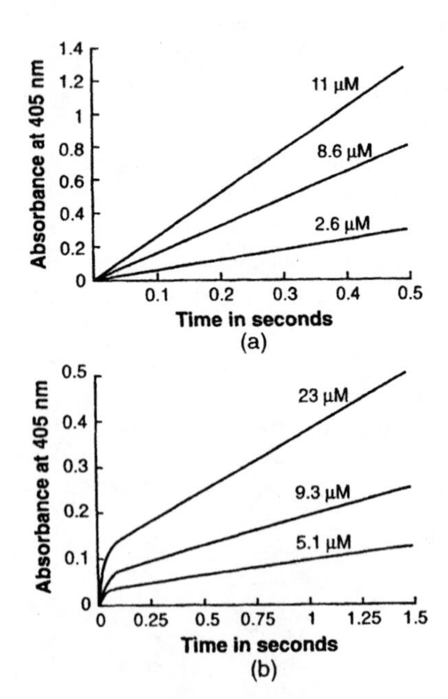

FIGURE 5-7. Transient kinetic time course for the hydrolysis of *p*-nitrophenyl phosphate by native (a) and mutant (b) protein tyrosine phosphatase. The point mutation substituted alanine for serine 222. The concentrations of substrate are shown next to the traces. Reprinted with permission from D. L. Lohse, J. M. Denu, N. Santoro, and J. E. Dixon, *Biochemistry* **36**, 4568 (1997). Copyright (c) 1997 American Chemical Society.

TABLE 5-2. Kinetic Parameters for Native and Mutated Tyrosine Phosphatases

Enzyme	k_{cat} (s^{-1})	k_2 (s^{-1})	k_3 (s^{-1})	K(μM)[a]
Native	20			
Serine 222 → alanine	1.3	34	1.4	0.46
Aspartate 181 → asparagine	0.27			
Serine 222 → alanine and Aspartate 181 → asparagine	0.055	1.9	0.057	0.81

Source: Adapted from D. L. Lohse, J. M. Denu, N. Santoro, and J. E. Dixon, *Biochemistry* **36**, 4568 (1997). [a]$(k_{-1} + k_2)/k_1$.

the transient kinetic experiments are presented in Table 5-2. For the native enzyme, only k_{cat} can be determined and is equal to k_2. For the serine to alanine mutant, the rate constant for hydrolysis of the intermediate is the same as k_{cat}, as expected. This mutation also suggests that serine plays a role in the hydrolysis mechanism.

The importance of the aspartic acid in the mechanism was inferred not only from its conservation in many different enzymes, but also by the pH dependence of the steady-state kinetic parameters, which suggested it functioned as a general acid in the phosphorylation of the enzyme. When the aspartic acid was mutated to asparagine, the enzyme still functioned, but k_{cat} was only about 1% of that of the native enzyme (Table 5-2). The transient kinetics did not show a burst phase, indicating that formation of the phosphoenzyme was still rate determining. As expected, the pH dependence of the steady-state kinetic parameters was altered by this mutation.

Finally, the double mutated enzyme was prepared in which the serine was changed to alanine and the aspartate to asparagine. The transient kinetics showed a burst so that the slow step was now the hydrolysis of the intermediate, as with the single mutant in which serine was changed to alanine. However, k_{cat} was reduced even further, and the rate constants for both phosphorylation of the enzyme and hydrolysis were greatly reduced (Table 5-2). On the basis of these and other data, the aspartate was postulated to serve as a general base in the hydrolysis of the intermediate. The intermediate is sufficiently stable in the double mutant so that it could be observed directly with phosphorus nuclear magnetic resonance (8). The mechanism proposed for this reaction has two distinct transition states, one for phosphorylation of the cysteine and another for hydrolysis of the phosphorylated enzyme. Based on the experiments discussed and other data, especially structural data, structures have been proposed for the two transition states as shown in Figure 5-8.

The mechanism of tyrosine phosphatases illustrates several important points. First, the general usefulness of the analysis developed for chymotrypsin is apparent. Both transient and steady-state kinetic experiments were important in postulating a mechanism. Second, the importance of site-specific mutations in helping to establish the mechanism is evident. However, this method is not without pitfalls. It is important to establish that mutations do not alter activity through structural changes in the molecule. In the present case, experiments were done to establish the structural integrity of the mutant enzymes. This ensures that the mutations are altering only the chemical aspects of the mechanism. Finally, this discussion again illus-

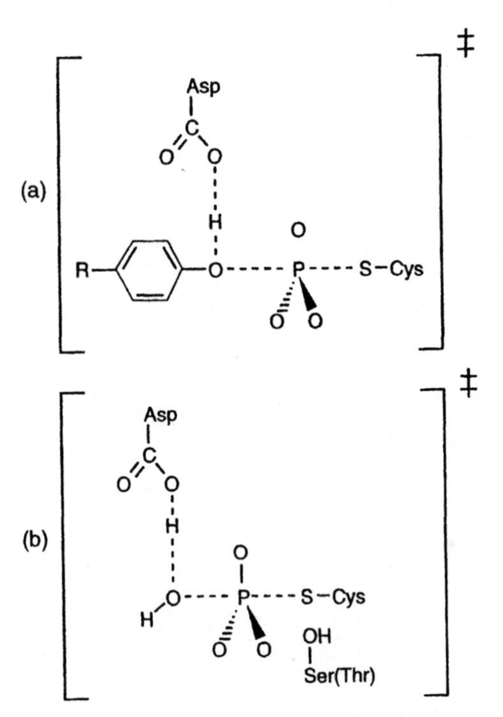

FIGURE 5-8. Proposed transition states for the protein tyrosine phosphatase reaction. (a) Formation of the cysteine phosphate intermediate with a tyrosine phosphate as substrate. (b) Hydrolysis of the cysteine phosphate intermediate. Adapted from J. M. Denu, D. L. Lohse, J. Vijayalakshmi, M. A. Saper, and J. E. Dixon, *Proc. Natl. Acad. Sci.* USA **93**, 2493 (1996). Republished with permission of the National Academy of Sciences USA.

trates the many different types of experiments that must be done in developing mechanisms. The kinetic results must be bolstered by structural and chemical information. Many other studies of enzymes could profitably be discussed, but instead we will turn our attention to kinetic studies in other types of biological systems.

5.5 RIBOZYMES

Enzymes are the most efficient and prevalent catalysts in physiological systems, but they are not the only catalysts of biological importance. RNA molecules have also been found to catalyze a wide range of reactions. Most of these reactions involve the processing of RNA, cutting RNA to the appropriate size or splicing RNA. RNA has also been implicated in peptide bond formation on the ribosome and has been shown to hydrolyze amino acid esters. These catalytic RNAs are called *ribozymes*. They are much less efficient than a typical protein enzyme and sometimes catalyze only a single event. In the latter case, this is not true catalysis. Ribozymes also may work in close collaboration with a protein in the catalytic event. We will consider an

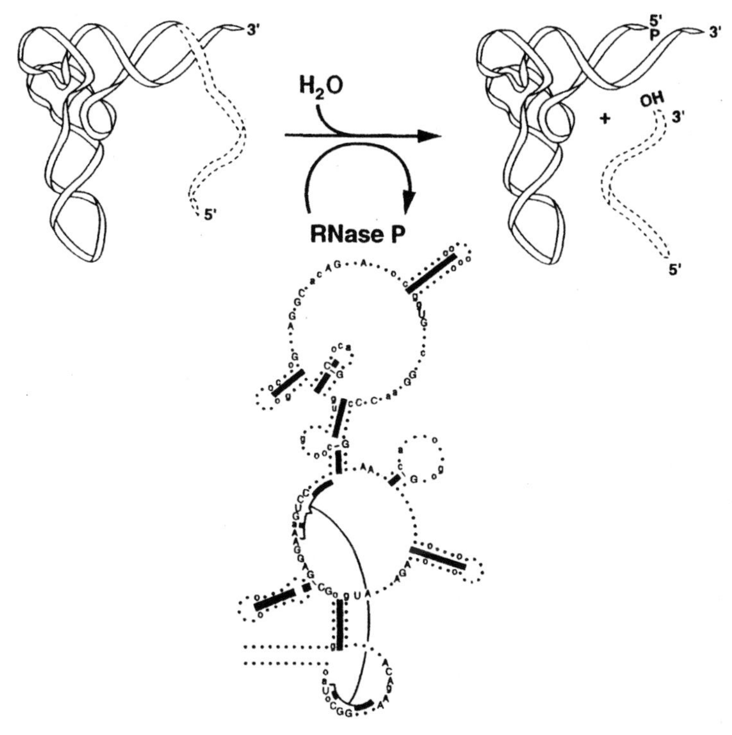

FIGURE 5-9. The reaction of pre-tRNA catalyzed by ribonuclease P is shown schematically at the top of the figure. The 5′ leader sequence of the pre-tRNA is removed in the reaction. The eubacterial consensus structure of ribonuclease P RNA is shown in the lower part of the figure. Proven helices are designated by filled rectangles, invariant nucleotides by uppercase letters, >90% conserved nucleotides by lowercase letters, and less conserved nucleotides by dots. Reproduced from J. W. Brown and N. R. Pace, *Nucleic Acids Res.* **20**, 1451 (1992). Reproduced by permission of Oxford University Press.

abbreviated discussion of a ribozyme, ribonuclease P. For more information, many reviews of ribozymes are available (10–12).

Ribonuclease P catalyzes an essential step in tRNA maturation, namely, the cleavage of the 5′ end of a precursor tRNA (pre-tRNA) to give an RNA fragment and the mature tRNA. This reaction is shown schematically in Figure 5-9. Also shown in this figure is a representation of the catalytic RNA. The catalytic RNA is about 400 nucleotides long and catalyzes the maturation reaction by itself *in vitro*. *In vivo*, a protein of about 120 amino acids also participates in the catalysis. The precise role of the protein is not known, but it appears to alter the conformation of the RNA. The protein–RNA enzyme has a broader selectivity for biological substrates and is a more efficient catalyst (13). Metal ions are also involved in this reaction, with Mg^{2+} being the most important physiologically. Metal ions probably play a role both in catalysis and in maintaining the active conformation of the RNA, but we will not consider this aspect of the reaction here. Transient kinetic studies of the ribozyme have been carried out, and the minimal mechanism consists of the binding of pre-tRNA to the ribozyme, cleavage of the phosphodiester bond, and independent dissociation of both products. We will present some of the results obtained with the RNA component of ribonuclease P.

If transient experiments are carried out with ribonuclease P with excess substrate, a "burst" of the tRNA product occurs at short times, followed by an increase in the product concentration that is linear with time. This behavior is familiar by now, as it has been observed for chymotrypsin and protein tyrosine phosphatase, as discussed above. In the case of ribonuclease P, a covalent intermediate is not formed with the ribozyme; instead, the dissociation of products is very slow relative to the hydrolytic reaction. The mechanism can be written as

$$\text{pre-tRNA} + \text{E} \rightleftharpoons \text{ES} \rightarrow \text{tRNA} + \text{P} + \text{E} \qquad (5\text{-}30)$$

where E is RNAase P RNA and P is the pre-tRNA fragment product. This is not exactly the same as the mechanism discussed for chymotrypsin, but the analysis is similar if the second step is assumed to be much slower than the first. (The detailed analysis is given in Ref. 14.) There is an important lesson to be learned here. "Burst" kinetics are observed whenever the product is formed in a rapid first step followed by regeneration of the enzyme in a slower second step. The apparent rate constants for these two steps can be determined.

What is desired, however, is to distinguish between the binding of enzyme and substrate and the hydrolysis; both of these reactions are aggregated into the first step in Eq. 5-30. In the case of chymotrypsin and protein tyrosine phosphatase, we assumed that the binding step was rapid relative to the formation of the enzyme–substrate intermediate. In this case, we cannot assume that binding is rapid relative to hydrolysis. The two steps were resolved by single turnover experiments: an excess of enzyme was used and because dissociation of products is slow, only a single turnover of substrate was observed (14). The mechanism can be written as

$$\text{E} + \text{pre-tRNA} \xrightarrow{k_1} \text{E} \cdot \text{pre-tRNA} \xrightarrow{k_2} \text{E} \cdot \text{tRNA} \cdot \text{P} \qquad (5\text{-}31)$$

The results obtained are shown in Figure 5-10 for two different concentrations of enzyme. At the lower concentration of enzyme, it is clear that the curve is sigmoidal, rather than hyperbolic. This is because the first step is sufficiently slow that it takes some time for the concentration of the first enzyme–substrate complex to build up. The data at the higher enzyme concentration are essentially hyperbolic because the rate of the first reaction is greater at higher enzyme concentration. These data can be explained by analyzing the mechanism in Eq. 5-31 as two consecutive irreversible first-order reactions. Why irreversible? Fortunately, the rate of dissociation of pre-tRNA from the ribozyme is sufficiently slow that it does not occur on the time scale of the experiment, and furthermore, the analysis of the product gives both E-tRNA and free tRNA so that if some dissociation of tRNA occurs it is not relevant. Why is the first step assumed to be first order? This is another example of a pseudo-first-order rate constant because the enzyme concentration

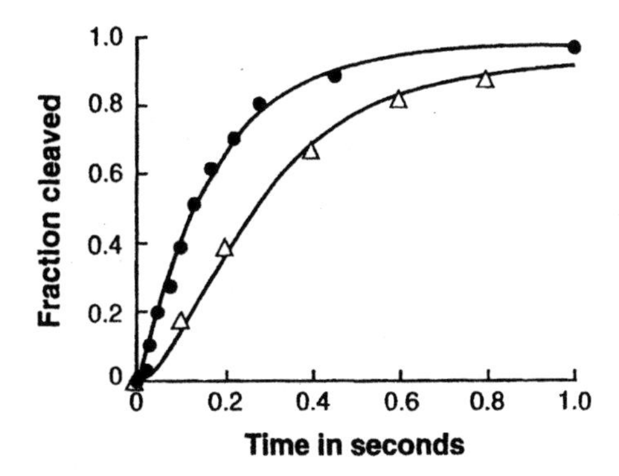

FIGURE 5-10. Single turnover measurements of the hydrolysis of pre-tRNA catalyzed by the RNA component of RNAase P. The fraction of pre-tRNA cleaved is plotted versus time. The pre-tRNA was mixed with excess concentrations of RNAase P RNA, 1.4 μM (Δ) or 19 μM (O). The data are the fit to a mechanism of two consecutive first-order reactions (Eq. 5-38). Reprinted with permission from J.A. Beebe and C.A. Fierke, *Biochemistry* **33**, 10294 (1994). Copyright 1994 American Chemical Society.

is effectively constant throughout the experiment. The pseudo-first-order rate constant is $k_1[E]$, where k_1 is the second-order rate constant for the reaction of the enzyme with the substrate.

To simplify the nomenclature, we will rewrite the mechanism as

$$A \rightarrow B \rightarrow C \tag{5-32}$$

The rate equation for the time dependence of A is

$$-\frac{d[A]}{dt} = k_1[A] \tag{5-33}$$

which is easily integrated to give

$$[A] = [A]_0 e^{-k_1 t} \tag{5-34}$$

where $[A]_0$ is the starting concentration, in this case of pre-tRNA. The rate equation for the time dependence of B is

$$-\frac{d[B]}{dt} = k_2[B] - k_1[A] \tag{5-35}$$

$$-\frac{d[B]}{dt} = k_2[B] - k_1[A]_0 e^{-k_1 t}$$

The solution to this differential equation is

$$[B] = \frac{[A]_0 k_1}{k_2 - k_1}(e^{-k_1 t} - e^{-k_2 t}) \tag{5-36}$$

This solution does not work if $k_1 = k_2$ as the denominator goes to zero. For this special case,

$$[\mathrm{B}] = k_1[\mathrm{A}]_0 t\mathrm{e}^{-k_1 t} \tag{5-37}$$

Finally, the time dependence of C can be obtained from mass balance since $[\mathrm{A}]_0 = [\mathrm{A}] + [\mathrm{B}] + [\mathrm{C}]$:

$$[\mathrm{C}] = [\mathrm{A}_0]\left(1 - \frac{k_2}{k_2 - k_1}\mathrm{e}^{-k_1 t} + \frac{k_1}{k_2 - k_1}\mathrm{e}^{-k_2 t}\right) \tag{5-38}$$

If Eq. 5-38 is used to analyze the data in Figure 5-10, it is found that $k_1 = 6 \times 10^6\ \mathrm{M}^{-1}\,\mathrm{s}^{-1}\ [\mathrm{RNA}]_0$ and $k_2 = 6\,\mathrm{s}^{-1}$.

Understanding how to analyze "burst" kinetics and consecutive first-order reactions is sufficient for the kinetic analysis of many enzymatic reactions, as the conditions can usually be adjusted to conform to these relatively simple mechanisms.

The mechanistic work carried out with RNAseP has permitted the establishment of a minimal mechanism for the RNA portion of the enzyme. However, much remains to be done. The roles of metals, specific groups on the RNA, and the protein remain to be delineated. Understanding how ribozymes function is at the forefront of modern biochemistry and has important implications for both physiology and the evolution of enzymes.

5.6 DNA MELTING AND RENATURATION

We will conclude this chapter with a discussion of the denaturation and renaturation of DNA. Understanding the dynamics of such processes is clearly of biological importance. At the outset, it must be stated that a detailed understanding of the kinetics and mechanisms has not been achieved. However, this is not for lack of effort, and a qualitative understanding of the mechanisms has been obtained.

We will begin with some of the elementary steps in the dynamics of the interactions between the two chains of helical DNA. As discussed in Chapter 3, the thermodynamics of hydrogen bonding between bases has been studied in nonaqueous solvents, where the dimers formed are reasonably stable. Kinetic studies of hydrogen-bonded dimers have also been carried out, and the reactions have been found to be extremely fast, occurring on the submicrosecond and nanosecond time scales. For example, the kinetics of formation of a hydrogen bonded dimer between 1-cyclohexyluracil and 9-ethyladenine has been studied in chloroform (15). (The hydrocarbon arms have been added to the bases to increase their solubility in chloroform.) The reaction is found to occur in a single step with a second-order rate constant for the formation of the dimer of $2.8 \times 10^9\ \mathrm{M}^{-1}\,\mathrm{s}^{-1}$ and a dissociation rate constant of $2.2 \times 10^7\ \mathrm{s}^{-1}$ at 20°C. The second-order rate constant is the maximum possible value; that is, it is the value expected if every collision between the reactants produced a hydrogen-bonded dimer. The upper limit for the rate constant of a bimolecular reaction can be calculated from the known rates of diffusion of the reactants in the solvent. In all cases where hydrogen-bonded dimer formation has been studied, the formation of the dimer has been found to be diffusion controlled.

What does this tell us about the rate of hydrogen bond formation that occurs after the two reactants have diffused together? To answer this question, we will postulate a very simple mechanism, namely, diffusion together of the reactants to form a dimer that is not hydrogen bonded, followed by the formation of hydrogen bonds. This can be written as

$$\mathrm{A} + \mathrm{B} \underset{k_{-1}}{\overset{k_1}{\rightleftharpoons}} [\mathrm{A}, \mathrm{B}] \underset{k_{-2}}{\overset{k_2}{\rightleftharpoons}} \mathrm{A\text{-}B} \tag{5-39}$$

In this mechanism, k_1 is the rate constant for the diffusion-controlled formation of the intermediate and k_{-1} is the rate constant for the diffusion-controlled dissociation of the intermediate. Since only a single step is observed in the experiments, assume that the initial complex formed is in a steady state:

$$[\mathrm{A}, \mathrm{B}] = k_1[\mathrm{A}][\mathrm{B}]/(k_{-1} + k_2) + k_{-2}[\mathrm{A\text{-}B}]/(k_{-1} + k_2) \tag{5-40}$$

The overall rate of the reaction is

$$\frac{\mathrm{d}[\mathrm{A\text{-}B}]}{\mathrm{d}t} = k_2[\mathrm{A}, \mathrm{B}] - k_{-2}[\mathrm{A\text{-}B}] \tag{5-41}$$

If Eq. 5-40 is substituted into Eq. 5-41, we obtain

$$\frac{\mathrm{d}[\mathrm{A\text{-}B}]}{\mathrm{d}t} = k_\mathrm{f}[\mathrm{A}][\mathrm{B}] - k_\mathrm{r}[\mathrm{A\text{-}B}] \tag{5-42}$$

with

$$k_\mathrm{f} = k_1 k_2/(k_{-1} + k_2)$$

and

$$k_\mathrm{r} = k_{-1} k_{-2}/(k_{-1} + k_2)$$

The experimental results indicate that $k_\mathrm{f} = k_1$. This is true if $k_2 \gg k_{-1}$ or formation of the hydrogen bonds is much faster than diffusion apart of the intermediate. However, we can calculate the value of k_{-1}; it is about $10^{10}\ \mathrm{s}^{-1}$. Therefore, the rate constant for formation of the hydrogen bonds, k_2, must be greater than about $10^{11}\ \mathrm{s}^{-1}$. A more exact analysis would put in a separate step for formation of each of the two hydrogen bonds, but this would not change the conclusion with regard to the rate of hydrogen bond formation. Hydrogen bond formation is very fast!

We know that hydrogen bonding alone cannot account for the stability of the DNA double helix: Base stacking and hydrophobic interactions also are important.

An estimate of the rate of base stacking formation and dissociation has been obtained in studies of polyA and polydA (16). These molecules undergo a transition from "stacked" to "unstacked" when the temperature is raised. This transition is accompanied by spectral changes and can easily be monitored. Conditions were adjusted so that the molecules were in the middle of the transition, and the temperature was raised very rapidly ($<1\ \mu s$) with a laser. This is an example of a kinetic study near equilibrium. As the system returns to equilibrium at the higher temperature, the relaxation time can be measured and is the sum of the rate constants for stacking and unstacking (Eq. 4-43). The results obtained indicate that the rate constants are in the range of 10^6–$10^7\ s^{-1}$. Although this is a very fast process, it is considerably slower than hydrogen bonding or simple rotation of the bases.

Moving up the ladder of complexity, we now consider the formation of hydrogen bonded dimers between oligonucleotides (17,18). Again, only a few prototype reactions will be considered, namely, the reaction of a series of A_n oligonucleotides with U_n oligonucleotides to form double helical dimers. These reactions occur in the millisecond to second time regime with $n = 8$–18. In all cases, only a single relaxation time was observed. The relaxation time was consistent with a simple bimolecular reaction (Eq. 4-47):

$$1/\tau = k_1([A_n]_e + [U_n]_e) + k_{-1} \tag{5-43}$$

A typical plot of $1/\tau$ versus the sum of the equilibrium concentration is given in Figure 5-11, and some of the rate constants obtained are given in Table 5-3. The bimolecular rate constants are all about $10^6\ M^{-1}\ s^{-1}$, whereas the dissociation rate constants vary widely, reflecting the stability of the double helix that is formed. Although the second-order rate constant is quite large, it is considerably below the value expected for a diffusion controlled reaction.

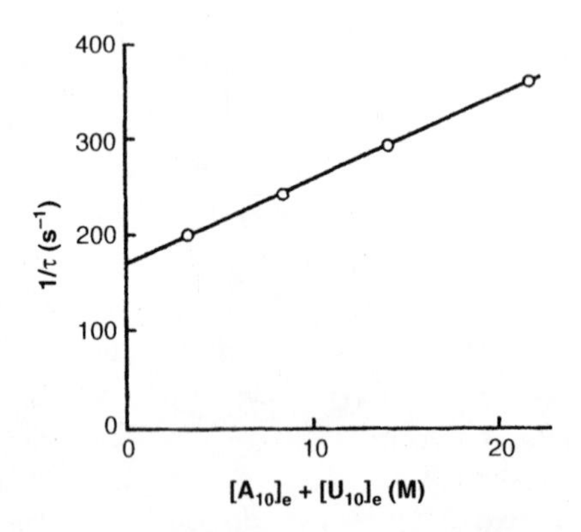

FIGURE 5-11. A plot of $1/\tau$ versus the sum of the equilibrium concentrations of A_{10} and U_{10} at 17°C (Eq. 5-43). The two reactants combine to form a double helix structure. Adapted from D. Pörschke and M. Eigen, Cooperative non-enzymic base recognition. III, *J. Mol. Biol.* **62**, 361 (1971). Copyright 1971, with permission from Elsevier.

TABLE 5-3. Rate Constants for the Reaction of $[A_n]$ with $[U_n]$ at 17°C.

n	$10^6 k_1$ ($M^{-1}\,s^{-1}$)	k_{-1} (s^{-1})
8	1.7	2400
9	7.1	640
10	0.83	175
11	0.58	28
14	1.44	1

Source: Adapted from D. Pörschke and M. Eigen, *J. Mol. Biol.* **62**, 361 (1971).

A detailed analysis of the mechanism can be made as above for the case of the formation of a hydrogen-bonded dimer. This case is considerably more complex, as a separate step is needed for the pairing of each of the n bases. The analysis carried out was bolstered by determination of the activation energies for the rate constants. The conclusion reached is that *nucleation* of the double helix after the molecules have diffused together is rate determining. The formation of the first base pair is rapid, but dissociation is even more rapid. This is also true for the formation of two base pairs. However, when the third base pair is formed, the structure is stabilized. After a nucleus of three base pairs is present, each later base pair forms with a specific rate constant of about $10^7\,s^{-1}$. Note that this is consistent with the results obtained for the "stacking" and "unstacking" of poly A. Altogether these results indicate that the elementary steps in the dynamics of DNA base pairing and stacking (and whatever other types of interactions are important for base pairing) are very rapid. We now turn to consideration of DNA itself.

As might be expected, the melting and reformation of DNA is very complex. DNA is very heterogeneous, and we have discussed previously (Chapter 3) that A–T-rich regions are much less thermodynamically stable than G–C-rich regions. In addition, DNA has a complex structure, with loops and bulges. Native DNA has a very precise alignment of base pairs. If extensive melting occurs, base pairings other than the native pairings may form when the temperature is lowered to form the native structure, thereby slowing down reformation of the native structure. The ultimate form of melting is separation of the individual strands. Because the strands are very long, the rate of melting could be limited by the hydrodynamics of moving the chains apart. Bearing these factors in mind, let us examine some experimental results. Fortunately, the melting and reformation can easily be followed because the spectrum of DNA changes when the bases are stacked. (This effect is discussed further in Chapter 8.) Furthermore, the melting/reformation processes are easy to initiate by changing the temperature.

Some typical results for the denaturation of a viral DNA are shown in Figure 5-12. The denaturation was initiated by the raising of the temperature to give a loss in the native spectrum of 25%, 53%, and 95%. In the first case, the loss in absorbance is approximately first order with a rate constant of about

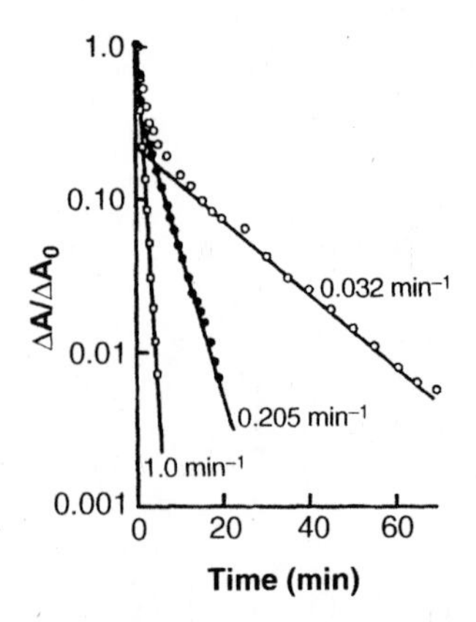

FIGURE 5-12. The kinetics of T2 DNA denaturation in 0.08 M NaCl, 80% formamide. The fraction of the absorbance at 259 nm associated with the native structure is plotted versus the time. The reaction was initiated by heating the three samples to a loss in the native ultraviolet spectrum of 25% (□), 53% (•), and 95% (○). Reproduced from M. T. Record and B. H. Zimm, *Biopolymers* 11, 1435 (1972). Copyright © 1972 *Biopolymers*. Reprinted by permission of John Wiley & Sons, Inc.

1 min^{-1}. The other two cases do not strictly follow first-order kinetics; that is, the logarithmic plot in Figure 5-12 is not linear, and the rates are clearly much slower. The approximate first-order rate constants at long times are shown in the figure. In contrast to the results in model systems, the melting kinetics cannot be fit to a single relaxation time, and the time scale is 100–1000 s. The viral DNA studied has about 2×10^5 base pairs. If only the elementary steps discussed above were involved, the DNA would be completely melted in less than 1 second. How can these results be explained? The rate of melting is governed by the rate of unwinding of the double helix, and the rate of unwinding varies as the melting proceeds. Melting will usually start in the interior of the DNA, and unwinding can only occur if one end of the loop rotates with respect to the other. Initially, this rotation is fast because the double helix is approximately a rod, but as melting proceeds, the bulky loops produced make it harder for one end of the helix to rotate with respect to the other. The rate then becomes slower, as observed. Thus, strand unwinding and separation become rate limiting. In support of this mechanism, if DNA is subjected to a very drastic denaturing force such as a very high pH, all of the base pairs rapidly break in times less than 0.01 s. This does not involve unwinding and strand separation as the native DNA can rapidly be formed by lowering the pH. A detailed description of this mechanism has been developed (19).

The slow unwinding time could be a severe problem for the replication of large DNAs. However, nature has solved this problem by creating enzymes that (1) can make breaks in the DNA chains, thereby reducing the length of DNA that must be rotated during the unwinding, and (2) can unwind DNA.

As a final example, we consider the renaturation of separated DNA strands. DNA can be denatured by raising the temperature significantly above the midpoint of the

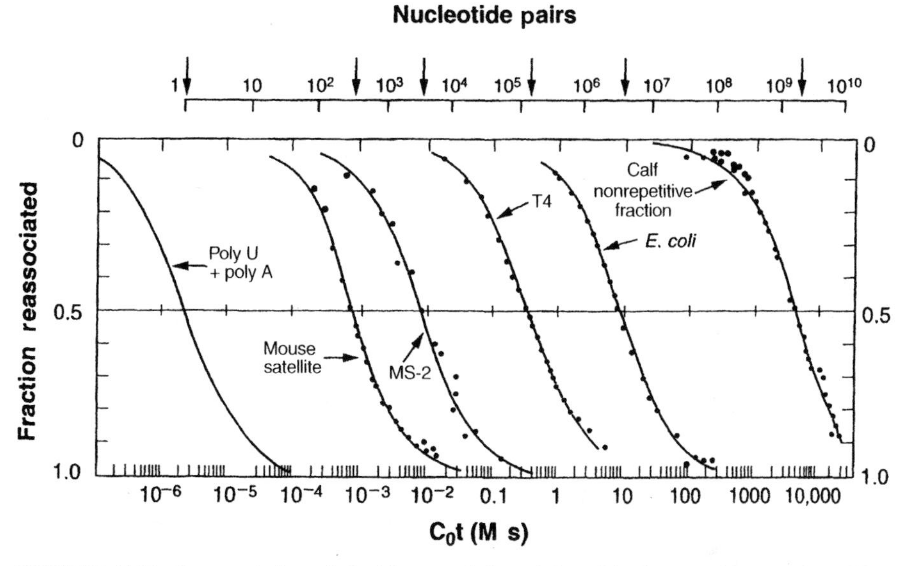

FIGURE 5-13. Reassociation of double-stranded nucleic acids from various sources. The genome size, which is a measure of the complexity of the DNA, is shown above the figure. The nucleic acids all have a single-stranded chain length of about 400 bases. Adapted with permission from R. J. Britten and D. E. Kohne, *Science* **161**,529 (1968). Copyright © 1968 American Association for the Advancement of Science.

melting transition or by raising the pH. After the strands have separated, native DNA can be formed by lowering the temperature or lowering the pH. As might be expected, the renaturation process is second order. The rate of reaction depends on the source of the DNA and its length. The dependence on length can be eliminated by first sonicating the DNA to form fragments of about the same size. If the rates of renaturation are measured at the same DNA concentration in moles of nucleotides per liter, they very by several orders of magnitude, as shown in Figure 5-13. The figure includes data for polyA–polyU and a double stranded RNA (MS-2 viral RNA). The scale at the top is the number of nucleotide pairs in the genome. The larger the genome, the smaller the number of complementary fragments. Only complementary strands can form native DNA, so that the smaller the number of complementary strands, the slower the rate. For polyU and polyA, every strand is complementary so that this system is assigned a value of 1 on this scale.

The renaturation process can be analyzed quantitatively. If the complementary strands are represented as A and A′, the reaction can be written as

$$\mathrm{A} + \mathrm{A}' \xrightarrow{k} \mathrm{AA}' \tag{5-44}$$

Since the concentrations of A and A′ are equal, the rate equation is

$$-\frac{\mathrm{d}[\mathrm{A}]}{\mathrm{d}t} = k[\mathrm{A}]^2 \tag{5-45}$$

which can be integrated to give

$$1/[\mathrm{A}] - 1/[\mathrm{A}]_0 = kt \tag{5-46}$$

The data were found to obey this equation. The half-time for the reaction, that is, when $[\mathrm{A}] = [\mathrm{A}]_0/2$, is given by

$$t_{1/2} = 1/(k[\mathrm{A}]_0) \tag{5-47}$$

In the above equations, the concentration should be equal to the concentration of complementary strands, which we generally do not know. If the total number of base pairs in the smallest repeating sequence is N, then the concentration of complementary strands is proportional to the total concentration of DNA, c_0, divided by N. If this relationship is substituted in Eq. 5-47, we find that $c_0 t_{1/2} \propto N$. Thus, the total concentration of DNA times the half-time for renaturation is a measure of N. A long half-time means that a given fragment will have to sample many other fragments before it finds its complement, so the sequence is complex. N is a measure of the "complexity" of the DNA, and the scale on the top of Figure 5-13 can be equated to N. On this basis, the calf DNA is the most complex DNA that was studied. The study of the kinetics of renaturation can be a useful probe of a gene. It has been used, for example, to determine if fragments of a viral gene were present in viral transformed cells (20).

This concludes the sampling of the application of kinetics to biological systems. Many interesting studies have not been discussed, and many more remain to be done. The study of the time dependence of biological processes is essential in the quest for the elucidation of molecular mechanisms.

REFERENCES

1. G. G. Hammes, *Enzyme Catalysis and Regulation*, Academic Press, New York, 1982.
2. A. Fersht, *Structure and Mechanism in Protein Science: A Guide to Enzyme Catalysis and Protein Folding*, Freeman, San Francisco, CA, 1999.
3. D. L. Purich (ed.), *Contemporary Enzyme Kinetics and Mechanism*, Academic Press, New York, 1996.
4. B. S. Hartley and B. A. Kilby, *Biochem. J.* **56**, 288 (1954).
5. G. P. Hess, in *The Enzymes*, 3rd edition, Vol. 3 (P. Boyer, ed.) Academic Press, New York, 1970 p. 213.
6. J. M. Denu, J. A. Stuckey, M. A. Saper, and J. E. Dixon, *Cell* **87**, 361 (1996).
7. S. Shenolikaar, *Annu. Rev. Cell Biol.* **10**, 55 (1994).
8. J. M. Denu, D. L. Lohse, J. Vijayalakshmi, M. A. Saper, and J. E. Dixon, *Proc. Natl. Acad. Sci. USA* **93**, 2493 (1996).
9. D. L. Lohse, J. M. Denu, N. Santoro, and J. E. Dixon, *Biochemistry* **36**, 4568 (1997).
10. T. R. Cech, in *The RNA World*, Cold Spring Harbor Laboratory Press, Cold Spring Harbor, NY, 1993, p. 239.

11. D. B. McKay and J. E. Wedekind, in *The RNA World*, 2nd edition, Cold Spring Harbor Laboratory Press, Cold Spring Harbor, NY, 1999, p. 265.
12. S. Altman and L. Kirsebom, in *The RNA World*, 2nd edition, Cold Spring Harbor Laboratory Press, Cold Spring Harbor, NY, 1999, p. 351.
13. A. Loria, S. Niranjanakumari, C. A. Fierke, and T. Pan, *Biochemistry* **37**, 15466 (1998).
14. J. A. Beebe and C. A. Fierke, *Biochemistry* **33**, 10294 (1994).
15. G. G. Hammes and A. C. Park, *J. Am. Chem. Soc.* **91**, 956 (1969).
16. T. G. Dewey and D. H. Turner, *Biochemistry* **18**, 5757 (1979).
17. D. Riesner and R. Romer, in *Physico-Chemical Properties of Nucleic Acids*, Vol. 2 (J. Duchesne, ed.), Academic Press, London, 1973.
18. D. Pörschke and M. Eigen, *J. Mol. Biol.* **62**, 361 (1971).
19. M. T. Record and B. H. Zimm, *Biopolymers* **11**, 1435 (1972).
20. S. J. Flint, P. H. Gallimore, and P. A. Sharp, *J. Mol. Biol.* **96**, 47 (1975).

PROBLEMS

5-1. The hydration of CO_2 is catalyzed by the enzyme carbonic anhydrase. The overall reaction at neutral pH can be written as

$$CO_2 + H_2O \overset{CA}{\rightleftharpoons} HCO_3^- + H^+$$

The steady-state kinetics of both hydration and dehydration have been studied at pH 7.1, 0.5°C. Some typical data are given below for an enzyme concentration of 2.8×10^{-9} M.

Hydration		Dehydration	
$10^3/v$ (M^{-1} s)	$10^3[CO_2]$ (M)	$10^3/v$ (M^{-1} s)	$10^3[CO_3{}^-]$ (M)
36	1.25	95	2.0
20	2.5	45	5.0
12	5.0	29	10
6.0	20	25	15

Calculate the steady-state parameters for the forward and reverse reactions.

5-2. Studies of the inhibition of enzymes by various compounds often provide information about the nature of the binding site and the mechanism. Competitive inhibition is when the inhibitor, I, competes with the substrate for the catalytic site. This mechanism can be written as

$$E + S \rightleftharpoons ES \rightarrow E + P$$
$$E + I \rightleftharpoons EI$$

Derive the steady-state rate law for this mechanism and show that it follows Michaelis-Menten kinetics when the inhibitor concentration is constant. Assume that the inhibitor concentration is much greater than the enzyme concentration.

5-3. In the text, the steady-state rate law was derived with the assumption that the reaction is irreversible and/or only the initial velocity was determined. Derive the steady-state rate law for the reversible enzyme reaction:

$$E + S \rightleftharpoons X \rightleftharpoons E + P$$

Show that the rate law can be put into the form

$$v = \frac{(V_S/K_S)[S] - (V_P/K_P)[P]}{1 + [S]/K_S + [P]/K_P}$$

where V_S and V_P are the maximum velocities for the forward and reverse reactions, and K_S and K_P are the Michaelis constants for the forward and reverse reactions.

When equilibrium is reached, $v = 0$. Calculate the ratio of the equilibrium concentrations of S and P, [P]/[S], in terms of the four steady-state parameters. This relationship is called the Haldane relationship and is a method for determining the equilibrium constant of the overall reaction.

5-4. The kinetics of the formation and breakdown of hydrogen bonded loops or hairpins within small RNA molecules can be studied with relaxation methods since the formation of the helical structures is accompanied by a spectral change. This reaction can be represented as

$$\text{hydrogen bounded} \underset{k_2}{\overset{k_1}{\rightleftharpoons}} \text{non-hydrogen bonded}$$

What is the relaxation time for this reaction?

The relaxation time determined at a specific concentration was found to be 10 μs. Calculate the individual rate constants if the equilibrium constant is 0.5.

If the concentration of the RNA is doubled, would the relaxation time get smaller, larger, or stay the same?

5-5. Consider the binding of a protein, P, to a DNA segment (gene regulation). Assume that only one binding site for P exists on the DNA and that the concentration of DNA binding sites is much less than the concentration of P. The reaction mechanism for binding can be represented as

$$P + DNA \rightleftharpoons P\text{-}DNA \rightarrow P\text{-}DNA'$$

where the second step represents a conformational change in the protein. Calculate the rate law for the appearance of P-DNA′ under the following conditions.

a. The first step in the mechanism equilibrates rapidly relative to the rate of the overall reaction and [P-DNA] $\ll$ [DNA].

b. The intermediate, P-DNA, is in a steady state.

c. The first step in the mechanism equilibrates rapidly relative to the rate of the overall reaction and the concentrations of DNA and P-DNA are comparable. Express the rate law in terms of the *total* concentration of DNA and P-DNA, that is, [DNA] + [P-DNA].

d. The following initial rates were measured with an initial DNA concentration of 1 μM.

[P] (μM)	10^4 Rate (M/s)
100	8.33
50	7.14
20	5.00
10	3.33

Which of the rate laws is consistent with the data?

5-6. Many biological reactions are very sensitive to pH. This can readily be incorporated into the rate laws because protolytic reactions can be assumed to be much faster than other rates in most cases. For example, in enzyme mechanisms the ionization states of a few key protein side chains are often critical. Suppose that two ionizable groups on the enzyme are critical for catalytic activity and that one of them needs to be protonated and the other deprotonated. The protolytic reactions can be written as

$$\mathrm{EH_2} \rightleftharpoons \mathrm{EH} + \mathrm{H^+} \rightleftharpoons \mathrm{E} + 2\,\mathrm{H^+}$$

If only the species EH is catalytically active and the protolytic reactions are much more rapid than the other steps in the reaction, all of the rate constants that multiply the free enzyme concentration in the rate law have to be multiplied by the fraction of enzyme present as EH.

a. Calculate the fraction of free enzyme present as EH at a given pH. Your answer should contain the concentration of $\mathrm{H^+}$ and the ionization constants of the two side chains, $K_{\mathrm{E1}} = [\mathrm{E}][\mathrm{H^+}]/[\mathrm{EH}]$ and $K_{\mathrm{E2}} = [\mathrm{EH}][\mathrm{H^+}]/[\mathrm{EH_2}]$.

b. Assume that ES in the Michaelis–Menten mechanism (Eq. 5-2) also exists in three protonation states, $\mathrm{ESH_2}$, ESH, and ES, with only ESH being catalytically active. Calculate the fraction of the enzyme–substrate complex present as EHS. Designate the ionization constants as K_{ES1} and K_{ES2}.

c. Use the results of parts A and B to derive equations for the pH dependence of V_{m},K_{M}, and $V_{\mathrm{m}}/K_{\mathrm{M}}$. Measurement of the pH dependence of the steady-state parameters permits determination of the ionization constants, and sometimes identification of the amino acid side chains.

SPECTROSCOPY

CHAPTER 6

Fundamentals of Spectroscopy

6.1 INTRODUCTION

Spectroscopy is a powerful tool for studying biological systems. It often provides a convenient method for the analysis of individual components in a biological system such as proteins, nucleic acids, and metabolites. It can also provide detailed information about the structure and mechanism of action of molecules. In order to obtain the maximum benefit from this tool and to use it properly, a basic understanding of spectroscopy is necessary. This includes a knowledge of the fundamentals of spectroscopic phenomena, as well as of the instrumentation currently available. A detailed understanding involves complex theory, but a grasp of the important concepts and their application can be obtained without resorting to advanced mathematics and theory. We will attempt to do this by emphasizing the physical ideas associated with spectral phenomena and utilizing a few of the concepts and results from molecular theory.

Very simply stated, spectroscopy is the study of the interaction of radiation with matter. Radiation is characterized by its energy, E, which is linked to the frequency, ν, or wavelength, λ, of the radiation by the familiar Planck relationship:

$$E = h\nu = hc/\lambda \tag{6-1}$$

where c is the speed of light, 2.998×10^{10} cm/s(2.998×10^{8} m/s), and h is Planck's constant, 6.625×10^{-27} erg-s(6.625×10^{-34} J-s). Note that $\lambda\nu = c$.

Radiation can be envisaged as an electromagnetic sine wave that contains both electric and magnetic components, as shown in Figure 6-1. As shown in the figure, the electric component of the wave is perpendicular to the magnetic component. Also shown is the relationship between the sine wave and the wavelength of the light. The useful wavelength of radiation for spectroscopy extends from x-rays, $\lambda \sim 1 - 100$ nm, to microwaves, $\lambda \sim 10^{5} - 10^{6}$ nm. For biology, the most useful radiation for spectroscopy is in the ultraviolet and visible region of the spectrum. The entire useful spectrum is shown in Figure 6-2, along with the common names for the various regions of the spectrum. If radiation is envisaged as both an electric

Physical Chemistry for the Biological Sciences by Gordon G. Hammes

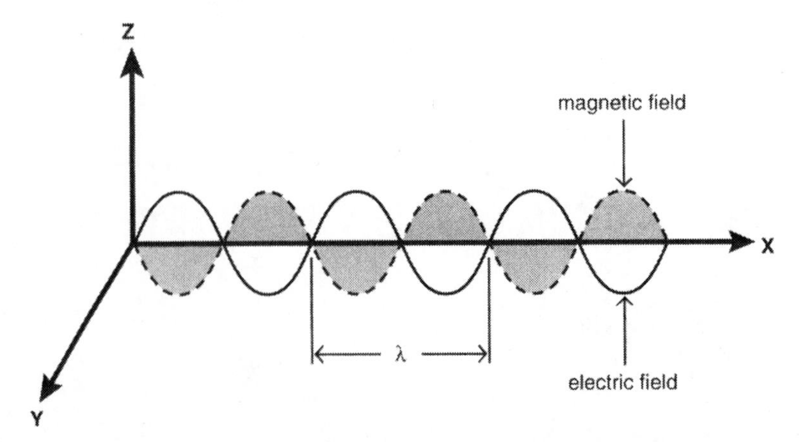

FIGURE 6-1. Schematic representation of an electromagnetic sine wave. The electric field is in the *xz* plane and the magnetic field in the *xy* plane. The electric and magnetic fields are perpendicular to each other at all times. The wavelength, λ, is the distance required for the wave to go through a complete cycle.

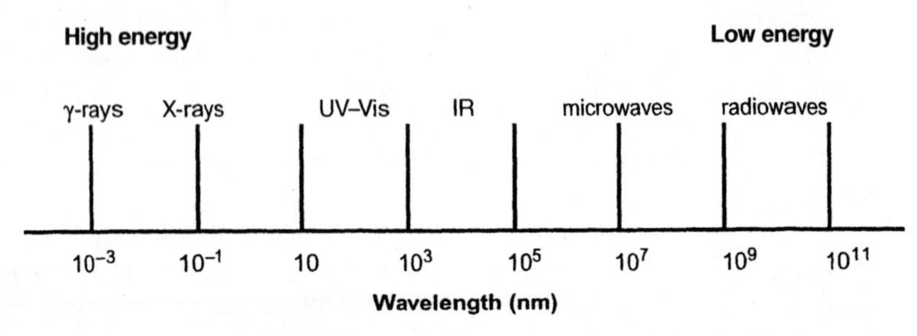

FIGURE 6-2. Schematic representation of the wavelengths associated with electromagnetic radiation. The wavelengths, in nanometers, span 14 orders of magnitude. The common names of the various regions also are indicated approximately (UV is ultraviolet, Vis is visible, and IR is infrared).

and a magnetic wave, then its interactions with matter can be considered as electromagnetic phenomena, due to the fact that matter is made up of positive and negative charges. We will not be concerned with the details of this interaction, which falls into the domain of quantum mechanics. However, a few of the basic concepts of quantum mechanics are essential for understanding spectroscopy.

6.2 QUANTUM MECHANICS

Quantum mechanics was developed because of the failure of Newtonian mechanics to explain experimental results that emerged at the beginning of the 20th century.

For example, for certain metals (e.g., Na), electrons are emitted when light is absorbed. This *photoelectric* effect has several nonclassical characteristics. First, for light of a given frequency, the kinetic energy of the electrons emitted is independent of the light intensity. The number of electrons produced is proportional to the light intensity, but all of the electrons have the same kinetic energy. Second, the kinetic energy of the photoelectron is zero until a threshold energy is reached, and then the kinetic energy becomes proportional to the frequency. This behavior is shown schematically in Figure 6-3, where the kinetic energy of the electrons is shown as a function of the frequency of the radiation. An explanation of these phenomena was proposed by Einstein, who, following Planck, postulated that energy is absorbed only in discrete amounts of energy, $h\nu$. A photon of energy $h\nu$ has the possibility of ejecting an electron, but a minimum energy is necessary. Therefore,

$$\text{kinetic energy} = h\nu - h\nu_0 \tag{6-2}$$

where $h\nu_0$ is the *work function* characteristic of the metal. This predicts that altering the light intensity would affect only the number of photoelectrons and not the kinetic energy. Furthermore, the slope of the experimental plot (Fig. 6-3) is h.

This explanation of the photoelectric effect postulates that light is corpuscular and consists of discrete photons characterized by a specific frequency. How can this be reconciled with the well-known wave description of light briefly discussed

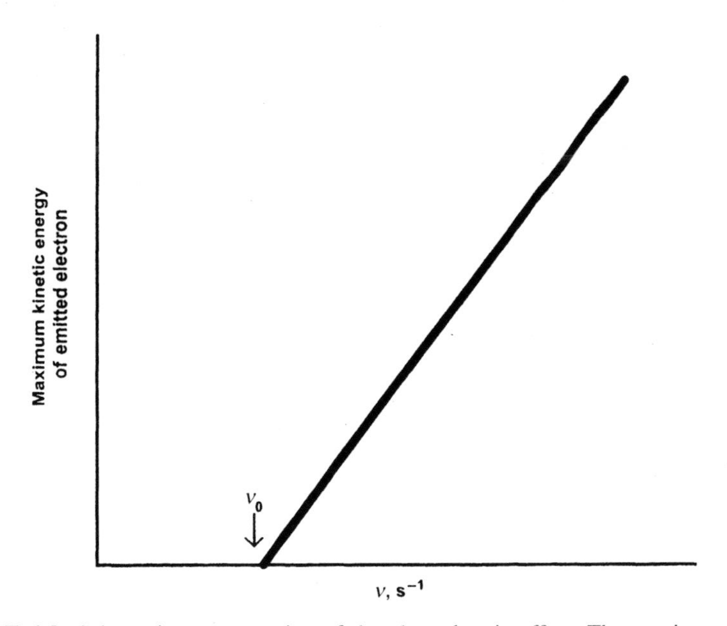

FIGURE 6-3. Schematic representation of the photoelectric effect. The maximum kinetic energy of an electron emitted from a metal surface when it is illuminated with light of frequency ν is shown. The frequency at which electrons are no longer emitted determines the work function, $h\nu_0$, and the slope of the line is Planck's constant (Eq. 6-2).

above? The answer is that both descriptions are correct—light can be envisioned either as discrete photons or a continuous wave. This wave–particle duality is a fundamental part of quantum mechanics. Both descriptions are correct, but one of them may more easily explain a given experimental situation.

About this point in history, de Broglie suggested that this duality is applicable to matter also, so that matter can be described as particles or waves. For light, the energy is equal to the momentum, p, times the velocity of light, and by Einstein's postulate is also equal to $h\nu$.

$$E = h\nu = pc \tag{6-3}$$

Furthermore, since $\lambda\nu = c$, $p = h/\lambda$. For macroscopic objects, $p = m\text{v}$, where v is the velocity and m is the mass. In this case, $\lambda = h/(m\text{v})$, the de Broglie wavelength. These fundamental relationships have been verified for matter by several experiments such as the diffraction of electrons by crystals. The postulate of de Broglie can be extended to derive an important result of quantum mechanics developed by Heisenberg in 1927, namely the uncertainty principle:

$$\Delta p \, \Delta x \geq h/(2\pi) \tag{6-4}$$

In this equation, Δp represents the uncertainty in the momentum and Δx the uncertainty in the position. The uncertainty principle means that it is not possible to determine the precise values of the momentum, p, and the position, x. The more precisely one of these variables is known, the less precisely the other variable is known. This has no practical consequences for macroscopic systems but is crucial for the consideration of systems at the atomic level. For example, if a ball weighing 100 grams moves at a velocity of 100 miles/hour (a good tennis serve), an uncertainty of 1 mile/hour in the speed gives $\Delta p \sim 4.4 \times 10^{-2}$ kg m/s and $\Delta x \sim 2 \times 10^{-33}$ m. We are unlikely to worry about this uncertainty! On the other hand, if an electron (mass $= 9 \times 10^{-28}$ g) has an uncertainty in its velocity of 1×10^{8} cm/s, the uncertainty in the position is about 1 Å, a large distance in terms of atomic dimensions. As we will see later, quantum mechanics has an alternative way of defining the position of an electron.

A second puzzling aspect of experimental physics in the late 1800s and early 1900s was found in the study of atomic spectra. Contrary to the predictions of classical mechanics, discrete lines at specific frequencies were observed when atomic gases at high temperatures emitted radiation. This can only be understood by the postulation of discrete energy levels for electrons. This was first explained by the famous Bohr atom, but this model was found to have shortcomings, and the final resolution of the problem occurred only when quantum mechanics was developed by Schrödinger and Heisenberg in the late 1920s. We will only consider the development by Schrödinger, which is somewhat less complex than that of Heisenberg.

Schrödinger postulated that all matter can be described as a wave and developed a differential equation that can be solved to determine the properties of a system. Basically, this differential equation contains two important variables, the kinetic energy and the potential energy. Both of these are well-known concepts from classical mechanics, but they are redefined in the development of quantum mechanics. If the wave equation is solved for specific systems, it fully explains the previously puzzling results. Energy is quantized, so discrete energy levels are obtained. Furthermore, a consequence of quantum mechanics is that the position of a particle can never be completely specified. Instead, the probability of finding a particle in a specific location can be determined, and the average position of a particle can be calculated. This probabilistic view of matter is in contrast to the deterministic character of Newtonian mechanics and has sparked considerable philosophic debate. In fact, Einstein apparently never fully accepted this probabilistic view of nature. In addition to the above concepts, quantum mechanics also permits quantitative calculations of the interaction of radiation with matter. The result is the specification of rules that ultimately determine what is observed experimentally. We will make use of these rules without considering the details of their origin, but it is important to remember that they stem from detailed quantum mechanical calculations.

6.3 PARTICLE IN A BOX

As an example of a simple quantum mechanical result that leads to quantization of energy levels, we consider a particle of mass m moving back and forth in a one-dimensional box of length L. This actually has some practical application. It is a good model for the movement of pi electrons that are delocalized over a large part of a molecule, for example, biological molecules such as carotenoids, hemes, and chlorophyll. This is not an ordinary box because inside the box, the potential energy of the system is 0, whereas outside of the box, the potential energy is infinite. This is depicted in Figure 6-4. The Schrödinger equation in one dimension is

$$-\frac{h^2}{8m\pi^2}\frac{d^2\psi_n}{dx^2} + U = E_n\psi_n \tag{6-5}$$

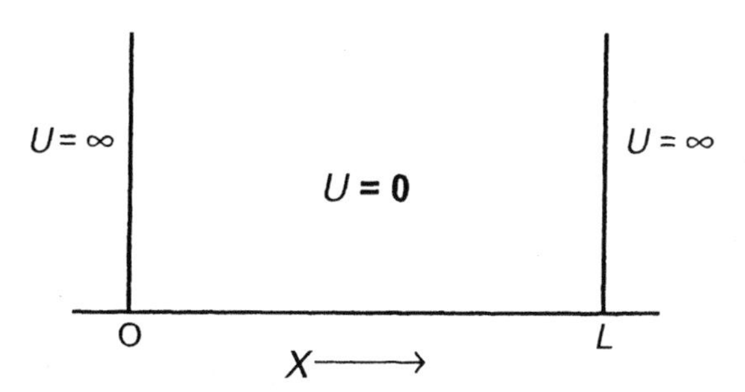

FIGURE 6-4. Quantum mechanical model for a particle in a one-dimensional box of length L. The particle is confined to the box by setting the potential energy equal to 0 inside the box and to ∞ outside of the box.

where Ψ_n is the wave function, x is the position coordinate, U is the potential energy, and E_n is the energy associated with the wave function Ψ_n. Since the potential walls are infinitely high, the solution to this equation outside of the box is easy—there is no chance the particle is outside the box, so the wave function must be 0. Inside the box, $U = 0$, and Eq. 6-5 can be easily solved. The solution is

$$\psi_n = A \sin bx \tag{6-6}$$

where A and b are constants. At the ends of the box, Ψ must be zero. This happens when $\sin n\pi = 0$ and n is an integer, so b must be equal to $n\pi/L$. This causes Ψ_n to be 0 when $x = 0$ and $x = L$ for all integral values of n. To evaluate A, we introduce another concept from quantum mechanics, namely that the probability of finding the particle in the interval between x and $x + dx$ is $\Psi^2 dx$. Since the particle must be in the box, the probability of finding the particle in the box is 1, or

$$\int_0^L \psi_n^2 dx = \int_0^L A^2 \sin^2(n\pi x/L) dx = 1 \tag{6-7}$$

Evaluation of this integral gives $A = \sqrt{2/L}$. Thus, the final result for the wave function is

$$\psi_n = \sqrt{2/L}\,\sin(n\pi x/L) \tag{6-8}$$

Obviously n cannot be 0, as this would predict that there is no probability of finding the particle in the box, but n can be any integer. The wave functions for a few values of n are shown in Figure 6-5. Basically Ψ_n is a sine wave, with the "wavelength"

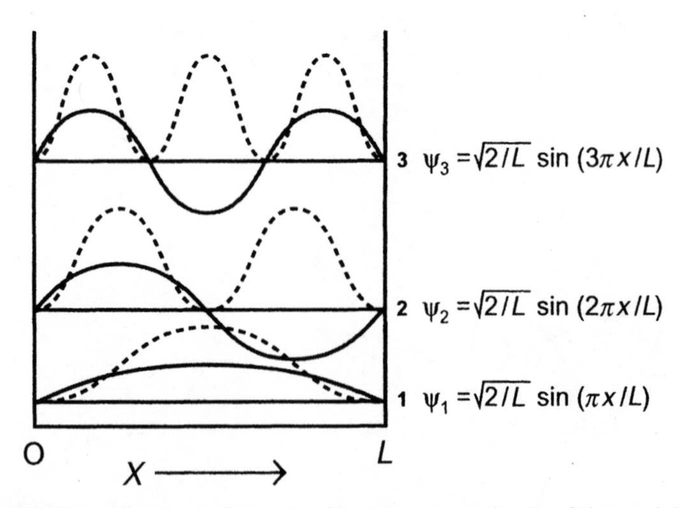

FIGURE 6-5. Wave functions, Ψ, for the first three energy levels of the particle in a box (Eq. 6-8). The dashed lines show the probability of the finding the particle at a given position x, Ψ^2.

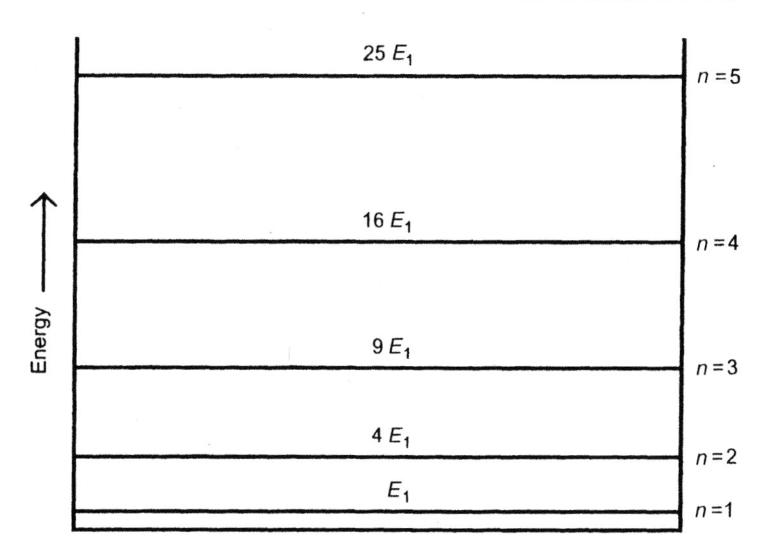

FIGURE 6-6. Energy levels for a particle in a box (Eq. 1-9). The energy levels are n^2E_1, where E_1 is the energy when $n = 1$.

decreasing as n increases. (More advanced treatments of quantum mechanics use the notation associated with complex numbers in discussing the wave equation and wave functions, but this is beyond the scope of this text.)

To determine the energy of the particle, all we have to do is put Eq. 6-8 back into Eq. 6-5 and solve for E_n. The result is

$$E_n = (h^2n^2)/(8\,mL^2) \tag{6-9}$$

Thus, we see that the energy is quantized, and the energy is characterized by a series of energy levels, as depicted in Figure 6-6. Each energy level, E_n, is associated with a specific wave function, Ψ_n. Notice that the energy levels would be very widely spaced for a very light particle such as an electron, but would be very closely spaced for a macroscopic particle. Similarly, the smaller the box, the more widely spaced the energy levels. For a tennis ball being hit on a tennis court, the ball is sufficiently heavy and the court (box) sufficiently big so that the energy levels would be a continuum for all practical purposes. The uncertainty in the momentum and position of the ball cannot be blamed on quantum mechanics in this case! The particle in a box illustrates how quantum mechanics can be used to calculate the properties of systems and how quantization of energy levels arises. The same calculation can be easily done for a three-dimensional box. In this case, the energy states are the sum of three terms identical to Eq. 6-9, but with each of the three terms having a different quantum number.

The quantum mechanical description of matter does not permit the determination of the precise position of the particle to be determined, a manifestation of the Heisenberg uncertainty principle. However, the probability of finding the particle within a given segment of the box can be calculated. For example, the probability of finding the particle in the middle of the box, that is, between $L/4$ and $3L/4$ for the lowest energy state is

$$\int_{L/4}^{3L/4} \psi_1^2 \mathrm{d}x = (2/L) \int_{L/4}^{3L/4} \sin^2(\pi x/L)\mathrm{d}x$$

Evaluation of this integral gives a probability of 0.82. The probability of finding the particle within the middle part of the box is independent of L, the size of the box, but does depend on the value of the quantum number, n. For the second energy level, $n = 2$, the probability is 0.50. The probability of finding the particle at position x in the box is shown as a dashed line for the first three energy levels in Figure 6-5.

An important result of quantum mechanics is that not only do molecules exist in different discrete energy levels, but the interaction of radiation with molecules causes shifts between these energy levels. If energy or radiation is *absorbed* by a molecule, the molecule can be raised to a higher energy state, whereas if a molecule loses energy, radiation can be *emitted*. For both cases, the change in energy is related to the radiation that is absorbed or emitted by a slight modification of Eq. 6-1, namely, the change in energy state of the molecule, ΔE, is

$$\Delta E = h\nu = hc/\lambda \tag{6-10}$$

The change in energy, ΔE, is the difference in energy between specific energy levels of the molecule, for example, $E_2 - E_1$ where 1 and 2 designate different energy levels. It is important to note that since the energy is quantized, the light emitted or absorbed is always a specific single frequency. Equation 6-10 can be applied to the particle in a box for the particle dropping from the $n + 1$ energy level to the n energy level:

$$\Delta E = \frac{h^2}{8mL^2}[(n+1)^2 - n^2] = \frac{h^2}{8mL^2}(2n+1) = hc/\lambda \tag{6-11}$$

If the particle is assumed to be an electron moving in a molecule 20 Å long and $n = 10$, then $\lambda \sim 600\,\mathrm{nm}$. This wavelength is in the visible region and has been observed for π electrons that are highly delocalized in molecules.

In practice, energy levels are sometimes so closely spaced that the frequencies of light emitted appear to create a continuum of frequencies. This is a shortcoming of the experimental method—in reality, the frequencies emitted are discrete entities. The particle in a box is a rather simple application of quantum mechanics, but it illustrates several important points that are also found in more complex calculations for molecular systems. First, the system can be described by a wave function. Second, this wave function permits determination of the probability of important characteristics of the system, such as positions. Finally, the energy of the system can be calculated and is found to be quantized. Moreover, the energy can only be absorbed or emitted in quantized packages characterized by specific frequencies. Quantum mechanical calculations also tell us what conditions are necessary for energy to be emitted or absorbed by a molecule. These calculations tell us *whether* radiation will be emitted or absorbed and what quantized packets of energy are available. We will only utilize the results of these calculations and will not be concerned with the details of the interactions between light and molecules, other than the above concepts.

6.4 PROPERTIES OF WAVES

It is useful to consider several additional aspects of light waves in order to understand better some of the experimental methods that will be discussed later. Thus far, we have considered light to be a periodic electromagnetic wave in space that could be characterized, for example, by a sine function:

$$I = I_0 \sin(2\pi x/\lambda) \tag{6-12}$$

Here I is the magnetic or electric field, I_0 is the maximum value of the electric or magnetic field, x is the distance along the x-axis, and λ is the wavelength. A light wave can also be periodic in time, as illustrated in Figure 6-7. In this case,

$$I = I_0 \sin 2\pi \nu t = I_0 \sin \omega t \tag{6-13}$$

Now, I is the light intensity, I_0 is the maximum light intensity, ν is the frequency in s^{-1}, as defined in Figure 6-7, ω is the frequency in radians ($\omega = 2\pi\nu$), and t is the time. The velocity of the propagating wave is $\lambda\nu$, which in the case of electromagnetic radiation is the speed of light, that is, $\lambda\nu = c$. If light of the same frequency and maximum amplitude from two sources is combined, the two sine functions will be added. If the two light waves start with zero intensity at the same time ($t = 0$), the two waves add and the intensity is doubled. This is called *constructive interference*. If the two waves are combined with one of the waves starting at zero intensity and proceeding to positive values of the sine function, whereas the other begins at zero intensity and proceeds to negative values, the two intensities cancel each other out. This is called *destructive interference*. Obviously, it is possible to have cases in between these two extremes. In such cases, a phase difference is said to exist between the two waves. Mathematically, this can be represented as

$$I = I_0 \sin(\omega t + \delta) \tag{6-14}$$

where δ is called a phase angle and can be either positive or negative. When many different waves of the same frequency are combined, the intensity can always be

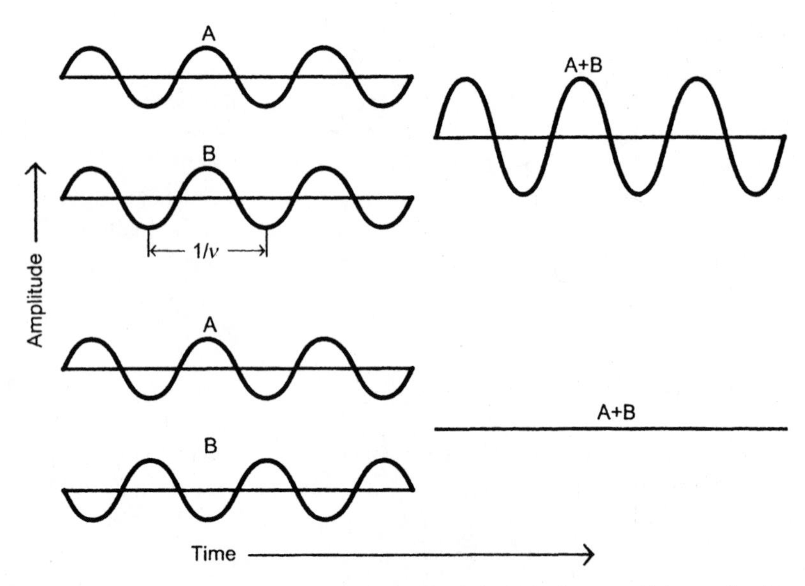

FIGURE 6-7. Examples of constructive and destructive interference. Constructive interference: when the upper two wave forms of equal amplitude and a phase angle of 0° (or integral multiples of 2π) are added (left), a sine wave with twice the amplitude and the same frequency results (right). Destructive interference: when the lower two wave forms are added (left), the amplitudes of the two waves cancel (right). The phase angle in this case is 90° (or odd integral multiples of $\pi/2$).

described by such a relationship. These phenomena are shown schematically in Figure 6-7.

A standard way of carrying out spectroscopy is to apply continuous radiation and then look at the intensity of the radiation after it has passed through the sample of interest. The intensity is then determined as a function of the frequency of the radiation, and the result is the absorption spectrum of the sample. The color of a material is determined by the wavelength of the light absorbed. For example, if white light shines on blood, blue/green light is absorbed so that the transmitted light is red. Several examples of absorption spectra are shown in Figure 6-8. We will consider why and how much the sample absorbs light a bit later, but you are undoubtedly already familiar with the concept of an absorption spectrum.

The use of continuous radiation is a useful way to carry out an experiment, but there is an interesting mathematical relationship that permits a different approach to the problem. This mathematical operation is the *Fourier transform*. The principle of a Fourier transform is that if the frequency dependence of the intensity, $I(\nu)$, can be determined, it can be transformed into a new function, $F(t)$, that is a function of the time, t. Conversely, $F(t)$ can also be converted to $I(\nu)$. Both of these functions contain the same information. Why then are these transformations advantageous? It can be quite time consuming to determine $I(\nu)$, but a short pulse of radiation can be applied very quickly. Basically, what this transformation means is that looking at

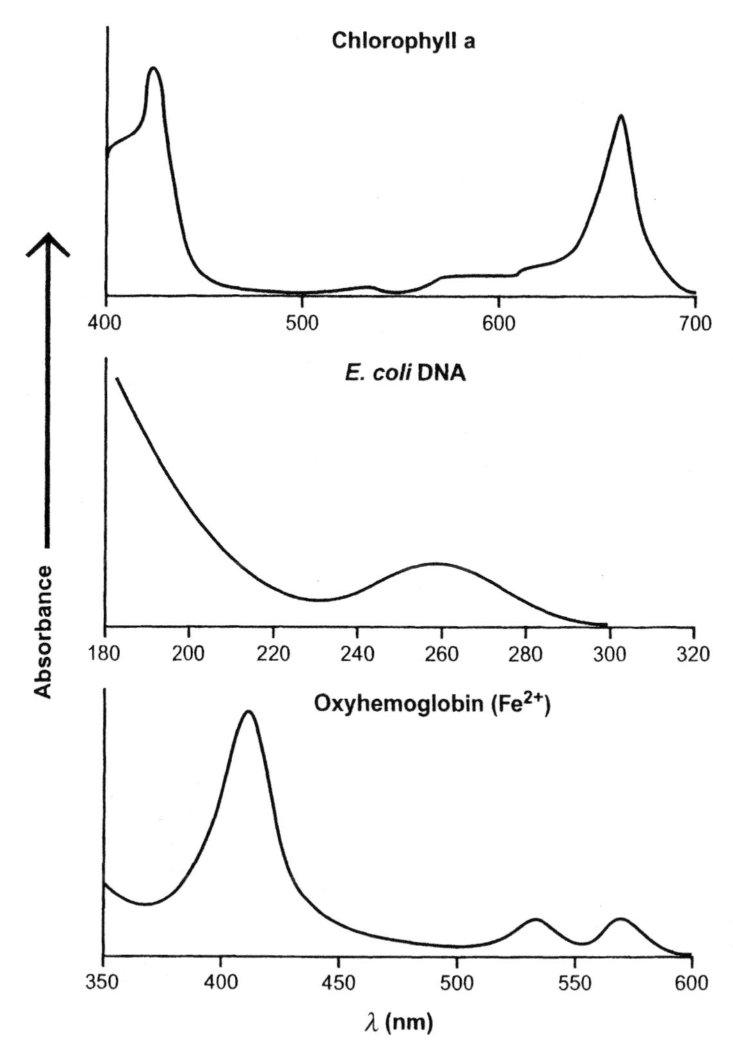

FIGURE 6-8. Absorption of light by biological molecules. The absorbance scale is arbitrary and the wavelength, λ, is in nanometers. Chlorophyll *a* solutions absorb blue and red light and are green in color. DNA solutions absorb light in the ultraviolet and are colorless. Oxyhemoglobin solutions absorb blue light and are red in color.

the response of the system to application of a pulse of radiation, such as shown as in Figure 6-9, is equivalent to looking at the response of the system to sine wave radiation at many different frequencies. In other words, a square wave is mathematically equivalent to adding up many sine waves of different frequencies, and vice versa. This is shown schematically in Figure 6-9 where the addition of sine waves with four different frequencies produces a periodic "square" wave. The larger the number of sine waves added, the more "square" the wave becomes. In mathematical

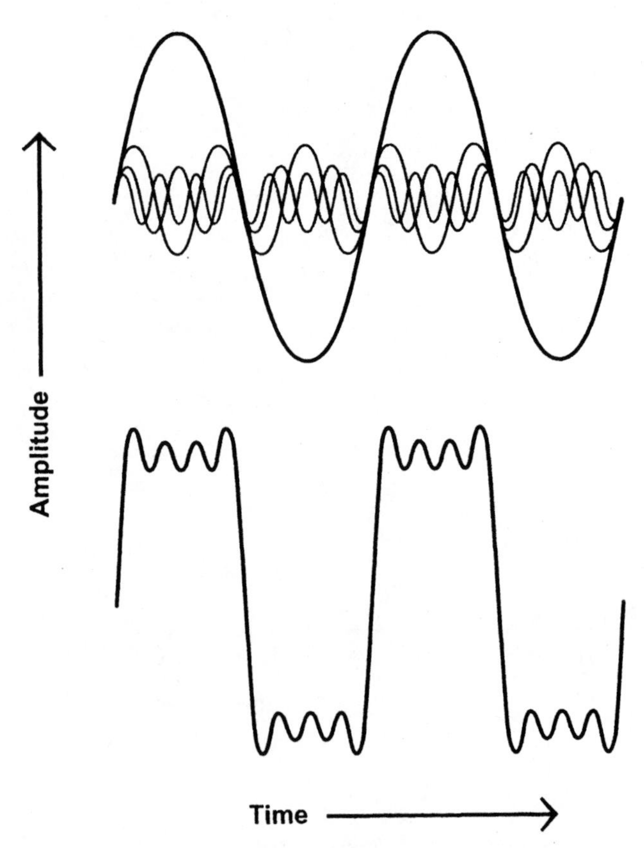

FIGURE 6-9. The upper part of the figure shows sine waves of four different frequencies, and the lower part of the figure is the sum of the sine waves, which approximates a square wave pulse of radiation. When sine waves of many more frequencies are included, the time dependence becomes a pulsed square wave. This figure illustrates that the superposition of multiple sine waves is equivalent to a square wave pulse, and vice versa. This equivalency is the essence of Fourier transform methods. Copyright by Professor T. G. Oas, Duke University. Reproduced with permission.

terms, a square wave can be represented as an infinite series of sine functions, a Fourier series.

The mathematical equivalence of timed pulses and continuous waves of many different frequencies has profound consequences in determining the spectroscopic properties of materials. In many cases, the use of pulses permits thousands of experiments to be done in a very short time. The results of these experiments can then be averaged, producing a far superior frequency spectrum in a much shorter time than could be determined by continuous wave methods. In later chapters, we will be dealing with both continuous wave spectroscopy and Fourier transform spectroscopy. It is important to remember that both methods give identical results. The method of choice is the one that produces the best data in the shortest time, and in some cases at the lowest cost.

With this brief introduction to the underlying theoretical principles of spectroscopy, we are ready to proceed with consideration of specific types of spectroscopy and their application to biological systems.

REFERENCES

The topics in this chapter are discussed in considerably more depth in a number of physical chemistry textbooks, such as those cited below.

1. I. Tinoco, Jr., K. Sauer, J. C. Wang, and J. D. Puglisi, *Physical Chemistry: Principles and Applications in Biological Sciences*, 4th edition, Prentice-Hall, Englewood Cliffs, NJ, 2002.
2. R. J. Silbey, R. A. Alberty, and M. G. Bawendi, *Physical Chemistry*, 4th edition, John Wiley & Sons, New York, 2004.
3. P. W. Atkins and J. de Paula, *Physical Chemistry*, 7th edition, Freeman, New York, 2001.
4. R. S. Berry, S. A. Rice, and J. Ross, *Physical Chemistry*, 2nd edition, Oxford University Press, New York, 2000.
5. D. A. McQuarrie and J. D. Simon, *Physical Chemistry: A Molecular Approach*, University Science Books, Sausalito, CA, 1997.

PROBLEMS

6.1. The energies required to break the C–C bond in ethane, the "triple bond" in CO, and a hydrogen bond are about 88, 257, and 4 kcal/mol, respectively. What wavelengths of radiation are required to break these bonds?

6.2. Calculate the energy and momentum of a photon with the following wavelengths: 150 pm (X-ray), 250 nm (ultraviolet), 500 nm (visible), and 1 cm (microwave).

6.3. The maximum kinetic energy of electrons emitted from Na at different wavelengths was measured with the following results.

λ (Å)	Maximum kinetic energy (eV)
4500	0.40
4000	0.76
3500	1.20
3000	1.79

Calculate Planck's constant and the value of the work function from these data. ($1\text{eV} = 1.602 \times 10^{-19}$ J.)

6.4. Calculate the de Broglie wavelength for the following cases:

a. An electron in an electron microscope accelerated with a potential of 100 kV.

b. A He atom moving at a speed of 1000 m/s.

c. A bullet weighing 1 g moving at a speed of 100 m/s.

Assume that the uncertainty in the speed is 10% and calculate the uncertainty in the position for each of the three cases.

6.5. The particle in a box is a useful model for electrons that can move relatively freely within a bonding system such as π electrons. Assume that an electron is moving in a "box" that is 50 Å long, that is, a potential well with infinitely high walls at the boundaries.

a. Calculate the energy levels for $n = 1$, 2, and 3.

b. What is the wavelength of light emitted when the electron moves from the energy level with $n = 2$ to the energy level with $n = 1$?

c. What is the probability of finding the electron between 12.5 and 37.5 Å for $n = 1$?

6.6. Sketch the graph of I versus t for sine wave radiation that obeys the relationship $I = I_0 \sin(\omega t + \delta)$ for $\delta = 0$, $\pi/4$, $\pi/2$ and π.

Plot the sum of the sine waves when the sine wave for $\delta = 0$ is added to that for $\delta = 0$ or $\pi/4$, or $\pi/2$, or π. This exercise should provide you with a good understanding of constructive and destructive interference.

Do your results depend on the value of ω? Briefly discuss what happens when waves of different frequency are added together.

CHAPTER 7

X-ray Crystallography

7.1 INTRODUCTION

The primary tool for determining the atomic structure of macromolecules is the scattering of X-ray radiation by crystals. Strictly speaking, this is not considered spectroscopy even though it involves the interaction of radiation with matter. Nevertheless, the importance of this tool in modern biology makes it mandatory to have some understanding of how macromolecular structures are determined with X-ray radiation. The basic principles of the methodology will be discussed without delving into the details of how structures are determined. We then consider some of the important results that have been obtained as well as how they can be used to understand the function of macromolecules in biological systems.

Although considerable progress has been made in determining the structures of macromolecules that are not strictly crystalline, we will concentrate on the structure of macromolecules from high-quality crystals. High-quality crystals have molecules arranged in a regular array, and this regular array, or lattice, serves as a scattering surface for X rays.

Why are X rays used to determine molecular structures? Ordinary objects can be easily seen with visible light, and very small objects can be seen in microscopes with a high-quality lens that converts the electromagnetic waves associated with light into an image. However, the resolution of a light microscope is limited by the wavelength of light that is used. Distances cannot be resolved that are significantly shorter than the wavelength of the light that illuminates the object. The wavelength of visible light is thousands of angstroms, whereas distances within molecules are approximately angstroms.

The obvious answer to this resolution problem is to use radiation that has a much shorter wavelength than visible light, namely, X rays which have wavelengths in the angstrom region. In principle, all that is needed is a lens that will convert the scattering from a molecule into an image. What does this entail? We have previously shown that light has an amplitude and a periodicity with respect to time and distance, or a phase. A lens takes both the amplitude and phase information and converts it into an image. Unfortunately, a lens does not exist that will carry out this

Physical Chemistry for the Biological Sciences by Gordon G. Hammes

function for X-ray radiation. Instead, what can be measured is the amplitude of the scattered radiation. Methods are then needed to obtain the phase and ultimately the molecular structure. The principles behind this methodology are given below, without presenting the underlying mathematical complications. More advanced texts should be consulted for the mathematical details (1,2).

7.2 SCATTERING OF X RAYS BY A CRYSTAL

X rays are produced by bombarding a target (typically a metal) with high-energy electrons. If the energy of the incoming electron is sufficiently large, it will eject an electron from an inner orbital of the target. A photon is emitted when an outer orbit electron moves into the vacated inner orbital. For typical targets, the wavelengths of the photons are tenths of angstrom to several angstroms. In a normal laboratory experiment, X rays are produced by special electronic tubes. However, more intense short-wavelength sources can be obtained using high-energy electron accelerators (synchrotrons), and these sources are often used to obtain high-resolution structures. The intensity of X rays from a synchrotron is thousands of times greater than a conventional source, and radiation of different wavelengths can be selected. Smaller crystals can be used for structure determinations with synchrotron radiation than with conventional sources, which is a considerable advantage.

Crystals can be considered as a regular array of molecules, or scattering elements. Only seven symmetry arrangements of crystals are possible. For example, one possibility is a simple cube. In this case, the three axes are equal in length and are 90° with respect to each other. Other lattices can be generated by having angles other than 90° and unequal sides of the geometric figure. A summary of the seven crystal systems is given in Table 7-1.

The crystal type does not give a complete picture of the possible arrangements of atoms within the crystal, that is, the lattice. The lattice is an infinite array of points (atoms) in space. A. Bravais showed that all lattices fall within 14 types, the *Bravais lattices*. For example, for a cubic crystal, three lattices are possible: one with an atom at each corner of the cube, one with an additional atom at the center of the cube (body-centered cube), and one with an additional atom in the center of

TABLE 7-1. The Seven Crystal Systems

System	Axes	Angles
Cubic	$a = b = c$	$\alpha = \beta = \gamma = 90°$
Rhombohedral	$a = b = c$	$\alpha = \beta = \gamma$
Tetragonal	$a = b; c$	$\alpha = \beta = \gamma = 90°$
Hexagonal	$a = b; c$	$\alpha = \beta = 90°; \gamma = 120°$
Orthorhombic	$a; b; c$	$\alpha = \beta = \gamma = 90°$
Monoclinic	$a; b; c$	$\alpha = \gamma = 90°; \beta$
Triclinic	$a; b; c$	$\alpha; \beta; \gamma$

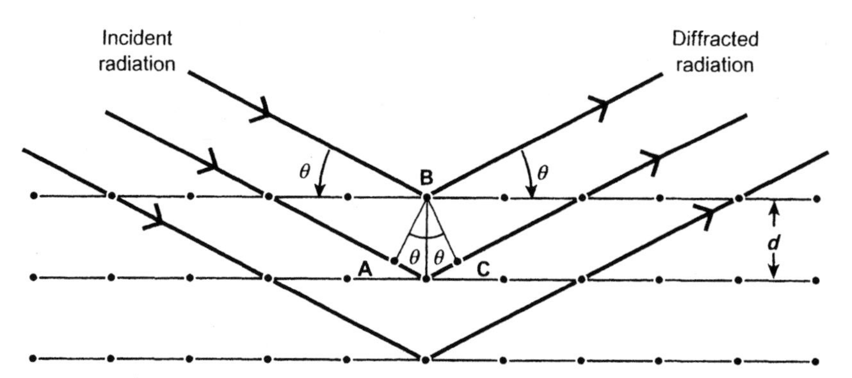

FIGURE 7-1. Diffraction of X-ray radiation by a crystal lattice. The parallel lines are planes of atoms separated by a distance, d, and the radiation impinges on the crystal at an angle θ. This diagram can be used to derive the Bragg condition for maximum constructive interference as described in the text (Eq. 7-1).

each face of the cube (face-centered cube). Determination of crystal type and lattice classification is important for the subsequent analysis of the X-ray scattering data.

W. L. Bragg showed that the scattering of X rays from a crystal can be described as the scattering from parallel planes of molecules, as illustrated in Figure 7-1. If the incident beam of X rays is at an angle θ with respect to the molecular plane, then it will be scattered at an angle θ. This is called *elastic* scattering and assumes that the radiation does not lose or absorb energy in the scattering process. In reality, some *inelastic* scattering occurs in which energy can be gained or lost, but this effect can be neglected. From Figure 7-1, it is clear that scattering will occur for each plane and scattering center with the same incident and exit angle. From a wave standpoint, the radiation will be scattered from each plane, and the radiation from each plane will have a different phase as it exits, since each wave would have traveled a different distance depending on the depth of the plane in Figure 7-1. However, if the wavelength of radiation is such that the difference in path length traveled by the beams from different planes is equal to the wavelength or an integer multiple of the wavelength, then the two waves will be in phase and constructive interference will occur, that is, the intensity of the radiation will be at a maximum. The phenomena of constructive and destructive interference have been discussed in Chapter 6. Figure 6-7 illustrates these phenomena when the time dependence of an electromagnetic wave is considered. The same analysis is applicable for the propagation of a wave in space (Fig. 6-1 and Eq. 6-12) as illustrated in Figure 7-2.

The condition for maximum constructive interference can be calculated by reference to Figure 7-1:

$$\mathrm{AB} + \mathrm{BC} = n\lambda$$

where n is an integer and λ is the wavelength. The distances AB and BC are equal to $d \sin \theta$, where d is the distance between planes. Thus, the Bragg equation is

$$n\lambda = 2d \sin \theta \tag{7-1}$$

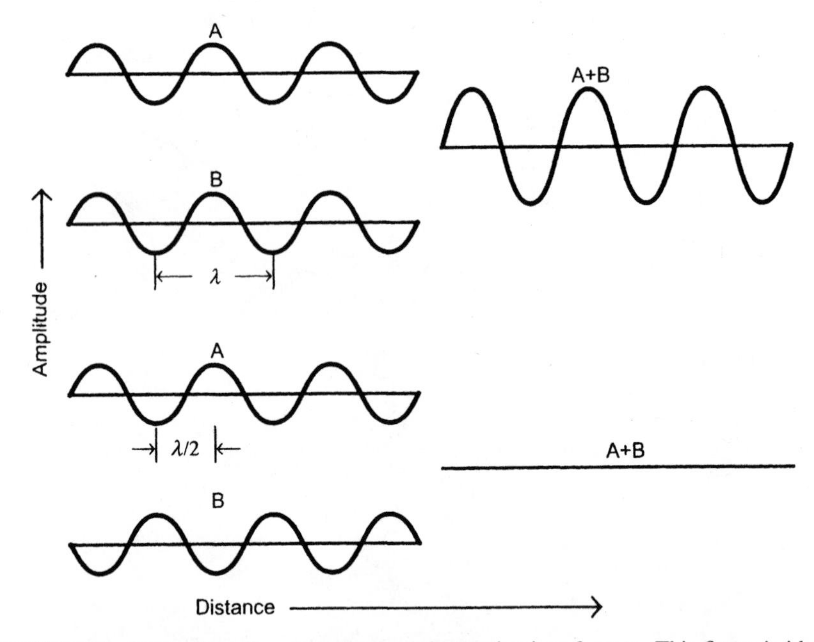

FIGURE 7-2. Illustration of constructive and destructive interference. This figure is identical to Figure 6-7, except that now the abscissa is distance rather than time.The upper two waves are in phase, which means they differ by integral multiples of the wavelength, λ, and constructive interference occurs. The bottom two waves differ by $\lambda/2$ in phase, and destructive interference occurs.

The Bragg equation gives the condition for diffraction so that if a crystal is rotated in a beam of X rays, the scattering pattern is a series of intensity maxima. A real crystal is more complex than a set of parallel planes of point-scattering sources. In fact, multiple planes exist, and a molecule is not a simple point scatterer. Electrons in atoms are the scatterers and each atom has a different effectiveness as a scatterer. Consequently, when an experiment is carried out, a set of diffraction maxima of different intensities is observed. A schematic representation of the experimental setup is shown in Figure 7-3. Either the crystal or detector, or both are rotated to obtain the scattering intensity at various angles. The diffraction pattern has a strong peak at the center, the unscattered beam, and a radial distribution around the center, corresponding to different planes and values of n in Eq. 7-1.

The Bragg equation can be used to derive the minimum spacing of planes that can be resolved for X rays of a given wavelength. Since the maximum value of $\sin\theta$ is 1, $d_{\text{min}} = \lambda/2$.

7.3 STRUCTURE DETERMINATION

The analysis of a diffraction pattern is considerably more difficult than suggested above. The scattering intensity depends on the scattering effectiveness of the

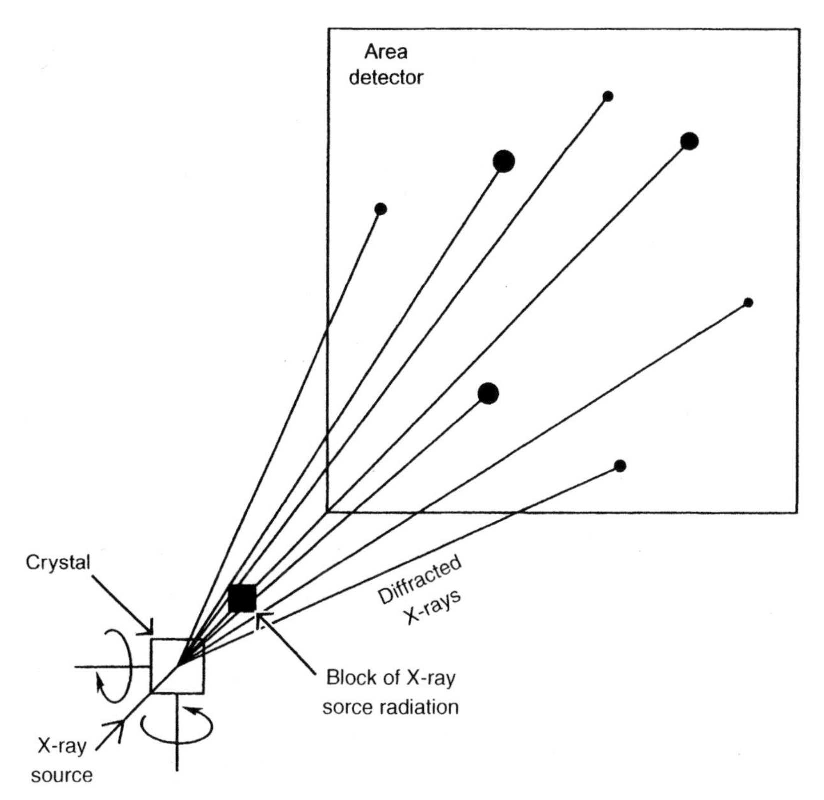

FIGURE 7-3. Experimental setup for measuring X-ray diffraction from crystals. The crystal can be rotated around both perpendicular axes, and the diffraction pattern is measured on an area detector or film. The incident X-ray radiation is prevented from hitting the area detector by insertion of a beam blocker between the crystal and the detector.

individual atoms and the phase of the wave from each scattering source. The structure factor, F, for each plane can be defined as the sum of the structure factors for individual atoms, f_i, times a phase factor, α_i, for each atom:

$$F = \sum_i f_j \alpha_i \tag{7-2}$$

The intensity of scattered radiation is proportional to the absolute value of the amplitude of the structure factor. The structure factor can be calculated if the atomic coordinates are known. Note that Eq. 7-2 is analogous to the general description of electromagnetic waves developed in Chapter 6, namely, an amplitude is multiplied by an angular dependence (sin and/or cos).

In proceeding to determination of the structure, it should be remembered that X rays are scattered by electrons in atoms so that representing a crystal as a series of point atoms is not a good picture of the real situation. Instead, a more realistic formulation is to represent the structure factor as an integral over a continuous distribution of electron density. The electron density is a function of the coordinates of the scattering centers, the atoms, and has a maximum around the position of each atom. What is desired in practice is to convert the measured structure factors into atomic coordinates. This is done by taking the Fourier transform of Eq. 7-2. In this case, the Fourier transform takes the structure factors, which are functions of the electron density, and inverts the functional dependence so that the electron density is expressed as a function of the structure factors. (This is analogous to the discussion in Chapter 6 in which the frequency and time dependences are interchanged by Fourier transforms.) This seems straightforward from the standpoint of the mathematics, but the problem is that the actual structure factors contain both amplitudes and phases. Only the amplitude, or to be more precise, its square, can be directly derived from the measured intensity of the diffracted beam. The phase factor must be determined before a structure can be calculated.

Two methods are commonly used to solve the phase problem for macromolecules. One of these is Multiple-wavelength Anomalous Dispersion or MAD. Thus far, we have discussed X-ray scattering in terms of *coherent* scattering where the frequency of the radiation is different than the frequency of oscillation of electrons in the atoms. However, X-ray frequencies are available that match the frequency of oscillation of the electrons in some atoms, giving rise to *anomalous* scattering that can be readily discerned from the wavelength dependence of the scattering. The feasibility of this approach is made possible by the use of synchrotron radiation since crystal monochromaters can be used to obtain high-intensity X rays at a variety of wavelengths. Anomalous scattering is usually observed with relatively heavy atoms (e.g., metals or iodine) that can be inserted into the macromolecular structure. The observation of anomalous scattering for specific atoms allows these atoms to be located in the structure, thus providing the phase information that is required to obtain the complete structure.

Another frequently used method for determining phases in macromolecules is *isomorphous replacement*. With this method a few heavy atoms are put into the structure, for example, metal ions. Since scattering is proportional to the square of the atomic number, the enhanced scattering due to the heavy atoms can be easily seen, and the positions of the heavy atoms determined. This is done by looking at the difference in structure factors between the native structure and its heavy atom derivative. Of course, it is essential that the isomorphous replacement not significantly alter the structure of the molecule being studied. We will not delve into the details of the methodology here.

Thus far, the assumption has been made that the X-ray radiation is monochromatic. An alternative is to use a broad spectrum of radiation ("white" radiation). The premise is that whatever the orientation of the crystal, one of the wavelengths would satisfy the Bragg condition. In fact, this was the basis for the first X-ray diffraction experiments carried out by Max von Laue in 1912, and this approach is called Laue diffraction. As might be expected, Laue diffraction patterns can be very complex and difficult to interpret as diffraction patterns from multiple wavelengths are observed simultaneously. The advantage is that a large amount

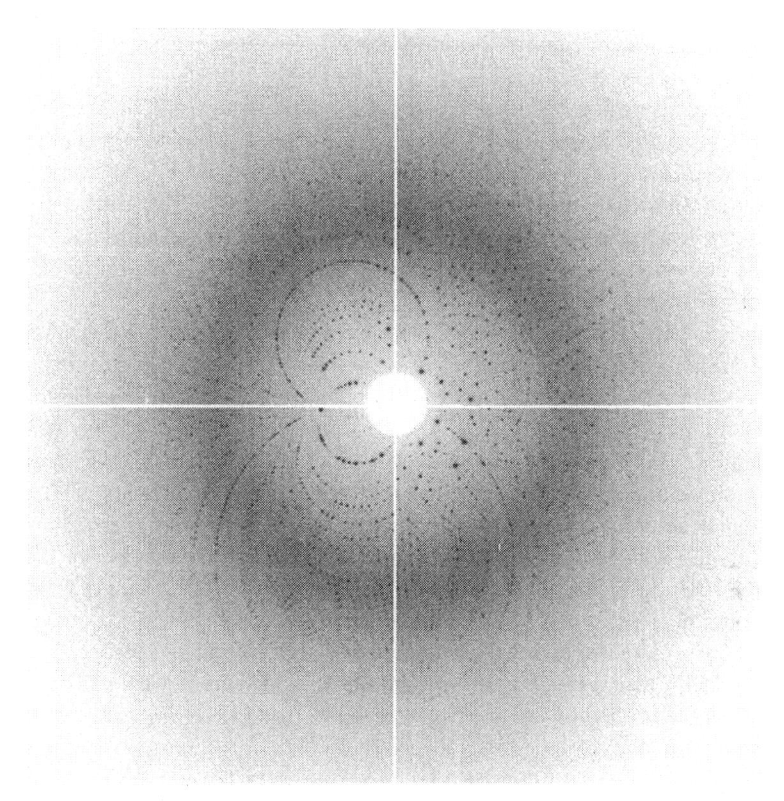

FIGURE 7-4. Laue diffraction pattern obtained with synchrotron radiation and an area detector. The protein crystal is DNA polymerase from *Bacillus stearothermophilus*. Copyright by Professor L. S. Beese, Duke University. Reproduced with permission.

of structural information can be obtained in a very short time. A Laue diffraction pattern of DNA polymerase from *Bacillus stearothermophilus* obtained with an area detector and synchrotron radiation is shown in Figure 7-4. Although most structural determinations use monochromatic radiation, the use of Laue diffraction has increased significantly in recent years.

What limits the precision of a structure determination? Real crystals are not infinite arrays of planes. They have imperfections so that the diffraction pattern is strong near the center and becomes weaker radially from the center, as the reflections from planes closer together become important. The precision of the distances in the final structure is limited by how many reflections are observed. The better the order in the crystal, the further from the center of the diffraction pattern that reflections can be seen. This problem can only be cured by obtaining better crystals. A minor problem is that atoms have thermal motion, so that an inherent uncertainty in the position exists. This can be helped by working at low temperatures, which sometimes will lock otherwise mobile structures into one conformation. Finally, it should be noted that in some cases, portions of the macromolecule may not be well defined because of intrinsic disorder, that is, more than one arrangement of the atoms occurs in the crystals.

The spatial resolution of a structure is usually given in terms of the distances that can be distinguished in the diffraction pattern. For good structures, this is typically 2 Å or better. However, this is not the uncertainty in the positions of atoms within the structure. In solving the structure of a macromolecule, other information is used from known structures of small molecules, for example, bond distances (C—C, etc.) and bond angles. For high-quality structures, the uncertainty in atomic positions is tenths of angstroms.

At the present time, the primary difficulty in determining the molecular structures of proteins and nucleic acids is obtaining good crystals, that is, crystals that give good diffraction patterns. If good diffraction patterns can be obtained, including isomorphous replacements or anomalous scattering atoms, computer programs are available to help derive the structure. However, as structures become larger and larger, both obtaining good crystals and the data analysis itself are a challenge.

Crystal structures represent a static picture of the equilibrium structure so that the correlation with biological function must be approached with caution. Furthermore, the packing of molecules within a crystal can alter the structure relative to that in solution. This is usually true only for atoms on the surface of the molecule so that the core structures, such as active sites, are normally an accurate reflection of the biologically active species. Recent applications of synchrotron radiation have permitted the time evolution of structures to be studied using Laue diffraction (3). Although such experiments are exceedingly difficult to carry out and interpret, in principle it is possible to observe the structure of biological molecules as they function.

7.4 NEUTRON DIFFRACTION

High-velocity thermal neutrons are generated by atomic reactors. If they are slowed down by collisions, typically with D_2O, then their energy corresponds to a wavelength of about 1 Å. The scattering of neutrons by atoms is quite different than the scattering of X rays. In the latter case, the scattering is by electrons, whereas in the former case, the scattering is by the nuclei. Moreover, the scattering of neutrons is not easily related to atomic number.

A major difference between neutron and X-ray scattering is that hydrogen is a very effective scattering center for neutrons, but is relatively ineffective for X rays. Consequently, the positions of hydrogen atoms are difficult to obtain from X-ray scattering but can be readily found through neutron scattering. D_2O scatters quite differently than H_2O: their scattering factors have opposite signs, that is, they are out of phase with respect to each other. The scattering factors for proteins and nucleic acids usually fall somewhere in between so that mixtures of D_2O and H_2O can be used as "contrast" agents. Appropriate mixtures can be used to effectively make certain macromolecules "invisible" to neutron scattering. For example, if a protein-nucleic acid interaction is being studied, the appropriate solvent mixture can make either the protein or the nucleic acid "invisible." Historically, this property was important for mapping the structure of the ribosome.

Neutron scattering studies are relatively rare because high-flux neutron sources are small in number. Nevertheless, neutron diffraction can be a useful tool in elucidating macromolecular structure.

7.5 NUCLEIC ACID STRUCTURE

Probably the most exciting structure determination of a biological molecule was that for B-DNA deduced by James Watson and Francis Crick. This structure was, in fact, based on diffraction patterns from fibers rather than crystals. The familiar right-handed double helix is shown in Figure 3-9 (see color plates). Thermodynamic aspects of this structure have been previously discussed in Chapter 3. In this structure, two polynucleotide chains with opposite orientations coil around an axis to form the double helix. The purine and pyrimidine bases from different chains hydrogen bond in the interior of the double helix, with adenine (A) hydrogen bonding to thymine (T) and guanine (G) hydrogen bonding to cytosine (C). The former pair has two hydrogen bonds and the latter has three hydrogen bonds. Hence, the G–C pair is significantly more stable than the A–T pair. Phosphate and deoxyribose are on the outside of the double helix and interact favorably with water. The double helical structure contains two obvious grooves, the major and minor grooves. The major groove is wider and deeper than the minor groove. These grooves arise because the glycosidic bonds of a base pair are not exactly opposite to each other.

We will not dwell on the biological function of DNA except to note that DNA must separate during the replication process, and the stability of various DNAs depends on the base composition of the DNA. At sufficiently high temperatures, all DNA structures are destroyed and the two polynucleotide chains separate.

Extensive X-ray studies have shown that other forms of helical DNA exist. If the relative humidity is reduced below about 75%, A-DNA forms. It is also a right-handed double helix of antiparallel strands, but a puckering of the sugar rings causes the bases to be tilted away from the normal of the axis. A-DNA is shorter and wider than B-DNA. This structure is found in biology for some double-stranded regions of RNA and in RNA–DNA hybrids.

A third type of DNA has been found that is a left-handed helix, Z-DNA. Z-DNA is elongated relative to B-DNA and has more unfavorable electrostatic interactions. This structure has been observed with specific short oligonucleotides at high salt concentrations. The biological significance of this structure is not clear, but its occurrence demonstrates that quite different structures can exist for DNA.

RNA has more diverse structures than DNA, in keeping with its diverse biological functions that include its role as messenger RNA (mRNA) during transcription, as transfer RNA (tRNA) in protein synthesis and as ribozymes in catalysis. In addition, RNA is found in ribosomes and other ribonucleic proteins. Typically, RNA does not form a double helix from two separate chains, as DNA does.

However, the same base pairing rules as found for DNA cause internal helices to be formed within an RNA molecule. In fact, the structures of many RNA molecules are inferred in this way. In RNA, uracil (U) is found rather than thymine, as in DNA.

The first structure of an RNA deduced by X-ray crystallography was that of yeast phenylalanine tRNA (4). This structure is shown in Figure 7-5 (see color plates). The L-shaped molecule is typical of tRNAs. The hydrogen bonding network is shown in this structure. The acceptor stem (upper right-hand corner) is where the amino acid is linked to form the aminoacyl-tRNA. The amino acid is transferred to the growing protein chain during protein synthesis. The anticodon, which specifies the amino acid to be added to a protein during synthesis, is at the end of the long arm of the L. It pairs with a specific mRNA that is the genetic information for the amino acid. The structures of many other tRNAs are now known and are quite similar.

For many years, a central dogma of biochemistry was that all physiological reactions are catalyzed by enzymes that are proteins. However, now many reactions are known that are catalyzed by RNA, as previously discussed in Chapter 5. In brief, these catalytic RNA (ribozymes) are particularly important in splicing and maturation of RNA. In some cases, the ribozyme cleaves other RNA, whereas in other cases it undergoes self-cleavage. Ribosomal RNA also plays a catalytic role in the formation of peptide bonds. Although there is little doubt that RNA functions as a catalyst physiologically, they are not as efficient as enzymes. In some cases, such as self-cleavage, the reaction is not truly catalytic as multiple turnovers cannot occur. Many ribozymes require a protein to function efficiently, and even in cases where true catalysis occurs, it is slow relative to typical enzymatic reactions. Kinetic aspects of ribonuclease P, a ribozyme important for the maturation of tRNA, were considered in Chapter 5.

The structures of several ribozymes have been determined (5). The first structure that was determined was that of "hammerhead" ribozyme (6). This ribozyme was first discovered as a self-cleaving RNA associated with plant viruses. The minimal structure necessary for catalysis contains three short helices and a universally conserved junction. The structure is shown in Figure 7-6, along with an indication of the cleavage site in the RNA chain. The three helices form a Y-shaped structure, with helices II and III essentially in line and helix I at a sharp angle to helix II. Distortions at the junction of the helices cause C_{17} to stack with helix I. This structure is somewhat misleading because multiple divalent metal ions are required in order for the RNA to fold and carry out catalysis. At least one of these metal ions is intimately involved in the catalytic process. The precise number of metal ions that are required for folding and/or catalysis is uncertain. Moreover, the two structures that have been determined are necessarily noncleavable variants: one has a DNA substrate strand and the other a 2′-*O*-methyl group at the cleavage site. The two structures are not identical and not entirely consistent with mutagenesis studies that place specific groups at the active catalytic site. This is undoubtedly a reflection of the flexibility of the structure. Nevertheless, knowledge of ribozyme structure is a requisite for understanding their mechanisms of action and for engineering ribozymes for enhanced catalysis and function.

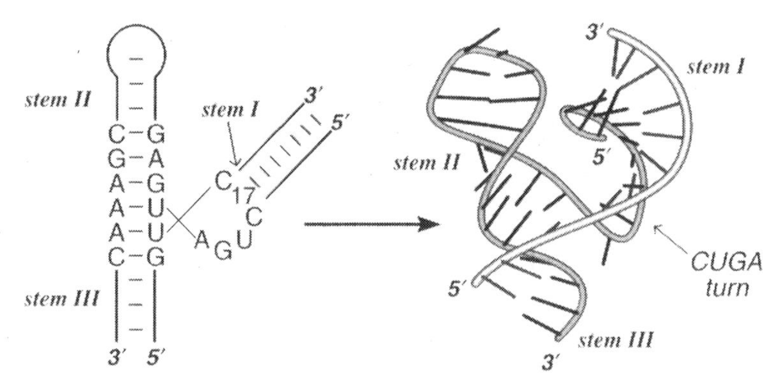

FIGURE 7-6. Secondary and tertiary structures of the hammerhead ribozyme. The capital letters refer to the bases of the nucleic acids, and the lines show hydrogen bonding interactions between the bases. The cleavage site is indicated by the arrow. Reprinted with permission from D. A. Doherty and J. A. Doudna, *Annu. Rev. Biochem.* **69**, 597 (2000). 2000 by Annual Reviews, www.annualreviews.org.

7.6 PROTEIN STRUCTURE

Hundreds of protein structures are known, with a great deal of variation in structure, as well as similarities, among them. Some of the thermodynamic aspects of protein structure as well as some of the basic structures have been discussed in Chapter 3. In this section, we again return to the common motifs, as well as a few specific examples. Fortunately, the coordinates for protein structures can be found in a single database, the Protein Data Bank, which can be accessed via the Internet. Free software is also available on the internet for viewing and manipulating structures with known coordinates, for example, Kinemage (7) and RasMol (8). This software can also be used for nucleic acid structures. Interested readers should explore this software and database, as their use is very important for understanding biological mechanisms on a molecular basis.

The linear sequence in which amino acids are arranged in a protein is termed the *primary* structure. The two most common long-range ordered structures in proteins are the α-helix and the β-sheet. Ordered structures within a protein are called *secondary* structures. Both structures are stabilized by hydrogen bonds between the NH and CO groups of the main polypeptide chain. In the case of the α-helix, depicted in Figure 3-4 (see color plates), a coiled rod-like structure is stabilized by hydrogen bonds between residues that are four amino acids apart. In principle, the screw sense of the helix can be right handed or left handed. However, the right-handed helices are much more stable because they avoid steric hindrance between the side chains and backbone. Helices with a different pitch, that is, stabilized by hydrogen bonding between residues other than those that are four amino acids apart are found, but the α-helix is the most common helical element found. The helical content of proteins ranges from almost 100% to very little. Helices are seldom more than about 50 Å long, but multiple helices can intertwine to give extended structures over 1000 Å long. A specific example is the interaction of myosin and tropomyosin in muscle.

Beta sheets are also stabilized by NH—CO hydrogen bonds, but in this case the hydrogen bonds are between adjacent chains, as depicted in Figure 3-5 (see color plates). The chains can be either in the same direction (parallel β-sheets) or in opposite directions (antiparallel β-sheets). These sheets can be relatively flat or twisted in protein structures. In schematic diagrams of proteins, α-helices are often represented as coiled ribbons and β-sheets as broad arrows pointing in the direction of the carboxyl terminus of the polypeptide chain (9).

Most proteins are compact globular structures, consisting of collections of α-helices and β-sheets. Obviously, if the structure is to be compact, it is necessary for the polypeptide chain to reverse itself. Many of these reversals are accomplished with a common structural element, a reverse turn or β-hairpin bend, that alters the hydrogen bonding pattern. In other cases, a more elaborate loop structure is used. These loops can be quite large. Although they do not have regular periodic structures, analogous to α-helices and β-sheets, they are still very rigid and well-defined structures.

The first protein whose structure was determined by X-ray crystallography to atomic resolution was myoglobin (10). Myoglobin serves as the oxygen transporter in muscle. In addition to the protein, it contains a *heme*, a protoporphyrin with a tightly bound iron. The iron is the locus where oxygen is bound. Myoglobin was a particularly fortuitous choice for structure determination. Of course, good crystals could be obtained. In addition, myoglobin is very compact and primarily consists of α-helices. Its structure is shown in Figure 7-7 (see color plates). The α-helices form a relatively compact electron density so that the polypeptide could be easily traced at low resolution. The structure has several turns that are necessary to maintain the compact structure. Overall, the myoglobin molecule is contained within a rectangular box roughly $45 \times 35 \times 25$ Å^3. The overall trajectory of the polypeptide chain (that is, the folding of the secondary structure) is called the *tertiary* structure.

For water-soluble proteins, including myoglobin, the interior is primarily hydrophobic or nonpolar amino acids such as leucine, pheylalanine, etc. Amino acids that have ionizable groups, such as glutamic acid, lysine, etc., are normally on the exterior of the protein. If a protein exists in a membrane, a hydrophobic environment, the situation is often reversed, with hydrophilic residues on the inside and nonpolar residues on the outside. In some cases, however, the structure is hydrophobic on both the "inside" and the "outside." Understanding how polypeptides fold into their native structures is an important subject under intense investigation. In an ideal world, the structure of a protein should be completely predictable from knowledge of its primary structure, that is, its amino acid sequence. However, we have not yet reached this goal.

As a second example of a protein structure, the structure of ribonuclease A is depicted in Figure 7-8 (see color plates) (11). This protein is quite compact and contains both α-helix and β-sheet modules with significant loops and connecting structures. In addition, four disulfide linkages are present. This enzyme is kidney shaped with the catalytic site tucked into the center of the kidney. The active site is shown by the placement of an inhibitor and two histidine residues that participate in the catalytic reaction. This structure proved much more elusive than myoglobin since a larger variety of structural elements are present.

Not all proteins are as compact as myoglobin and ribonuclease. In some cases, a single polypeptide chain may fold into two or more separate structural domains that are linked to each other, and many proteins exist as oligomers of polypeptide chains in their biologically active form. The arrangement of oligomers within a protein is called the *quaternary* structure. The first well-studied example of the latter structure was hemoglobin. Hemoglobin is a tetramer of polypeptide chains. Two different types of polypeptides are present: they are very similar but not identical. The typical structure of hemoglobin is $\alpha_2\beta_2$, with α and β designating the two types of quite similar polypeptide chains. Hemoglobin, of course, is used to transport oxygen in blood. In a sense, it is four myoglobins in a single molecule. Four porphyrins and irons are present, and each complex is associated with a single polypeptide chain. However, the interactions between the polypeptide chains are extremely important in the function of hemoglobin and influence how oxygen is released and taken up. In brief, the uptake and release are highly cooperative so that they occur over a very narrow range of oxygen concentration (cf. Ref. 12). This aspect of hemoglobin is discussed in Chapter 13.

Hemoglobin is known to exist in two distinct conformations that differ primarily in the subunit interactions and the details of the porphyrin binding site. The structures of both conformations of hemoglobin are shown in Figure 7-9 (see color plates). One of the conformations binds oxygen much better than the other, and it is the switching between these conformations that is primarily responsible for the cooperative uptake and release of oxygen. A complete discussion of hemoglobin structure is beyond the scope of this presentation, which is focused on illustrating some of the major features of protein structures.

7.7 ENZYME CATALYSIS

As final illustration of the application of X-ray crystallography to biology, we consider the enzyme DNA polymerase. Elucidating the mechanism by which enzymes catalyze physiological reactions has been a long-standing goal of biochemistry. X-ray crystallography has been used to probe many enzyme mechanisms. The choice of DNA polymerase is arbitrary, but it is clearly an enzyme at the core of biology and requires the knowledge of both protein and nucleic acid structures. DNA polymerase is the enzyme responsible for replication of new DNA. The reaction proceeds by adding one nucleotide at a time to a growing polynucleotide chain. This addition can only occur in the presence of a *DNA template* that directs which nucleotide is to be added by forming the correct hydrogen bonds between the incoming nucleotide and the template.

The molecular structures of a wide variety of DNA polymerases are known (cf. Ref. 13). The general structure of the catalytic site is similar in all cases and has been described as a right hand with three domains, fingers, a thumb, and a palm. The structure of a catalytic fragment of the thermostable *Bacillus stearothermophi-*

lus enzyme is shown in Figure 7-10 (see color plates) (14,15). The fingers and thumb wrap around the DNA and hold it in position for the catalytic reaction that occurs in the palm. To initiate the reactions, a primer DNA strand is required which has a free 3′-hydroxyl group on a nucleotide already paired to the template. The enzyme was crystallized with primer templates, which are included in Figure 7-10. The reaction occurs via a nucleophilic attack of this hydroxyl group on the α-phosphate of the incoming nucleotide, assisted by protein side chains on the protein and two metal ions, usually Mg^{2+}. We will not be concerned with the details of the chemical reaction.

The DNA interacts with the protein through a very extensive network of noncovalent interactions that include hydrogen bonding, electrostatics, and direct contacts. More than 40 amino acid residues are involved that are highly conserved in all DNA polymerases. The interactions occur with the minor groove of the DNA substrate and require significant unwinding of the DNA. The major groove does not appear to interact significantly with the protein and is exposed to the solvent.

The synthesis of DNA is a highly processive reaction, that is, many nucleotides are added to the growing chain without the enzyme and growing chain separating. In the case of the polymerase under discussion, more than 100 nucleotides are added before dissociation occurs. This requires the DNA chain to be translocated through the catalytic site. This translocation has been observed to occur in the crystals by soaking the crystals with the appropriate nucleotides and determining the structures of the products of the reactions. Remarkably, the crystals are quite active, with up to six nucleotides being added and a translocation of the DNA chain of 20 Å.

The enzyme exists in at least two conformations. The initial binding of the free nucleotide occurs in the open conformation, but the catalytic step is proposed to occur in a closed conformation in which the fingers and thumb clamp onto the DNA and close around the substrates. The fidelity of DNA polymerase is remarkable, with only approximately one error per 10^5 nucleotides incorporated (16). This is accomplished primarily by the very specific hydrogen bonds occurring between the template and the incoming nucleotide, but interactions with the minor groove are also important. The conformational change accompanying binding of the nucleotide to be added probably also plays a role in the fidelity by its sensitivity to the overall shapes of the reactants. Also, the template strand is postulated to move from a "preinsertion" site to an "insertion" site. The acceptor template base that interacts with the incoming nucleotide occupies the preinsertion site in the open conformation and the insertion site in the closed conformation where interaction with the incoming nucleotide occurs. Thus, the addition of each nucleotide is accompanied by a series of conformational changes that translocate the template into position for the next step in the reaction. A movie of this process is available on the Internet (15).

Even the high fidelity of the polymerase reaction is not sufficient for the reliable duplication of genes that is required in biological systems. Consequently, the enzyme has a built-in proofreading mechanism, an exonuclease enzyme activity that is located 35 Å or more from the polymerase catalytic site. If an incorrect nucleotide is incorporated, the match within the catalytic site is not perfect, and the polymerase reaction stalls. This brief pause is sufficiently long so that the offending nucleotide base can migrate to the exonuclease site and be eliminated. Unfortunately, the proofreading process itself is not perfect so that every 20 incorporations or so a correct base is chopped off. This wasteful process, however, increases the fidelity of replication by a factor of about 1000.

This very brief discussion of DNA polymerase indicates the great insight into biological reactions that knowledge of macromolecular structures can bring. If the crystals are biologically active, multiple structures with various substrates and inhibitors can provide a detailed mechanistic proposal. This paves the way for additional experiments that will shed even more light on the molecular mechanism.

REFERENCES

1. I. Tinoco, Jr., K. Sauer, J. C. Wang, and J. D. Puglisi, *Physical Chemistry: Principles and Applications in Biological Sciences*, 4th edition, Prentice Hall, Englewood Cliffs, NJ, 2002, pp. 667–711.
2. A. McPherson, *Introduction to Macromolecular Crystallography*, Wiley–Liss, Hoboken, NJ, 2003.
3. K. Moffat, *Chem. Rev.* **101**, 1569 (2001).
4. A. Rich and S. H. Kim, *Sci. Am.* **238**, 52 (1978).
5. E. A. Doherty and J. A. Doudna, *Annu. Rev. Biochem.* **69**, 597 (2000).
6. W. G. Scott, J. T. Finch, and A. Klug, *Cell* **81**, 991 (1995).
7. D. C. Richardson and J. S. Richardson, *Trends Biochem. Sci.* **19**, 135 (1994).
8. R. A. Sayle and E. J. Milner-White, *Trends Biochem. Sci.* **20**, 374 (1995).
9. J. S. Richardson, D. C. Richardson, N. B. Tweedy, K. M. Gernert, T. P. Quinn, M. H. Hecht, B. W. Erickson, Y. Yan, R. D. McClain, M. E. Donlan, and M. C. Suries, *Biophys. J.* **63**, 1186 (1992).
10. C. L. Nobbs, H. C. Watson, and J. C. Kendrew, *Nature* **209**, 339 (1966).
11. G. Kartha, J. Bello, and D. Harker, *Nature* **213**, 862 (1967).
12. M. F. Perutz, A. J. Wilkinson, M. Paoli, and G. G. Dodson, *Annu. Rev. Biophys. Biomol. Struct.* **27**, 1 (1998).
13. C. A. Brautigam and T. A. Steitz, *Curr. Opin. Struct. Biol.* **8**, 54 (1998).
14. J. R. Kiefer, C. Mao, J. C. Braman, and L. S. Beese, *Nature* **391**, 304 (1998).
15. S. J. Johnson, J. S. Taylor, and L. S. Beese, *Proc. Natl. Acad. Sci. USA* **100**, 3895 (2003).
16. T. A. Kunkel and K. Bebenek, *Annu. Rev. Biochem.* **69**, 497 (2000).

PROBLEMS

7.1. Assume a lattice of atoms equidistant from each other in all directions (a cubic lattice) with a distance between atoms of 2.86 Å. If a crystal of this material is irradiated with X rays with a wavelength of 0.585 Å, at what angles are Bragg reflections seen for the planes that are 2.86 Å apart? (More than one set of reflections will be seen, but we will not deal with that complexity here.) If the X rays have a wavelength of 6.00 Å, what would be observed?

7.2. The bond energy for a C—C bond is about 340 kJ/mol, 600 kJ/mol for C=C, and 400 kJ/mol for C—H. Calculate the energy/mol associated with X-ray radiation with wavelengths of 0.585 and 1.54 Å. What does this imply about the stability of biomolecules in an X-ray beam?

7.3. When a molybdenum target is used as an X-ray source, X rays of wavelength 0.710 Å are produced. For a particular crystal, a reflection was observed at 4° 48′. What are the possible distances between Bragg planes? The same reflection observed when a copper target is used as an X-ray source is found at 10° 27′. What is the wavelength of the X ray produced by the copper target?

7.4. A crystal structure consists of identical units called unit cells. The unit cells are arranged in order, much like a brick wall, to form the crystal. The unit cell contains multiple atoms, and often multiple molecules. Consider one particular plane of a unit cell that extends indefinitely in the X direction. The following structure factors, F_i, were obtained for the planes (Data from P. W. Atkins, *Physical Chemistry*, 3rd edition, Freeman, New York, 1986, p. 561.):

Plane	0	1	2	3	4	5	6	7	8	9	10	11	12	13	14	15
F_i	16	−10	2	−1	7	−10	8	−3	2	−3	6	−5	3	−2	2	−3

Assume that the electron density, $\rho(x)$, can be calculated from the structure factors by the relationship

$$\rho(x) = F_0 + 2\sum_{i>0} F_i \cos(2\pi i x)$$

Evaluate this sum for $x = 0$, 0.1, 0.2, 0.3, 0.4, 0.5, 0.6, 0.7, 0.8, 0.9, and 1.0, and construct a plot of $\rho(x)$ versus x. (This is best done with a computer program.) The maxima in the plot correspond to positions of atoms along the x axis. This exercise should give you a good idea of how electron density maps are obtained. The equation for $\rho(x)$ is an example of a Fourier transformation.

7.5. Go to the Internet address www.proteinexplorer.org and do the 1-h tour. This will allow you to make molecular drawings of proteins and nucleic acids whose structures are known.

CHAPTER 8

Electronic Spectra

8.1 INTRODUCTION

The most common type of spectroscopy involves light in the visible and ultraviolet regions of the spectrum interacting with molecules. This interaction causes electrons to shift between their allowed energy levels. If light shines on a colored sample, some of the radiation is absorbed by the sample. This absorption can heat the sample, it can cause photons of the same or lower energy to be emitted, or photochemical reactions can occur. We will not be concerned with the latter phenomenon. Of course, in all cases energy must be conserved, that is, the total energy of all of the three processes must be equal to the energy of the photons that are absorbed.

The absorption of light is due to the interaction of the oscillating electromagnetic field of the radiation with the charged particles in the molecule. If the electromagnetic force is sufficiently large, the electrons in the molecule are rearranged, that is, shifted to higher energy levels. The absorption process is extremely fast, occurring in about 10^{-15} s. The excited state equilibrates with its surrounding and returns to the ground electronic state either by the production of heat or by the emission of radiation with an energy equal to or less than that of the absorbed radiation. The latter process is called *fluorescence* and typically occurs in nanoseconds. The absorption process is shown schematically in Figure 8-1; a more exact description is discussed later (Fig. 8-11).

The details of light absorption by a molecule can be determined directly from quantum mechanical calculations. A transition of electrons from one energy level to another occurs when the wave functions for the two electronic states in question are electrically coupled. The detailed calculation tells us what transitions between the energy levels are allowed, for example, from energy level 1 to energy level 2. Electrons can only move between certain energy states; not all transitions are allowed. These *selection rules* are important because they allow us to predict what wavelengths of light will be absorbed and emitted. Although this facet of spectroscopy will not be of great concern here, it is important to remember that transitions of electrons between energy levels are governed by a set of rules. Exceptions to

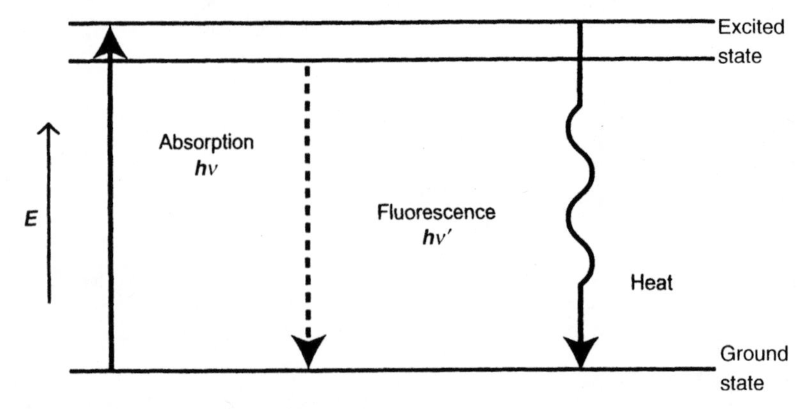

FIGURE 8-1. Simplified schematic diagram of electron excitation by absorbance of radiation. An electron is excited to a higher energy level by radiation of frequency ν. It returns to its ground state through fluorescence at a frequency ν' and/or by dissipation of heat. A more complete diagram is presented in Figure 8-11.

selection rules sometimes occur, but with a much lower probability than the favored event. This is primarily due to the fact that the potential energy functions used in the quantum mechanical calculations are not perfect.

8.2 ABSORPTION SPECTRA

From a practical standpoint, the most important aspect of the quantum mechanical calculation is the determination of how much light is absorbed by the sample. This is embodied in the Beer–Lambert law, which gives the relationship between the light intensity entering the solution, I_0, and the light intensity leaving the solution, I:

$$\log(I_0/I) = A = \varepsilon c l \tag{8-1}$$

Here A is defined as the absorbance, ε is the molar absorbtivity or extinction coefficient, c is the concentration of the absorbing material, and l is the thickness of the sample through which the light passes. In visible–ultraviolet spectrophotometers, a solution of the sample is put into a cuvette of calibrated dimensions, and visible or ultraviolet light is passed through the sample. The intensity of the light before and after passing through the solution is determined through the use of a variety of detectors, such as photomultipliers and photodiodes. A schematic diagram of a typical spectrophotometer is shown in Figure 8-2. The extinction coefficient is a different constant for each wavelength and is characteristic of the molecule under investigation. In principle, it can be calculated through quantum mechanics, but more practically it can be determined experimentally. A spectrophotometer also must account for light lost by reflection, absorption by the cuvette, and various other factors.

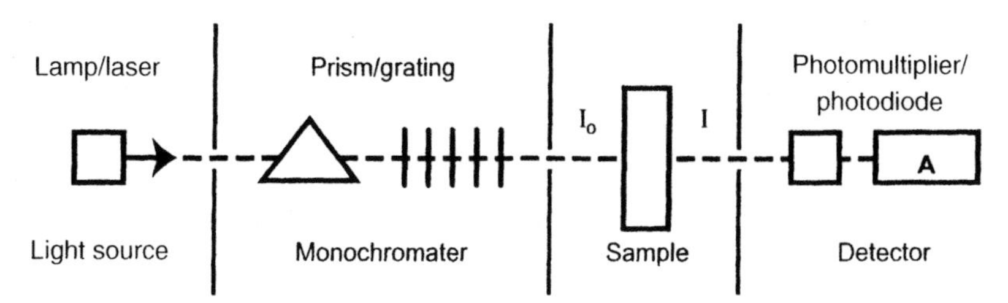

FIGURE 8-2. Schematic representation of an absorption spectrophotometer. Common light sources are tungsten and arc lamps and lasers. Monochromatic light of intensity I_0 is obtained with a prism or diffraction grating. The detection of the transmitted light of intensity I is with a photomultiplier or photodiode, with the electronic signal converted to the absorbance, A.

Deviations from the Beer–Lambert law can be observed in the laboratory. For example, the light might not be monochromatic, the sample might be inhomogeneous, light scattering could occur, the photodetectors might exhibit nonlinear behavior, etc. These are instrumental and sample artifacts. However, apparent deviations also occur that are due to the molecular properties of the sample: for example, sample aggregation or complex formation as the concentration is changed. In this case, the deviations can be used to study these properties. Experimental studies must carefully verify the validity of the Beer–Lambert law over the concentration range being studied.

A very important property of light absorbing solutions is that if multiple absorbing species are present, the total absorbance is simply the sum of the absorbance of the individual species. For example, for species X and Y, the absorbance is

$$A_\lambda = A_\lambda^{\mathrm{X}} + A_\lambda^{\mathrm{Y}} = \varepsilon_\lambda^{\mathrm{X}} l(\mathrm{X}) + \varepsilon_\lambda^{\mathrm{Y}} l(\mathrm{Y}) \tag{8-2}$$

If measurements are made at two different wavelengths where the extinction coefficients are sufficiently different, it is possible to determine the concentrations of the individual species if the extinction coefficients are known. Equation 8-2 generates two equations with different numerical values of the extinction coefficients at the two wavelengths. These two equations can be solved simultaneously to give the two concentrations. This is a very common method for determining the concentrations of individual species in solutions. In fact, this approach can be extended to more complicated situations. For example, if three species are present, absorbance measurements at three different wavelengths can be used to determine the concentrations of the individual components.

The wavelength at which two components have exactly the same extinction coefficient is called the *isobestic* wavelength. In this case, it is obvious that measurement of the absorbance determines the total concentration of the two components:

$$A_{\mathrm{isobestic}} = \varepsilon_{\mathrm{isobestic}} l[(\mathrm{X}) + (\mathrm{Y})] \tag{8-3}$$

Although one or more isobestic wavelengths may be found for any pair of components, this is not necessarily the case, as the two components may not have significant absorbance in the same wavelength region. A useful rule to remember is that if two or more isobestic wavelengths are seen at a series of different concentrations and/or varying conditions such as pH, then only two components are present. (This is not entirely rigorous, but the probability of this not being the case is exceedingly small.) This is especially useful when examining samples in which the individual extinction coefficients are not known.

8.3 ULTRAVIOLET SPECTRA OF PROTEINS

Proteins that do not contain strongly absorbing extrinsic cofactors such as hemes have a very characteristic ultraviolet spectrum, whereas they are essentially transparent in the visible region of the spectrum. The spectrum of serum albumin is shown in Figure 8-3 as a plot of the absorbance versus the wavelength. It has two absorption maxima, one at about 280 nm, the other, much stronger absorption, occurs around 190–210 nm. The absorption at 280 nm is frequently used to measure protein concentrations. Although the absorption is much stronger at 200 nm, it is harder to make measurements at this wavelength, so absorption at 200 nm is not routinely used as an analytical tool. The ultraviolet spectrum of a protein is sufficiently characteristic to use as a preliminary indication that a protein is present.

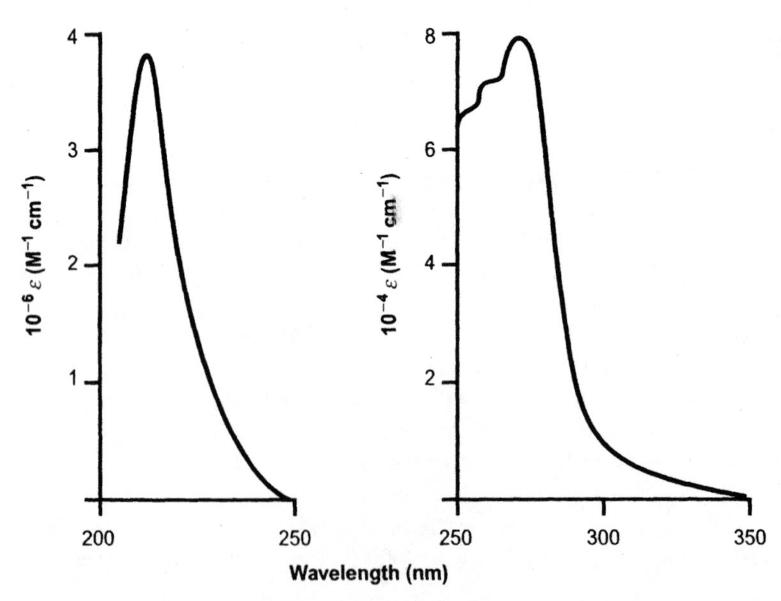

FIGURE 8-3. Ultraviolet absorption spectrum of bovine serum albumin at pH 7.0. The extinction coefficient is plotted versus the wavelength in two panels because of the large difference in the ordinate scale as the wavelength changes. Data from E. Yang in I. Tinoco, K. Sauer, J. C. Wang, and J. D. Puglisi, *Physical Chemistry*, 4th edition, Prentice-Hall, Englewood Cliffs, NJ, 2002, p. 548.

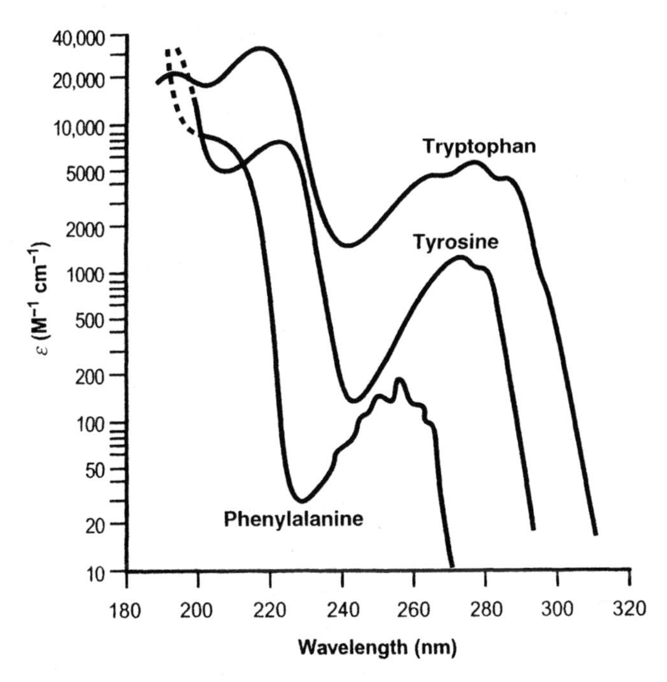

FIGURE 8-4. Absorption spectra of phenylalanine, tyrosine, and tryptophan at pH 6.0. Note the logarithmic scale on the ordinate. Adapted from S. Malik in D. B. Wetlaufer, Ultraviolet spectra of proteins and amino acids, *Adv. Protein Chem.* **17**, 303 (1962). © 1962, with permission from Elsevier.

The spectra of the aromatic amino acids phenylalanine, tyrosine, and tryptophan are shown in Figure 8-4. It can be readily seen that electronic transitions in tyrosine and tryptophan are primarily responsible for the absorption peak at around 280 nm. It can also be seen that all three amino acids absorb strongly at shorter wavelengths. Measurements of the spectra of polypeptides not containing aromatic amino acids show strong absorbance at around 192 nm due to electronic transitions associated with the peptide bond. Thus, the absorbance maximum in the 190–210 nm range is due to both aromatic amino acids and peptide bonds. The absorption due to side chains of other amino acids is typically at least an order of magnitude less in this wavelength region.

The spectra of aromatic amino acids and peptide bonds in proteins are significantly influenced by their local environments. This is because the spectra are fundamentally due to electrical phenomena so that structural features that determine the local charge distribution and the local dielectric constant strongly influence the spectra. Thus, the spectra of peptide bonds and aromatic amino acids buried inside the protein are somewhat different than those on the exterior of the protein. Hydrogen bonding and ion pair interactions also influence the spectra. This environmental sensitivity of protein spectra can be used to obtain information about the structure of proteins. For example, the amount of α-helix, β-sheet, and random coil can be estimated from the ultraviolet spectrum (1). A denatured protein usually has a somewhat different ultraviolet spectrum than the native protein so that protein folding reactions can be monitored by measuring spectral changes. However, the change in the ultraviolet spectrum is usually rather small so that other methods have proved more useful for assessing secondary structure and monitoring protein folding (e.g., circular dichroism, which is discussed later).

8.4 NUCLEIC ACID SPECTRA

In contrast to proteins, all of the common nucleotide bases have quite similar absorption spectra, with an absorption peak at around 260 nm. The spectra of adenine, thymine, cytosine, uracil, and guanine are shown in Figure 8-5. Nucleic acid solutions also are essentially transparent in the visible region of the spectrum. The intense absorption at 260 nm is frequently used as an indication of the presence of nucleic acids and can be used to determine their concentrations by use of the Beer–Lambert law.

Again, the local environment of individual nucleotide bases can influence the spectrum. Generally, nucleic acids have a lower absorbance than the sum of the absorbance of the individual nucleotides. This phenomenon is called *hypochromicity* and is due to the "stacking" or lining up of parallel planes of the bases. An increase in absorption is called *hyperchromicity*. These phenomena are very useful for determining whether nucleic acids are structured. For example, native (double-helix) DNA displays significant hypochromicity, and if the DNA is unwound, the absorbance at 260 nm increases.

Although the stable structure of DNA is the well-known double helix, the double-helix structure is lost as the temperature increases, and at a sufficiently

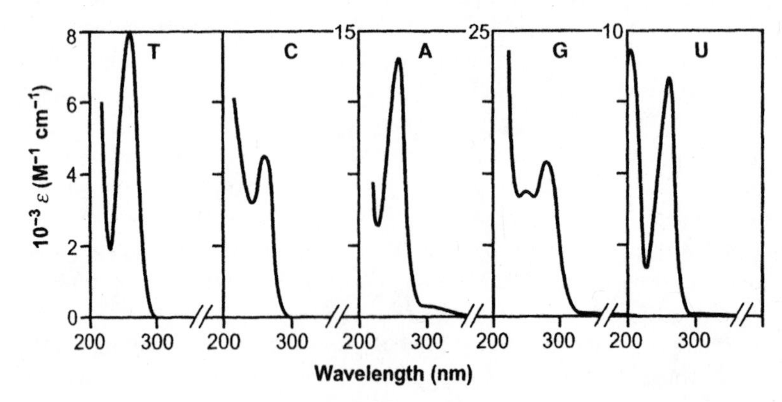

FIGURE 8-5. Absorption spectra in water of the four bases commonly found in nucleic acids: thymine (T), cytosine (C), adenine (A), guanine (G), and uracil (U). Data from H. Du, R. A. Fuh, J. Li, A. Corkan, and J. S. Lindsey, *Photochem. Photobiol.* **68**, 141 (1998) [http://omlc.ogi.edu/spectra/PhotochemCAD].

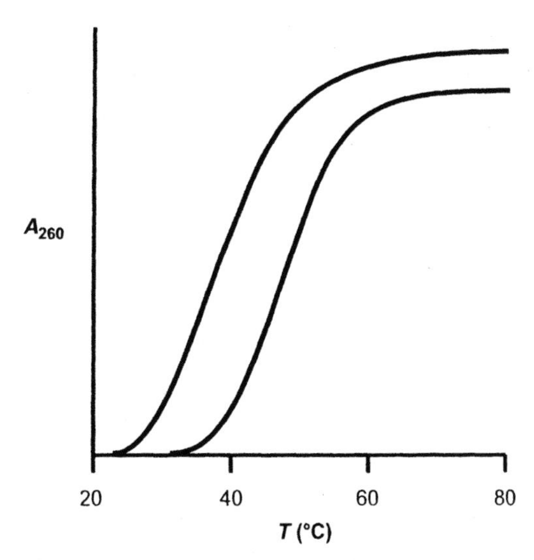

FIGURE 8-6. Schematic representation of DNA melting for two different DNA. The absorbance at 260 nm increases as the temperature is raised and the double-stranded structure breaks down. The temperature at which the melting occurs increases as the amount of G–C base pairs increases.

high temperature, the two strands of DNA separate. This loss of structure can be monitored by the increase in absorbance at 260 nm, as shown schematically in Figure 8-6. Obviously, the more stable the DNA, the higher the temperature that is needed to break down the structure so that doing a temperature "melt" of DNA is a useful tool for determining the stability of a specific DNA. With this technique, for example, it is easy to show that G-C base pairs are more stable than A-T pairs. By studying the stability of a series of nucleic acid oligomers, it is possible to predict the stability of DNA and, less reliably, RNA structures (cf. Chapter 3). Also, when DNA is synthesized, the double-helix structure must be broken in order to provide a single-strand template for the polymerization reaction. The kinetics of DNA structure formation/breakdown has been discussed in Chapter 5. The unwinding of DNA *in vivo* is done by helicase enzymes. Obviously, monitoring the DNA structure is important for understanding DNA synthesis, in general, and the mechanism of action of helicases, in particular. This subject is considered further later in this chapter.

8.5 PROSTHETIC GROUPS

Some proteins contain tightly bound organic molecules and/or metal ions. These are usually referred to as *prosthetic groups* and are essential for the biological activity of the macromolecules. The spectra of these prosthetic groups are often in the visible region of the spectrum, easily separable from the absorption due to the protein, and can be monitored to follow the reactions undergone by the protein.

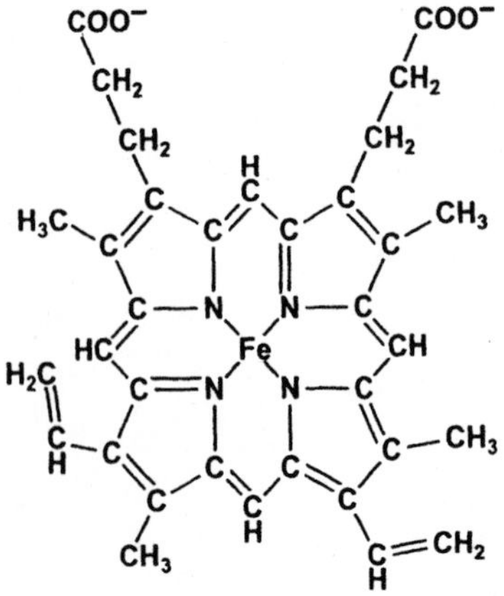

FIGURE 8-7. Structure of the heme found in hemoglobin. The Fe is in the oxidation state +2 when oxygen is bound.

One of the most common prosthetic groups is *heme*, the structure of which is shown in Figure 8-7. The heme binds iron tightly, either as Fe^{2+} or as Fe^{3+}. Probably, the most well-known heme protein is hemoglobin, mentioned in Chapter 7. Hemoglobin has a molecular weight of about 64,000. It has four polypeptide chains and contains four hemes. It is, of course, responsible for the transport of oxygen in the body. The structure of the protein is well known (Fig. 7-9, see color plates) and its function has been extensively studied (cf. Ref.2). For the purposes of this discussion, we consider only the visible spectra of the molecule. The active molecule contains Fe^{2+}: when oxygen is bound, it is coordinated directly to the iron. In the absence of oxygen, this iron coordination site is occupied by water. The visible spectra of oxy- and deoxyhemoglobin are quite distinct, as shown in Figure 8-8. This is because the binding of oxygen alters the environment of the iron/heme, thereby altering the electronic energy levels. Thus, the visible spectrum can be used to monitor the binding of oxygen. Oxyhemoglobin is responsible for the bright red color of oxygenated blood. The ferrous iron can be oxidized to the ferric form, which also alters the spectrum of the molecule. This form of hemoglobin is not involved in oxygen transport. The molecular mechanism for the binding of oxygen to hemoglobin is now understood in molecular detail. It is a highly cooperative process with alterations in the position of the iron in the molecule and significant conformational changes within the protein accompanying binding (cf. Chapter 13).

Cytochromes also contain hemes. For these molecules the oxidation–reduction of iron is crucial to their physiological function. The progress of the oxidation–reduction reactions can be readily monitored through changes in the visible spectrum.

Flavins constitute another type of common prosthetic group. Flavin mononucleotide (FMN) and flavin adenine dinucleotide (FAD) are important constituents in many enzymes that participate in redox reactions. Their structures are shown

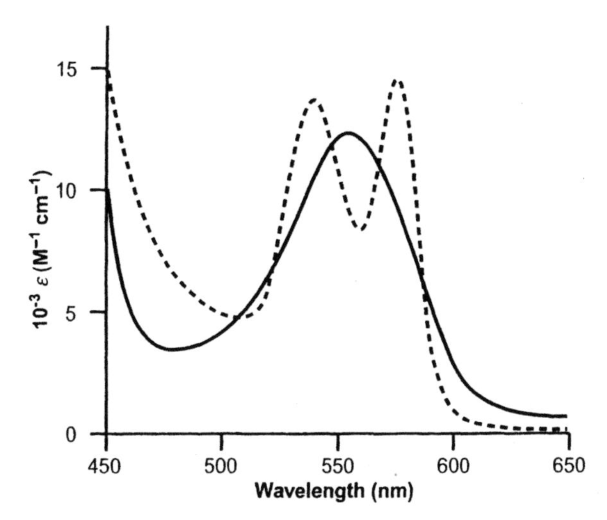

FIGURE 8-8. Visible absorption spectra of oxyhemoglobin (dashed line) and deoxyhemoglobin (solid line). These two forms of hemoglobin can be readily distinguished with absorption spectroscopy. Adapted from L. Stryer, *Biochemistry*, 2nd edition, Freeman, San Francisco, CA, 1981, p. 52. © 1976, 1981 by Lubert Stryer. Used with permission of Freeman.

in Figure 8-9. These flavin prosthetic groups can exist in three oxidation states: oxidized, semiquinone radical, and reduced. (A radical is an organic molecule with an unpaired electron.) All three of these oxidation states have different absorption peaks in the visible. The three oxidation states and their absorption maxima are included in Figure 8-9. The semiquinone can also lose a hydrogen to give a diradical (two unpaired electrons) with yet a different absorption maximum. The p*K* for the loss of this hydrogen is about 8.4. Flavin enzymes can undergo both one- and two-electron oxidation–reduction reactions and can be readily reoxidized by oxygen. The visible spectrum serves as a useful indicator of the oxidation state of the flavin in an enzyme. It can also be used to monitor the reaction progress of the enzymatic reactions.

Why do these prosthetic groups have electronic transitions at longer wavelengths than proteins? This can be readily understood in molecular and quantum mechanical terms. Both hemes and flavins have highly aromatic structures, that is, extensive conjugation occurs. As a result, the electrons are delocalized around the ring structures (π electrons). If the electrons are considered as "particles in a box," as discussed in Chapter 6, the size of the box is increased relative to less conjugated structures such as tyrosine, phenylalanine, and tryptophan. An increase in the size of the box increases the wavelengths that characterize the electronic transitions (Eq. 6-11).

8.6 DIFFERENCE SPECTROSCOPY

Because the absorption of light by proteins and nucleic acids depends on the local environment of the chromophores, changes in light absorption often occur when

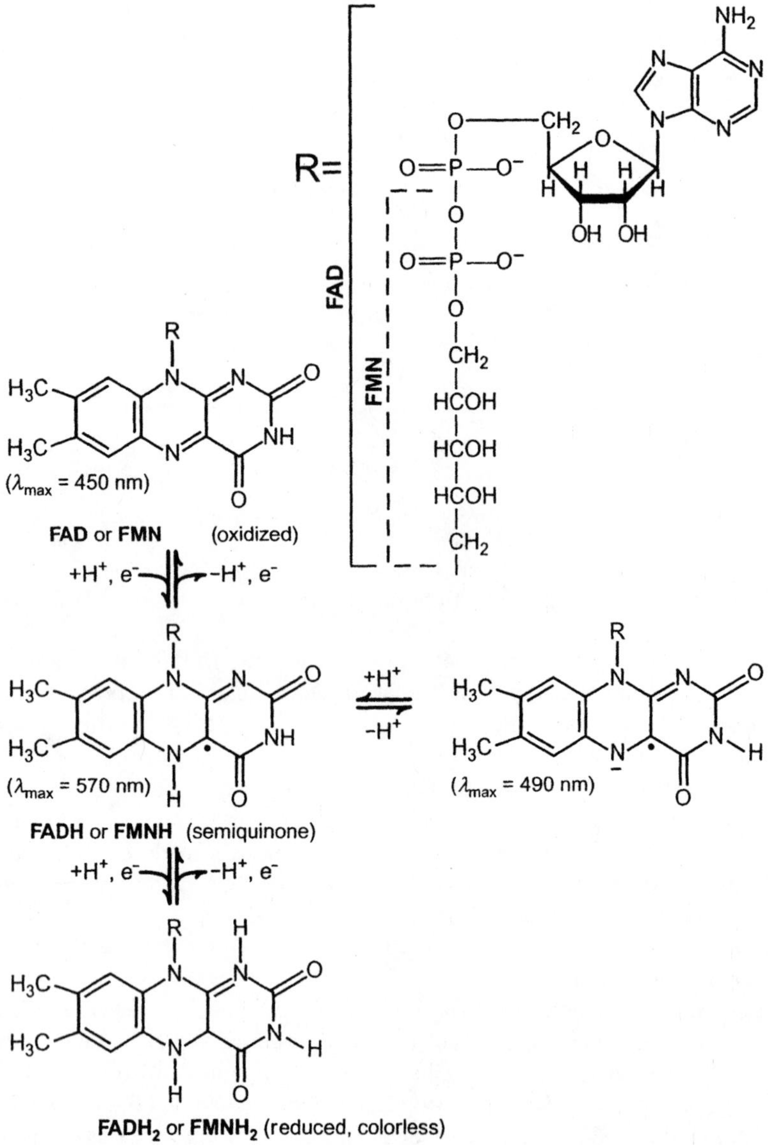

FAD or FMN (oxidized) (λ_{max} = 450 nm)

FADH or FMNH (semiquinone) (λ_{max} = 570 nm) ⇌ (λ_{max} = 490 nm)

$FADH_2$ or $FMNH_2$ (reduced, colorless)

FIGURE 8-9. Oxidation states of FAD and FMN with the absorption maxima indicated. These oxidation states are important for the biological function of these molecules and can be readily distinguished with absorption spectroscopy. The p*K* of the semiquinone is about 8.4.

small molecules bind to proteins or nucleic acids and when macromolecules interact. As an example, we consider the binding of nucleotides to the enzyme ribonuclease A. As the name implies, ribonuclease A hydrolyzes RNA. It is a small protein with a molecular weight of approximately 14,000 and contains tyrosine, tryptophan, and phenylalanine so that it has a typical protein ultraviolet spectrum. Its structure is shown in Figure 7-8 (see color plates). It binds nucleic acids and has a great preference for cleavage adjacent to a pyrimidine. When nucleotides bind to the enzyme, the ultraviolet spectrum is changed. However, the changes are very small so that a technique called difference spectroscopy must be used.

With difference spectroscopy, a double-beam spectrophotometer is used. Each beam contains two cuvettes. For one of the beams, one cuvette contains the nucleotide and the other the protein. In the other beam, one of the cuvettes contains a mixture of the nucleotide and the protein and the other contains buffer. The concentrations of nucleotide and the enzyme are exactly the same in each beam. Consequently, any difference in absorbance that is measured must be due to the interaction of the nucleotide and the protein. The absorbance in the beam containing the mixture of enzyme and nucleotide (assuming a path length of 1 cm) is

$$\begin{aligned} A_1 &= \varepsilon_{\mathrm{EL}}(\mathrm{EL}) + \varepsilon_{\mathrm{E}}(\mathrm{E}) + \varepsilon_{\mathrm{L}}(\mathrm{L}) \\ &= \varepsilon_{\mathrm{EL}}(\mathrm{EL}) + \varepsilon_{\mathrm{E}}[(\mathrm{E}_0) - (\mathrm{EL})] + \varepsilon_{\mathrm{L}}[(\mathrm{L}_0) - (\mathrm{EL})] \end{aligned} \tag{8-4}$$

where ε's are the extinction coefficients for the enzyme, E, the ligand, L, and the enzyme–ligand complex, EL. The study of ligand binding showed that only one nucleotide binds per enzyme molecule. The absorbance in the other beam is given by

$$A_2 = \varepsilon_{\mathrm{E}}(\mathrm{E}_0) + \varepsilon_{\mathrm{L}}(\mathrm{L}_0) \tag{8-5}$$

Subtraction of Eq. 8-5 from Eq. 8-4 gives the difference absorbance

$$\Delta A = (\varepsilon_{\mathrm{EL}} - \varepsilon_{\mathrm{E}} - \varepsilon_{\mathrm{L}})(\mathrm{EL}) = \Delta\varepsilon(\mathrm{EL}) \tag{8-6}$$

Thus, measuring the difference absorbance is a direct measure of the concentration of the enzyme–nucleotide complex.

Titration of the enzyme with the nucleotide and extrapolation to a concentration where all of the ligand is bound to the enzyme permits the determination of the difference extinction coefficient, $\Delta\varepsilon$. The difference extinction coefficient for the binding of 2′-cytidine monophosphate to ribonuclease is shown in Figure 8-10 at different wavelengths. Note that significant changes are found at wavelengths of 260 and 280 nm, indicating that the extinction coefficients of both the enzyme and the nucleotide are altered in the complex. Note also the isobestic point that is found as the pH is varied, demonstrating the presence of only two absorbing species, bound and unbound. Since the concentration of all species can be determined during the titration, the binding constant can be calculated and was found to be $1.8 \times 10^5\ \mathrm{M}^{-1}$ at pH 5.5. An extensive study carried out with various ligands over a range of pH provided valuable information about the molecular details of the interaction between the ligands and enzyme (3).

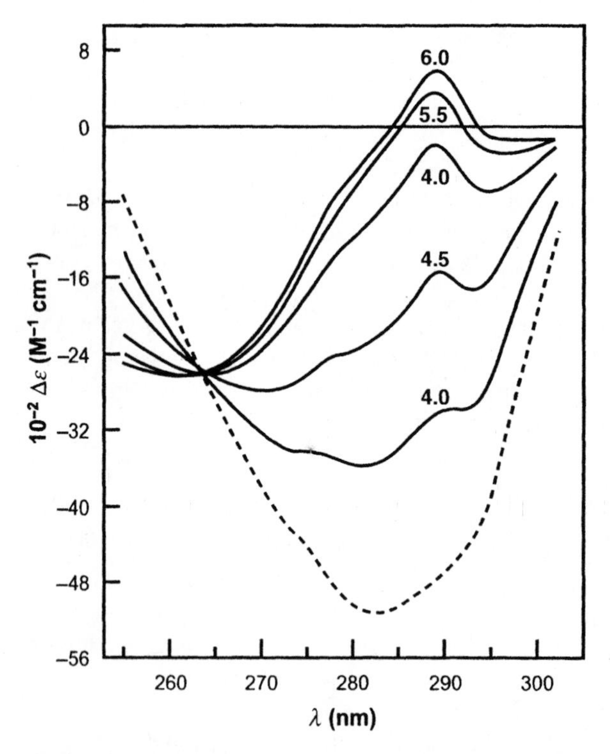

FIGURE 8-10. Difference absorption spectra of ribonuclease A-2′-CMP as a function of pH. The pH is indicated next to each difference spectrum. The dashed line is the calculated difference spectra for the protonated ring species of 2′-CMP. Reprinted in part with permission from D. G. Anderson, G. G. Hammes, and F. G. Walz, Jr., *Biochemistry* **7**, 1637 (1968). © 1968 by American Chemical Society.

8.7 X-RAY ABSORPTION SPECTROSCOPY

X rays can also be absorbed by materials, and the Beer–Lambert law (Eq. 8-1) is applicable. The energy associated with an X ray of 1 Å wavelength is about 10^6 kJ/mol ($E = hc/\lambda$). This energy is sufficiently large that the absorption of X rays by atoms results in electrons being ejected from inner orbits (1s, 2s, 2p). The X-ray absorption spectra of atoms consist of sharp peaks, the peaks occurring at different wavelengths for every element in the periodic table. If the atom is in a molecule, the ejected electron interacts with the environment, resulting in small oscillations around the main peak in the absorption spectra due to back scattering. Analysis of this fine structure provides information about the nature of the atoms surrounding the atom that has absorbed the X rays. This is particularly useful for probing the environment around heavy metals in biological structures.

8.8 FLUORESCENCE AND PHOSPHORESCENCE

Fluorescence is a very common phenomenon in biology and elsewhere. As previously discussed, fluorescence is the emission of light associated with electrons moving from an excited state to the ground state. Fluorescence and phosphorescence are often more useful tools for studying biological process than absorbance. They can be considerably more sensitive than absorbance so that much lower concentrations can be detected. They are also often extremely sensitive to the environment and thus are extremely good probes of the local-environment. Some other useful properties will be discussed a bit later.

In order to understand fluorescence and phosphorescene, a somewhat more complex energy diagram such as Figure 8-11 is useful. In this diagram, two electronic energy levels are shown, each with a manifold of vibrational energy levels. The vibrational energy levels are much more closely spaced than electronic energy levels and are due to the vibrations of atoms within a molecule. Different modes of vibration have different energies so that manifolds of quantized energy levels exist.

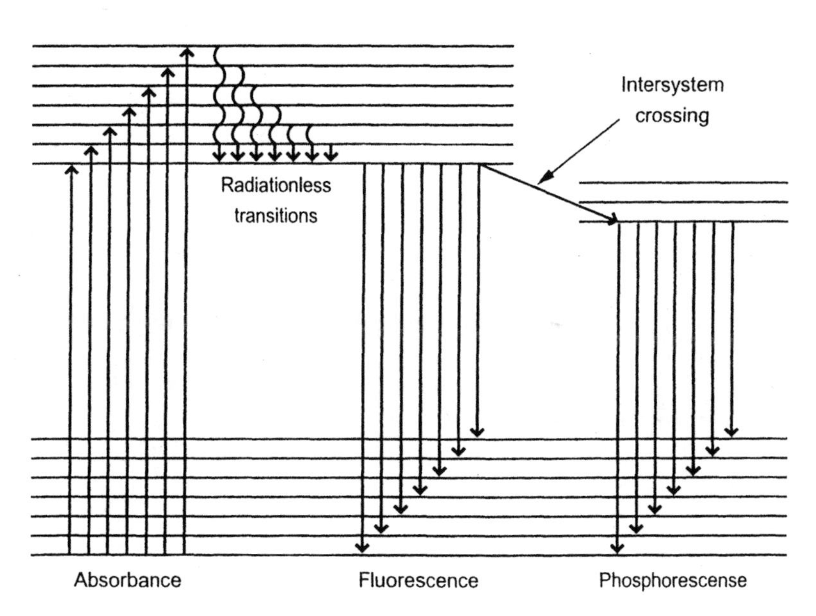

FIGURE 8-11. Schematic energy level diagram for absorption, fluorescence, and phosphorescence. This is often referred to as a Jablonski diagram. Electrons are excited from a ground-state singlet energy level, S_0, to various vibrational states of a higher electronic energy level, S_1. All electrons are paired in singlet states, that is, their spins are in opposite directions. Radiationless decay occurs to the ground vibrational energy level of S_1. As the electrons return to their ground electronic state, fluorescence occurs. In some cases, the excited electrons can move to a triplet state, T_1, in a radiationless transition. Triplet states have two unpaired electrons. The electrons in the triplet state can then return to the ground electronic state with phosphorescence occurring.

The properties of vibrational energy levels can be calculated through quantum mechanics. Generally, electrons are in their ground electronic energy state at room temperature, and the molecules are also in their lowest vibrational energy level. They can be excited by light to the next electronic energy level, but can be in many different vibrational energy levels in the excited state. After this excitation, the molecule will return to the lowest vibrational energy level of the second electronic energy level very rapidly, in 10^{-13} s or less. This generally occurs through a radiationless transition, that is, through the production of heat. Electrons in the second energy level can then decay to the electronic ground state, again in various vibrational energy levels. The emission produced is fluorescence. The decay to the ground electronic state can also occur without the emission of light, that is, by producing heat.

From Figure 8-11, it is clear that the emitted light must be at a longer wavelength than the absorbed light since the energy change associated with emission can never be greater than the energy change associated with absorption. Also from this diagram it is clear that the number of photons emitted can never exceed the number of photons absorbed. The efficiency of fluorescence is characterized by the *quantum yield*, Q, which is the fraction of photons absorbed that are eventually emitted:

$$Q = \frac{\text{number of photons emitted}}{\text{number of photons absorbed}} \tag{8-7}$$

In practice, the quantum yield can be quite close to 1, but for most molecules that fluoresce well, it is usually in the range of 0.3–0.7. Not all molecules that absorb light produce measurable fluorescence; in fact, most do not. Instead, decay to the ground state is radiationless, via heat production, and the quantum yield is essentially zero. The only common amino acid that has a useful fluorescence spectrum is tryptophan, and none of the common nucleic acids display significant fluorescence.

The measurement of fluorescence (and phosphorescence) is more complex than the measurement of absorption since the sample must be excited at a specific wavelength and the fluorescence observed at a different wavelength. Typically, the emission is detected perpendicular to the excitation beam. The emission spectrum is obtained by keeping the excitation wavelength constant and observing the emission over a range of wavelengths. The fluorescence of tryptophan, an amino acid found in many proteins, is shown in Figure 8-12. Also shown is the absorption spectrum. The emission spectrum in the figure was determined by exciting tryptophan with light at the absorption maximum of 275 nm. The excitation spectrum is obtained by observing the emission at a given wavelength and varying the excitation wavelength. The quantum yield is different for each pair of excitation and emission wavelengths. An absolute measurement of fluorescence would require knowledge of the number of photons absorbed at the excitation wavelength and the number of photons emitted at the emission wavelength. Such measurements are very difficult to make so that in practice the fluorescence intensity is usually expressed in relative terms or in comparison with a standard solution where the quantum yields are well known.

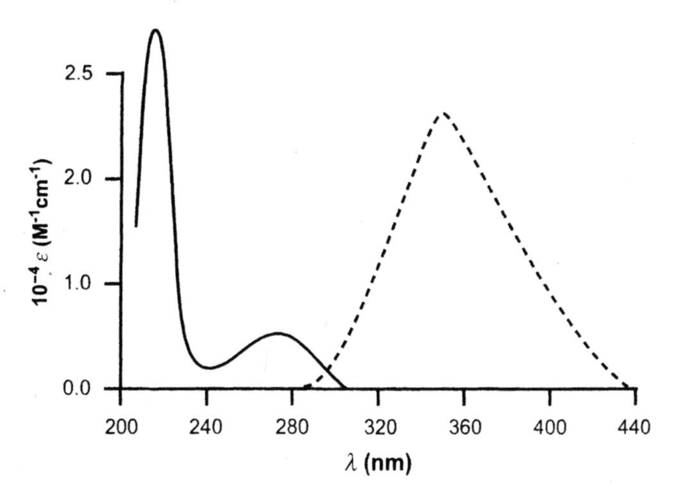

FIGURE 8-12. Absorption and fluorescence spectrum of tryptophan. The solid line is the extinction coefficient, and the dashed line is the fluorescence emission in arbitrary units with excitation at approximately 275 nm. D. Freifelder, *Physical Biochemistry*, 2nd edition, Freeman, New York, 1982, p. 539. © 1976, 1982 by Freeman. Used with permission.

A more quantitative description of fluorescence is useful for understanding the experimental measurements. The number of excited molecules is proportional to the amount of light absorbed, which can be determined from the Beer–Lambert law:

$$I_t = I_0 e^{-2.3\varepsilon(\lambda)cl} \qquad (8\text{-}8)$$

Here, I_t is the intensity of the light of incident intensity I_0 after it has passed a distance l through a solution of concentration c with an extinction coefficient of $\varepsilon(\lambda)$ at wavelength λ. Fluorescence measurements are made with solutions of low absorbance so that the exponential can be expanded to give

$$I_t = I_0[1 - 2.3\varepsilon(\lambda)cl] \qquad (8\text{-}9)$$

Thus, the concentration of excited molecules is proportional to the light absorbed,

$$I_0 - I_t = 2.3\varepsilon(\lambda)clI_0 \qquad (8\text{-}10)$$

The fluorescence emitted, F, is the product of the concentration of excited molecules times the quantum yield times the fraction of emitted photons collected by the instrument,

$$F = (\text{constant})Qc \qquad (8\text{-}11)$$

From the above discussion it is clear that the constant contains several proportionality relationships that are not easily determined, thus necessitating the determination of relative fluorescent intensities rather than absolute intensities. From a practical standpoint, it is important to note that the fluorescence is proportional to the concentration *if* the absorbance of the solution is sufficiently low to permit the expansion of the exponential. Fortunately, this situation is usually quite easy to obtain experimentally.

The discussion thus far is valid for *singlet* electronic states, that is, states in which all of the electron spins are paired. Many complex molecules have electronic states in which two unpaired electrons exist, a *triplet* electronic state. These are at a higher energy state than the ground singlet state and can be at a lower energy than the first excited singlet state. This is shown schematically in Figure 8-11. In principle, the molecule can move from the excited singlet state to the triplet state, and then decay can occur to the ground state via light emission or the production of heat. This is termed *phosphorescence*. As with fluorescence, the emitted light must be at a longer wavelength than the absorbed light.

What makes phosphorescence different than fluorescence, and how can they be distinguished? In the case of fluorescence, quantum mechanics tells us that the transitions between energy states are allowed, and they occur very rapidly, typically in nanoseconds. In the case of phosphorescence, quantum mechanics tells us that the interconversion from a single state to a triplet state is not allowed, that is, it will not occur readily. The result is that phosphorescence generally takes place in milliseconds or longer. Although this qualitative distinction between fluorescence and phosphorescence is usually sufficient to distinguish between the two, the proof of what is occurring requires determination of the magnetic properties of the excited state. Because the triplet state has unpaired electron spins, its energy will be altered in a magnetic field, whereas this will not happen for fluorescence. At room temperature, DNA does not phosphoresce, but in frozen solutions, thymidine will phosphoresce. We will not explicitly consider phosphorescence further, but it should be remembered that it is very similar to fluorescence; moreover, many biological systems display phosphorescence, perhaps most notable are plant and animal life in the ocean.

Figure 8-11 is a considerable simplification of the real situation where many different electronic states are possible so that multiple electronic energy levels must be taken into account. In addition, many different vibrational modes are possible, so multiple manifolds of vibrational energy levels are generally present. However, the principles enunciated are equally valid for these more complex situations.

8.9 RecBCD: HELICASE ACTIVITY MONITORED BY FLUORESCENCE

As mentioned above, part of the mechanism of DNA synthesis involves unwinding DNA. RecBCD is a protein from *Escherichia coli* that functions as a helicase (cf. Ref. 4). It also has an endo- and exonuclease activity, that is, it hydrolyzes DNA. The helicase activity requires the hydrolysis of MgATP to furnish the free energy necessary for the unwinding process. Monitoring absorbance is not a useful assay for following the unwinding process because of the small changes in absorbance accompanying unwinding relative to the total absorbance. However, a fluorescence assay was developed by Roman and Kowalczykowski (4) that provides a continuous assay of the helicase activity. This assay makes use of a protein that binds specifically to single-stranded DNA, SSB. The binding process stabilizes the single-stranded DNA. Fortunately, when SSB binds to single-stranded DNA, the fluorescence of SSB due to tryptophan is quenched. If SSB is put into an assay mixture containing double-stranded DNA, RecBCD, and MgATP, the SSB will bind to the single-stranded DNA as it is produced. The result is a decrease in the tryptophan fluorescence as more and more single-stranded DNA is formed. This change in fluorescence is a direct monitor of the rate of DNA unwinding since the binding of SSB to single-stranded DNA is fast relative to the rate of the unwinding reaction. The apparent turnover number for the enzyme determined by this method ranged from 14 to 56 μM base pairs s^{-1} (μM recBCD)$^{-1}$, depending on the experimental conditions.

The mechanism of action of the recBCD helicase is postulated to proceed by the enzyme binding to DNA and unwinding many base pairs before it dissociates from the DNA. An important question to ask is how much unwinding occurs before the helicase dissociates. When an enzyme catalyzes many events during a single binding, the reaction is called *processive*: the greater the amount of unwinding/binding event, the greater the processivity of the enzyme activity. With the assay described above, it was possible to quantitate the processivity (5). The processivity increases in a hyperbolic manner as the MgATP concentration increases, reaching a limiting value of 32 kilobases at saturating concentrations of MgATP. The apparent dissociation constant for MgATP deduced from the hyperbolic reaction isotherm is about 40 μM. The processivity was found to be quite sensitive to salt concentration. Thus, a substantial amount of DNA is unwound during each event, ultimately resulting in the synthesis of long stretches of DNA by the polymerase enzyme.

This is but one example of the use of fluorescence to develop an assay for an important biological process, with the subsequent determination of mechanistic aspects of the process.

8.10 FLUORESCENCE ENERGY TRANSFER: A MOLECULAR RULER

Thus far, we have considered only what happens within a single molecule when light is absorbed. However, it is also possible for a molecule in an excited electronic state to transfer the energy above the ground state to another molecule. This phenomenon is termed *Förster energy transfer* (6). This can be readily understood if a simple kinetic scheme is considered. If the energy donor, D, is considered, fluorescence can be described by the following sequence of events:

$$\begin{aligned} &\mathrm{D} \xrightarrow{h\nu} \mathrm{D}^* \text{ (light absorption)} \\ &\mathrm{D} \xrightarrow{k_f} \mathrm{D} + h\nu' \text{ (fluorescence of donor)} \\ &\mathrm{D}^* \xrightarrow{k_r} \mathrm{D} \text{ (radiationless de-excitation)} \end{aligned} \tag{8-12}$$

The k's are rate constants in this scheme, ν is the light frequency, and * designates the excited electronic state. The "natural" fluorescence lifetime is defined as

$$\tau_0 = 1/k_f \tag{8-13}$$

This lifetime cannot be measured experimentally but is a useful theoretical concept.

The fluorescence lifetime is measured by exciting the molecule with short pulses of light and following the decay of fluorescence. According to Eq. 8-12,

$$-d(D^*/dt = (k_f + k_r)(D^*) = (D^*)/\tau_D \tag{8-14}$$

so that the lifetime determined experimentally is

$$\tau_D = 1/(k_f + k_r) \tag{8-15}$$

Eq. 8-14 can be easily integrated with the boundary conditions that $(D^*) = (D_0^*)$ when $t = 0$ and $(D^*) = 0$ when $t = \infty$. The result is

$$(D^*) = (D_0^*)\exp(-t/\tau_D) \tag{8-16}$$

Thus, a plot of the natural logarithm of the fluorescence intensity versus time is a straight line with a slope of $-(1/\tau_D)$.

The quantum yield for this sequence of events is simply the fraction of molecules fluorescing after excitation, or

$$Q_D = k_f/(k_f + k_r) = \tau_D/\tau_0 \tag{8-17}$$

If the frequency of the fluorescence happens to overlap the absorption of a second molecule, the acceptor (A), then some of the energy of de-excitation can be transferred directly to A, which in turn can fluoresce. The scheme in Eq. 8-12 can be expanded to include this possibility:

$$\begin{aligned} &D^* + A \xrightarrow{k_T} D + A^* \text{ (energy transfer)} \\ &A^* \longrightarrow A + h\nu'' \end{aligned} \tag{8-18}$$

It should be noted that this energy transfer can occur even if A does not fluoresce, as A could return to its ground state without producing light. The quantum yield of D in the presence of A is now given by

$$Q_{DA} = k_f/(k_f + k_r + k_T) \tag{8-19}$$

The efficiency of energy transfer, E, is

$$E = k_T/(k_f + k_r + k_T) = 1 - Q_{DA}/Q_D = 1 - \tau_{DA}/\tau_D \tag{8-20}$$

where τ_{DA} is the fluorescence lifetime in the presence of the acceptor. In this case,

$$-d(D^*)/dt = (k_f + k_r + k_T)(D^*) \tag{8-21}$$

Thus, the efficiency of energy transfer can be measured quite readily.

This efficiency depends very strongly on the physical separation of the acceptor and donor, and therefore can be used to calculate the distance between A and D. The dependence of the efficiency on the distance between A and D, R, is given by

$$E = R_0^6/(R_0^6 + R^6) \tag{8-22}$$

where

$$R_0 \text{ (in nm)} = 8.79 \times 10^{-6} \, (J\kappa^2 n^{-4} Q_D)^{1/6} \tag{8-23}$$

In Eq. 8-23, J is a measure of the spectral overlap of the fluorescence emission of the donor with the absorption of the acceptor, κ is a measure of the relative orientations of the donor and acceptor, and n is the refractive index. Equation 8-22 was first enunciated by Förster, hence the name Förster energy transfer. Because R_0 can be readily calculated, R can be determined directly from measurements of E. Since E depends on the sixth power of the distance, it is a very sensitive measure of the distance. Typical values of R_0 are 3–6 nm so that distances in the range of 1–10 nm are readily accessible to this method. The measurement of energy transfer has proved very useful for providing structural information about biological systems, especially in cases where the molecular structures are not well known.

8.11 APPLICATION OF ENERGY TRANSFER TO BIOLOGICAL SYSTEMS

The use of energy transfer as a molecular ruler in macromolecular systems was elegantly tested by Stryer and Haugland (7). They synthesized a series of proline polymers with a fluorescent donor, dansyl, at one end and an energy acceptor, naphthyl, at the other end. Proline was selected because the polymer is a rigid cylinder whose length can be easily calculated. The number of prolines in the polymer varied from 1 to 12, and R varied from about 1.2 to 4.6 nm. The energy transfer efficiency obeyed Eq. 8-22 exceedingly well, as shown in Figure 8-13, and the experimental values of R agreed well with those determined from molecular models. They also demonstrated that the dependence of R_0 on the degree of overlap of the donor emission with the acceptor absorbance was quantitatively correct.

With the establishment of the methodology in model systems, energy transfer measurements were made in many biological systems. The most critical part of the application is to label a protein or nucleic acid with a fluorescent probe at a specific site, with essentially no labeling at nonspecific sites. This is essential if a reliable interpretation of the data is to be made. As a specific example, consider the

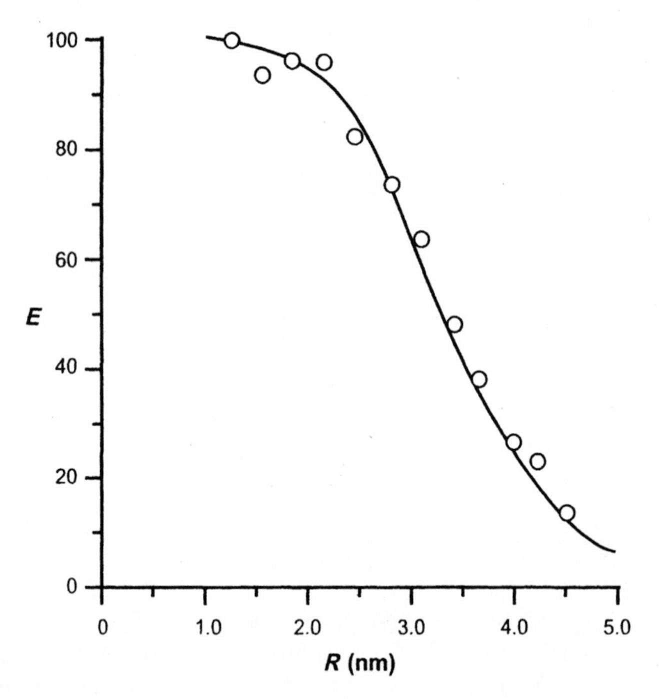

FIGURE 8-13. Percent efficiency of energy transfer, E, in proline polymers of various lengths, R, with a fluorescence donor at one end of the polymer and a fluorescence acceptor at the other end. The solid line corresponds to Eq. 8-22. Reproduced with permission from L. Stryer and R. P. Haugland, *Proc. Natl. Acad. Sci. USA* **58**, 719 (1967).

extensive studies on chloroplast coupling factor 1, CF_1 (8). CF_1 is the soluble part of an enzyme that is critical for ATP synthesis in chloroplasts. Very similar enzymes are in all organisms. In animals, the enzyme is located in the mitochondria. The enzyme is responsible for the synthesis of ATP from ADP and inorganic phosphate, an important physiological reaction. The synthesis occurs when a pH gradient is established across the membrane, and protons are pumped across the membrane as the synthesis proceeds. The coupling of a proton gradient and ATP synthesis is discussed in Chapter 3. Of course, an enzyme must catalyze the chemical reaction in both directions. Thus, the soluble portion of the enzyme catalyzes the reverse reaction, namely the hydrolysis of ATP. It cannot catalyze the synthetic reaction since a proton gradient cannot be created when the enzyme is not on the membrane.

CF_1 is a very complex protein and has five different subunits, with a subunit structure of $\alpha_3\beta_3\gamma\delta\varepsilon$. It binds three molecules of ATP or ADP, and the molecule is intrinsically asymmetric. In a series of experiments, fluorescent probes were put in a variety of positions on the molecule, and the distances between the probes were measured. Altogether, more than 30 distances were measured, including the distance of CF_1 sites from the surface of the membrane. These measurements were used to construct a model of the enzymes. The model is shown in Figure 8-14.

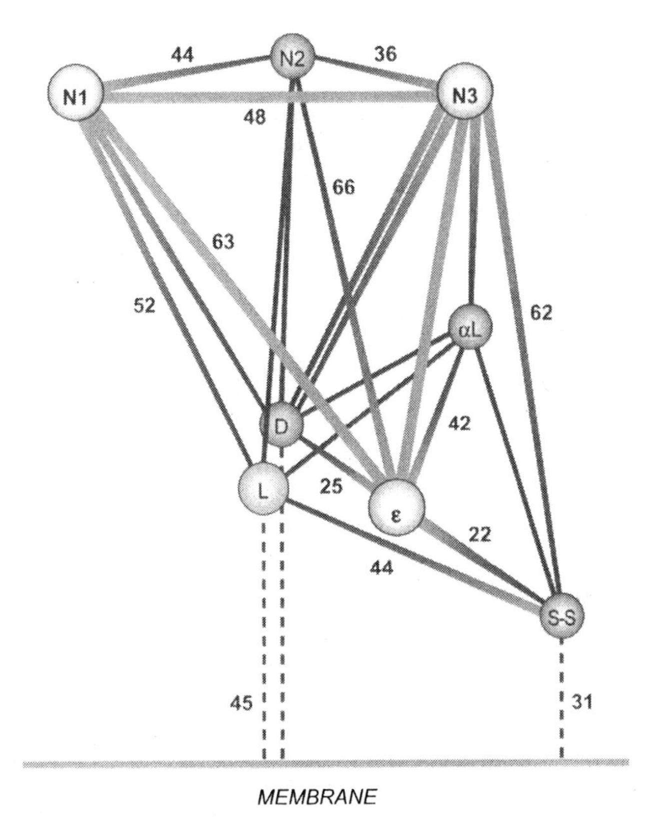

FIGURE 8-14. Spatial arrangement of specific sites on CF_1 determined by fluorescence energy transfer measurements. The sites shown are nucleotide binding sites (N_1, N_2, N_3), a disulfide on the γ polypeptide chain (S-S), specific sulfhydryl groups on the γ polypeptide chain (D, L), a sulfhydryl on the ε polypeptide chain (ε), and an amino group on an α polypeptide chain (αL). A few of the distances, in Å, are shown to establish the scale. The distances to the membrane are maximum values since the lines need not be perpendicular to the membrane surface. Reprinted from R. E. McCarty and G. G. Hammes, Molecular Architecture of Chloroplast Coupling Factor 1, *TIBS* **12**, 234 (1987). © 1987, with permission from Elsevier.

At the time this work was done, the structure of the enzyme was unknown, and the model summarized some important features of the enzyme, including the distance between nucleotide binding sites and the intrinsically asymmetric nature of the structure. The asymmetric structure is critical for the coordination of the chemical reaction and the pumping of protons across the membrane. Some time later, the structure of the mitochondrial enzyme was determined by X-ray crystallography (9). The distances determined by energy transfer proved to be consistent with the crystal structure. In fact, the elucidation of structural features by energy transfer measurements is generally useful for complex structures for which the molecular structure has not yet been established.

8.12 DIHYDROFOLATE REDUCTASE

As an example of the use of fluorescence to study ligand binding to a protein and enzyme catalysis, we consider the enzyme dihydrofolate reductase (DHFR). DHFR catalyzes the reduction of 7,8-dihydrofolate by NADPH to give 5,6,7,8-tetrahydrofolate and $NADP^+$. Tetrahydrofolate is essential for the biosynthesis of purines, as well as several other important biological molecules. Because of its central position in metabolism, it has been extensively studied. It is also a target for anticancer drugs such as methotrexate. The structures of the enzyme and enzyme–substrate/inhibitor complexes are known and many mechanistic studies of the enzyme have been carried out (cf. Refs. 10,11). This discussion will be limited to a few facets of the very extensive literature available for this enzyme.

NADPH is fluorescent, with an absorption maximum at about 340 nm and an emission maximum at about 450 nm. In addition, the enzyme contains tryptophan so that if the enzyme is excited at 290 nm, fluorescence is observed with a maximum intensity at 340–350 nm. If NADPH is added to the enzyme, this fluorescence is quenched about 65%, indicating the tryptophan environment is significantly altered when NADPH binds (12). Moreover, because NADPH has a maximum absorbance around 340 nm, significant energy transfer occurs from the excited tryptophan to bound NADPH. This results in a new emission maximum at 420–450 nm. The extent of NADPH binding to DHFR can be quantitatively assessed by measuring the quenching of tryptophan fluorescence. This can also be done by following the fluorescence at 450 nm. Quantitative analysis of the data permits the calculation of the equilibrium dissociation constant, which is 1.1 μM at pH 7.0.

The kinetics of ligand binding can also be measured by mixing NADPH and DHFR rapidly in a stopped flow apparatus and following the quenching of tryptophan fluorescence or the emission at 450 nm as a function of time. The time course of the reaction indicates a very rapid quenching, occurring in less than a second, followed by a relatively slower reaction, occurring in seconds. The first reaction is the binding of NADPH to DHFR, whereas the second reaction is the interconversion of the free enzyme from a form that does not bind NADPH to a form that does. The overall reaction can be represented by the following mechanism:

$$\begin{array}{l}\mathrm{DHFR} + \mathrm{NADPH} \underset{k_{-1}}{\overset{k_1}{\rightleftharpoons}} \mathrm{DHFR{-}NADPH} \\ k_{-2} \upharpoonleft\downharpoonright k_2 \\ \mathrm{DHFR}'\end{array}$$

A quantitative analysis of the data at pH 7.0 gives $k_1 = 1.7 \times 10^6\,M^{-1}\,s^{-1}$, $k_1 = 2.4\,s^{-1}$. The equilibrium constant for the transition from DHFR to DHFR' was found to be about 1, with rate constants of approximately $2.5 \times 10^{-2}\,s^{-1}$ (k_2 and k_{-2}; 11). Note that for this mechanism, the overall equilibrium dissociation

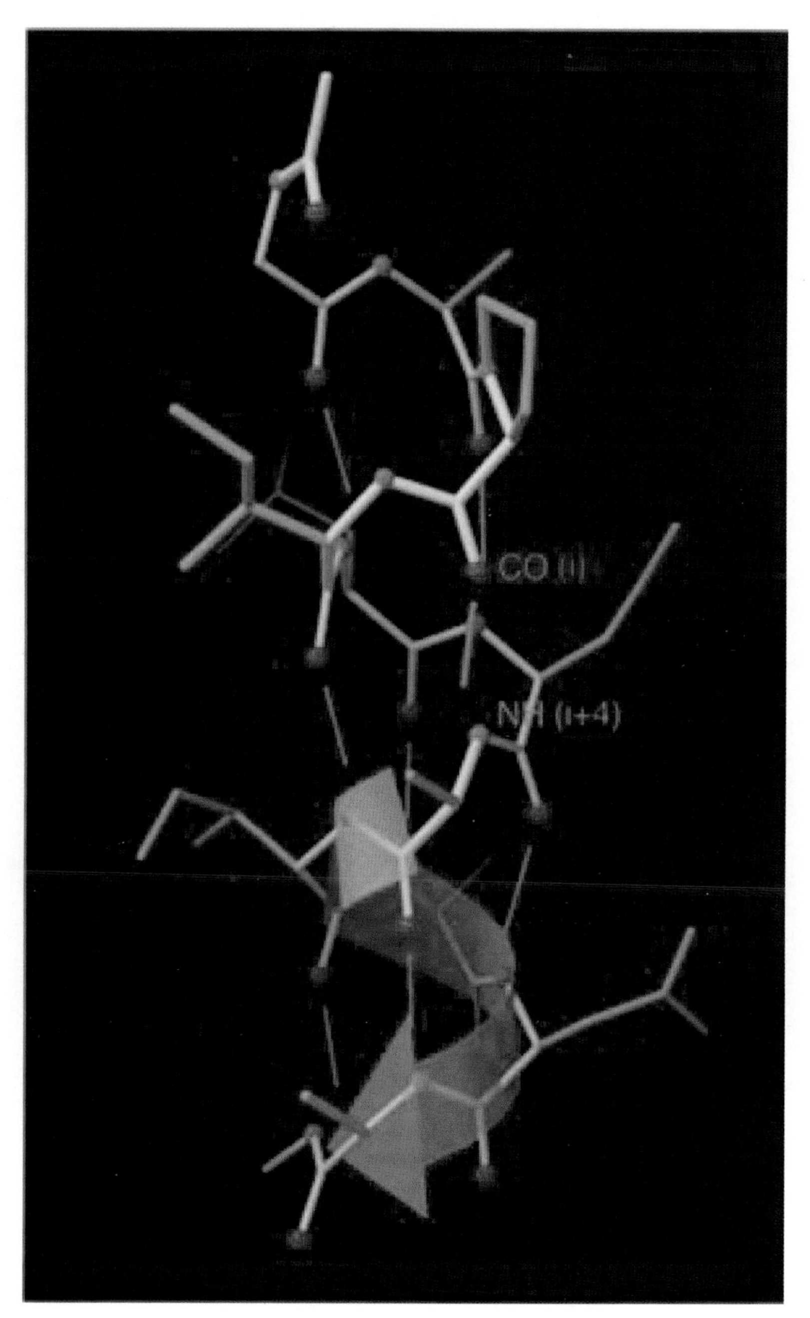

FIGURE 3-4. The α-helix found in many proteins. The yellow arrow follows the right-handed spiral of one helical turn. The hydrogen bonds between backbone peptide bonds are brown lines, the oxygens are red, and the nitrogens blue. The hydrogen bonds are formed between the *i*th carbonyl and $i + 4$ NH in the peptide backbone. Copyright by Professor Jane Richardson. Reprinted with permission.

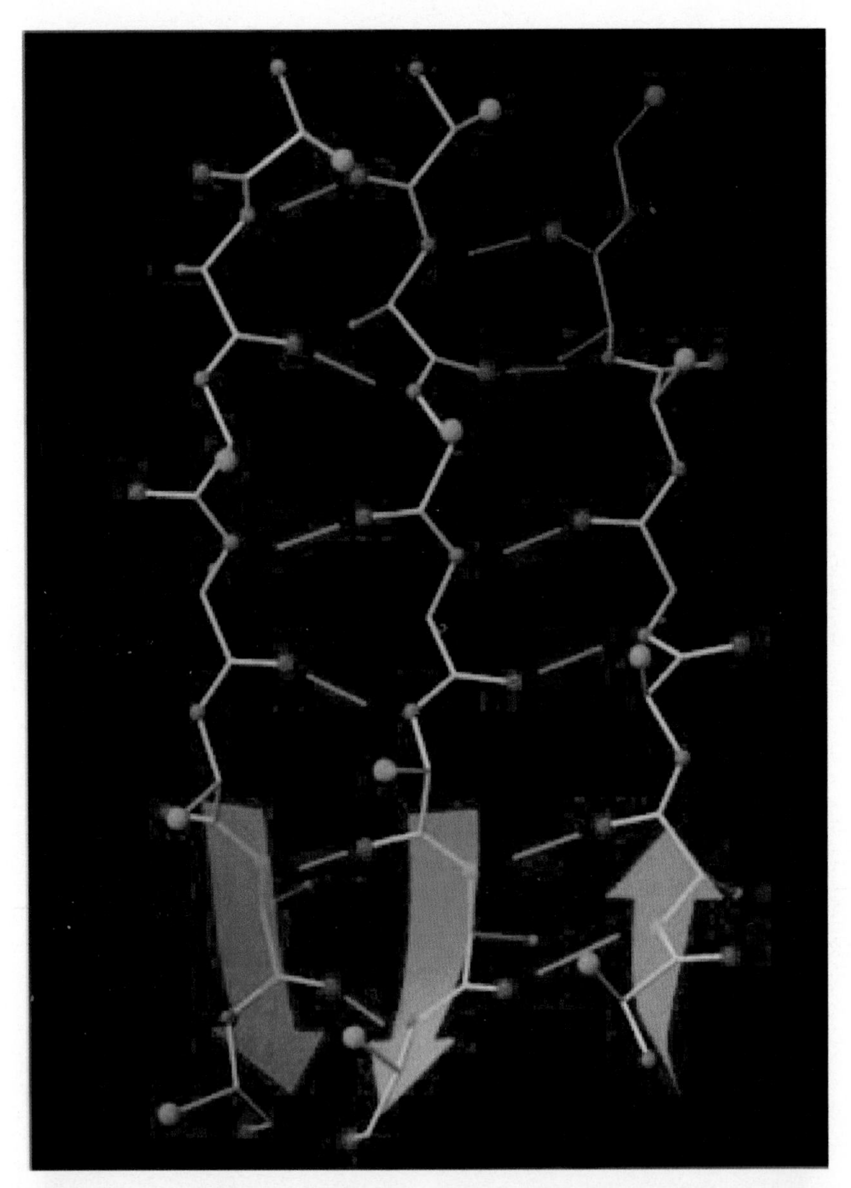

FIGURE 3-5. The β-pleated sheets found in many proteins. Both parallel and antiparallel strand-strand interactions are shown, as indicated by the yellow arrows. Again, the hydrogen bonds between backbone peptide bonds are shown as brown lines between the oxygens and nitrogens. Copyright by Professor Jane Richardson. Reprinted with permission.

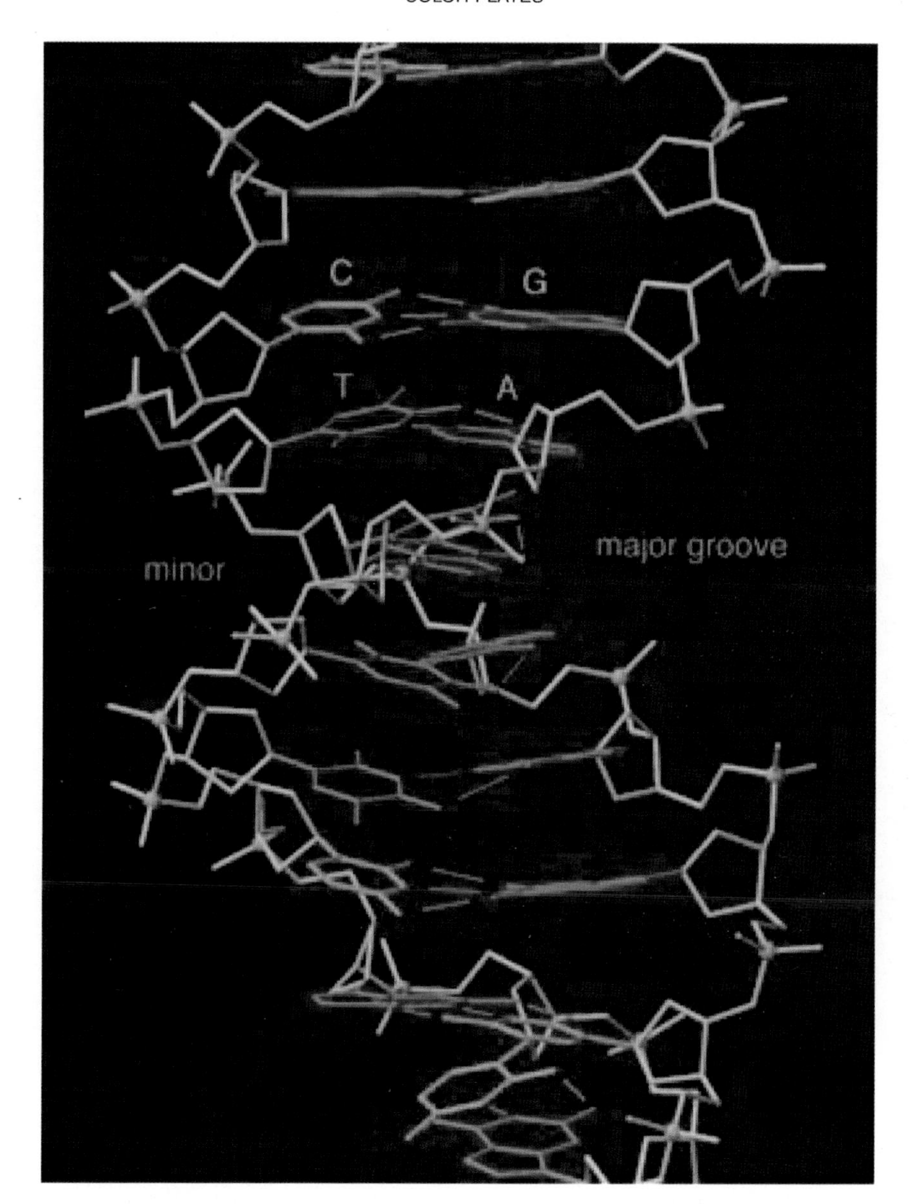

FIGURE 3-9. Stick-figure representation of the B form of the DNA double helix. The planes of the hydrogen-bonded bases can be seen, as well as the twisting of the chains to form a double helix. The pale yellow spheres are the phosphorous atoms of the sugar phosphate backbone on the outside of the helix. The bases are color coded, and the two grooves in the structure are labeled. Copyright by Professor Jane Richardson. Reprinted with permission.

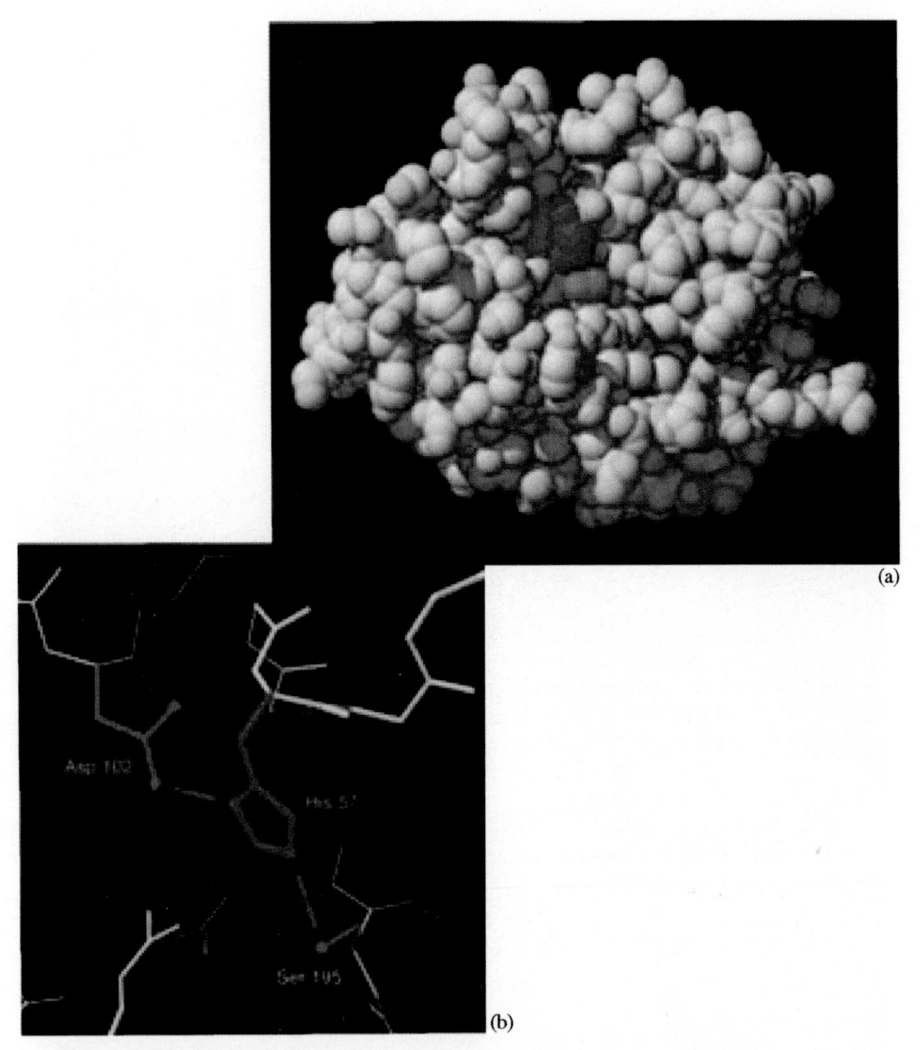

FIGURE 5-4. (a) Space filling representation of chymotrypsin, showing the active site pocket where substrate binding occurs. The blue at the active site is imidazole, the green is serine, and the red is aspartate. (b) Arrangement of the protein residues at the active site of chymotrypsin. The serine is the residue acylated; the imidazole of the histidine serves as a general base catalyst; and the aspartate carboxyl group is hydrogen bonded to the imidazole. Copyright by Professor Jane Richardson. Reprinted with permission.

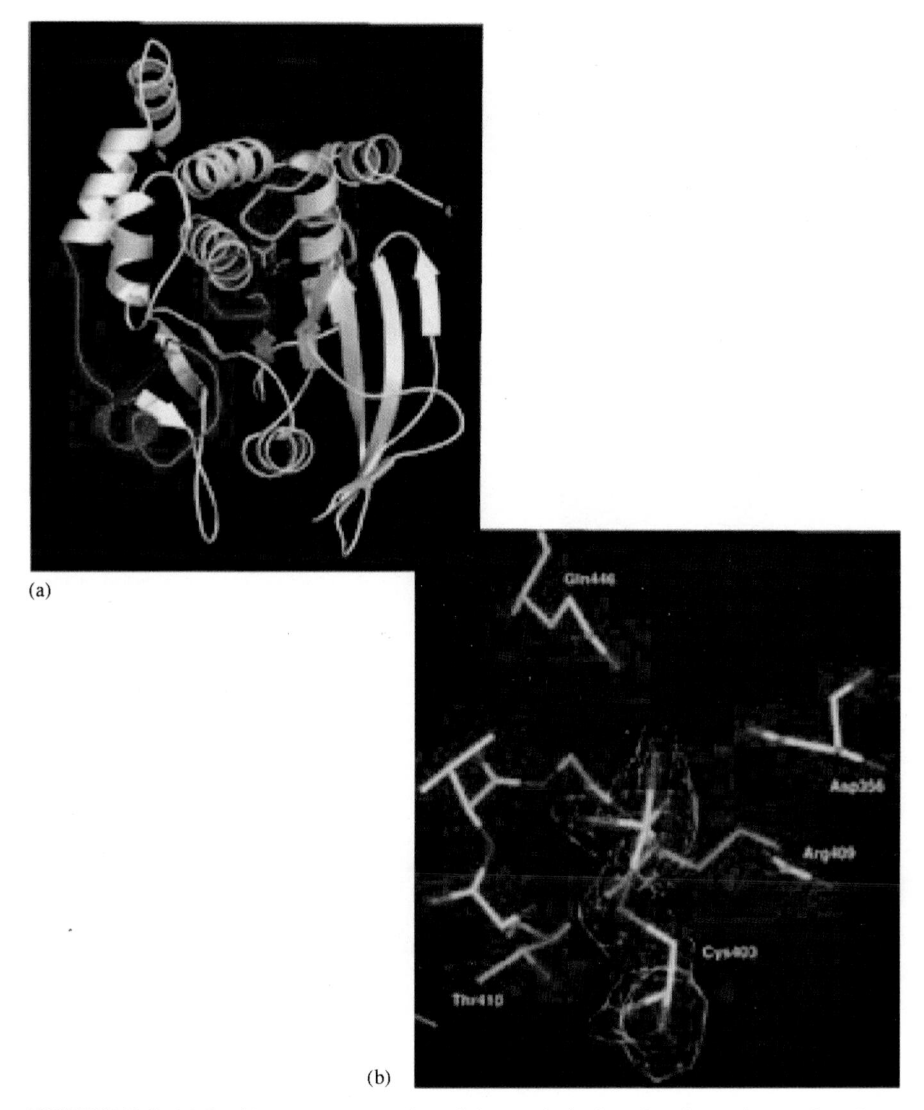

FIGURE 5-6. (a) Backbone representation of the catalytic domain of protein tyrosine phosphatase. Coils and arrows represent α-helices and β-strands, respectively. The cysteine (green), arginine (red), and aspartate (gold) are also shown. Reproduced from J. M. Denu, J. A. Stuckey, M. A. Saper, and J. E. Dixon, *Cell* **87**, 316 (1996). Copyright © 1996 with permission from Elsevier. (b) Vanadate bound at the active-site cysteine of protein tyrosine phosphatase. Other amino acids at the catalytic site are indicated, including the threonine and aspartate that participate in the catalytic mechanism. Reproduced from J. M. Denu, D. L. Lohse, J. Vijayalakshmi, M. A. Saper, and J. E. Dixon, *Proc. Natl. Acad. Sci. USA* **93**, 2493 (1996). Reprinted with permission of the Proceedings of the National Academy of Sciences USA. Reproduced by permission of the publisher via Copyright Clearance Center, Inc.

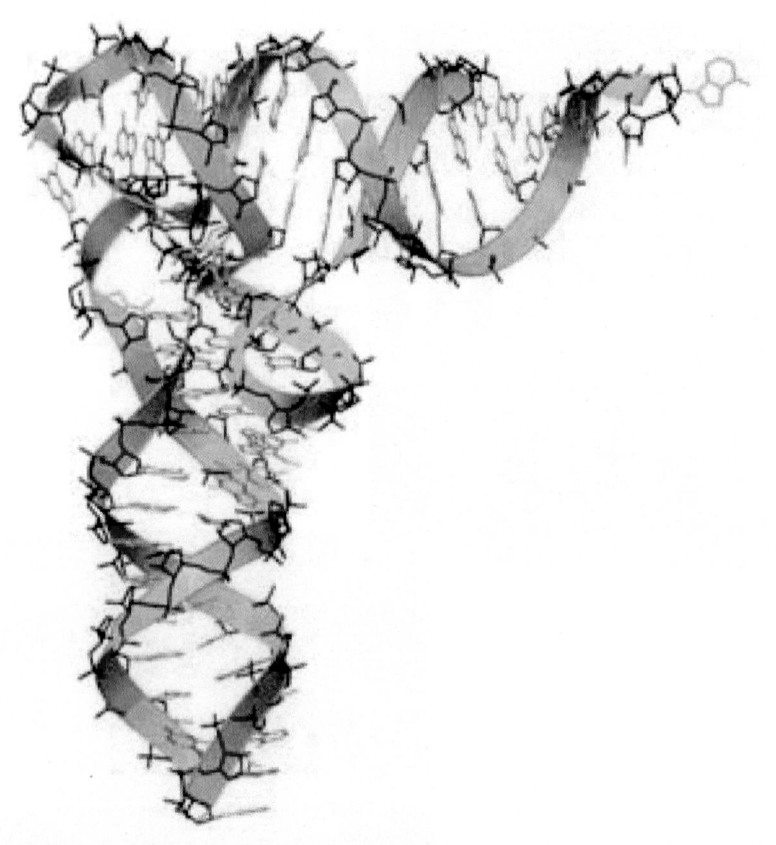

FIGURE 7-5. Structure of yeast phenylalanine t-RNA with the backbone in cyan (Protein Data Bank entry 1EHZ). The L-shaped structure can be seen and the hydrogen bonding of the side chains to form helical structures. The anticodon loop is at the bottom of the long arm of the L. Copyright by Professor David C. Richardson. Reprinted with permission. Kinemage graphics, then rendered in Raster3D.

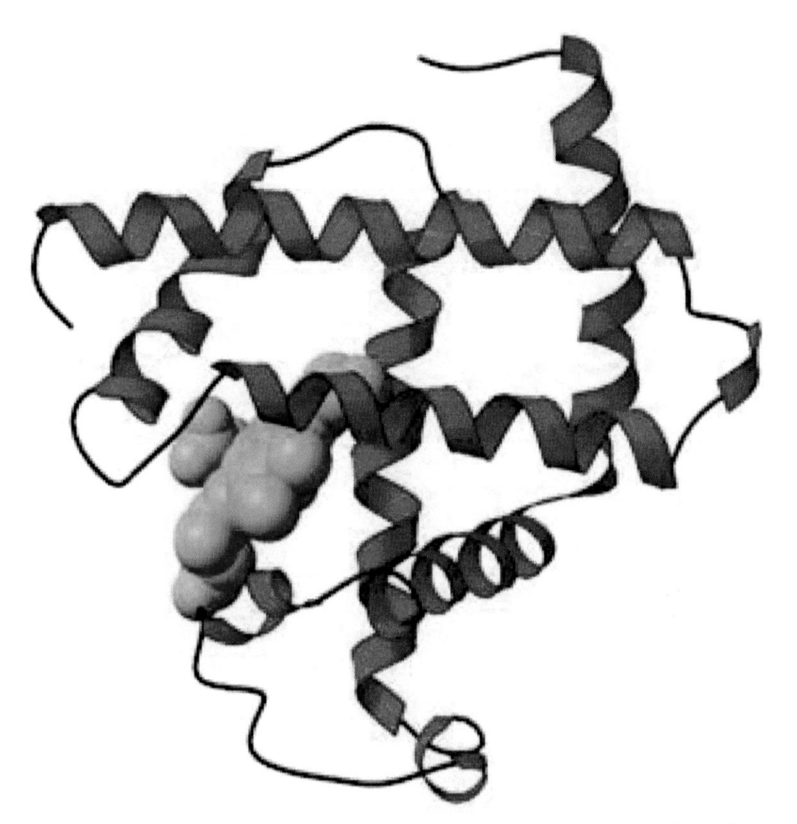

FIGURE 7-7. Ribbon structure of sperm whale myoglobin (Protein Data Bank entry 1A6M). The large amount of α-helical structure is apparent. The space filling structure in cyan is the heme. Copyright by Professor David C. Richardson. Reprinted with permission. Kinemage graphics, then rendered in Raster3D.

FIGURE 7-8. Schematic structure of ribonuclease A with uridine vanadate (magenta) bound to the active site (Protein Data Bank entry 1RUV). Both α-helical and β-sheet structure can be seen, as well as structure that does not have a regular array of hydrogen bonds. The disulfide linkages are in yellow, and the green residues are histidines 12 and 119, which are essential for catalytic activity. Copyright by Professor David C. Richardson. Reprinted with permission. Kinemage graphics, then rendered in Raster3D.

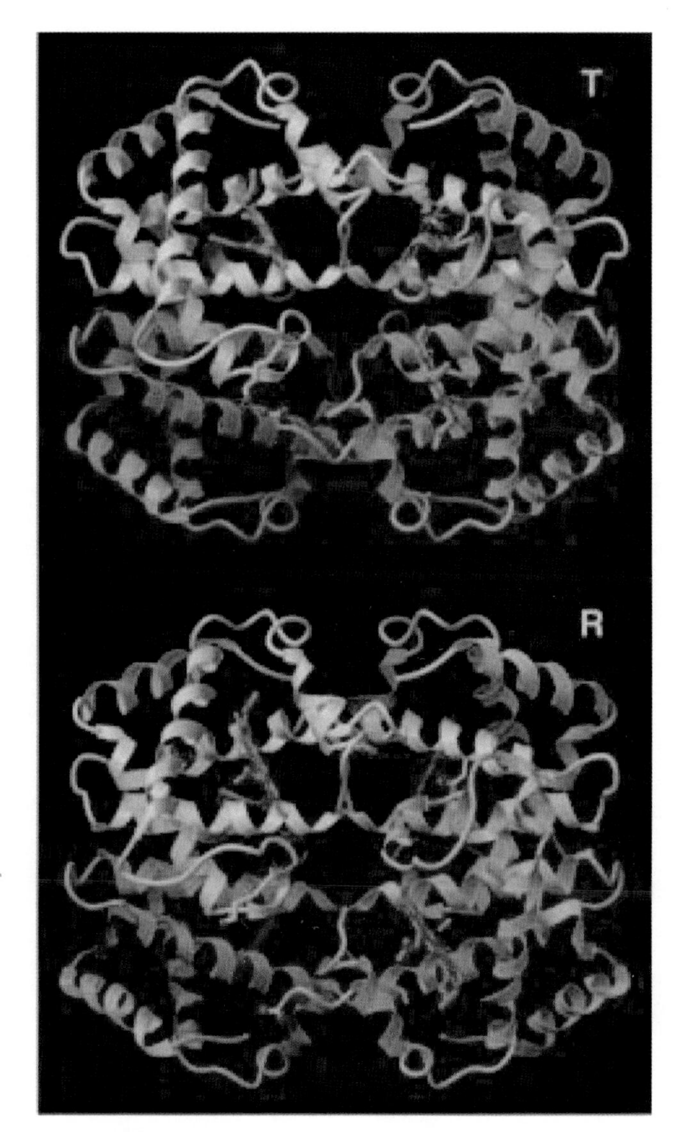

FIGURE 7-9. Schematic representation of the R (*bottom*, park) and T (*top*, blue) forms of the hemoglobin $\alpha_2\beta_2$ tetramer. The hemes where the oxygen binds can be seen in the structure. The yellow side chains form salt bonds in the T structure that are broken in the R structure. One pair of the α-β subunits also rotates with respect to the other by about 15° in the interconversion of R and T forms. Copyright by Professor Jane Richardon. Reprinted with permission.

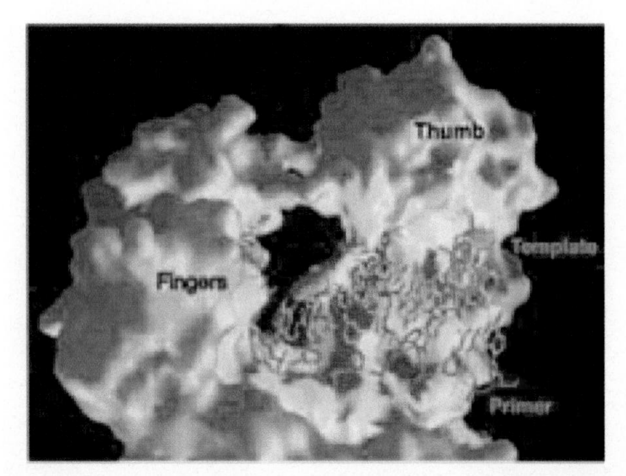

FIGURE 7-10. Structure of a catalytic fragment of DNA polymerase from *Bacillus stearothermophilus*. The "thumb" and "fingers" of the structure are labeled, and the template and primers can be seen as stick structures. Copyright by Professor L. S. Beese, Duke University. Reproduced with permission.

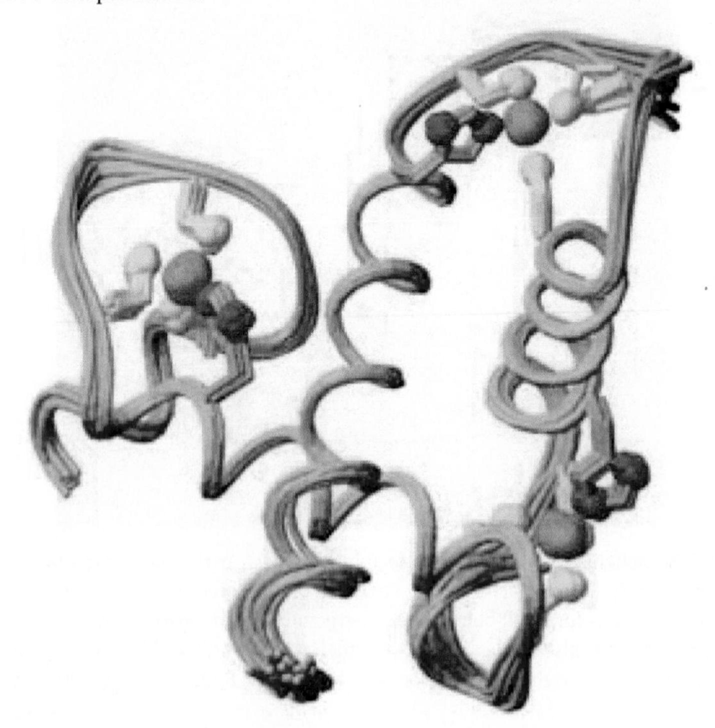

FIGURE 12-2. NMR structure of the TAZ2 (CH3) domain of CBP. A superposition of 20 structures are shown. The orange balls are Zn^{2+}, the yellow ball-and-stick representations are cysteines, and the purple and blue structures are imidazoles from histidines. The four α-helices can be easily discerned. PDB entry 1F81. Copyright by Professor David C. Richardson. Reprinted with permission. Kinemage graphics, then rendered in Raster3D.

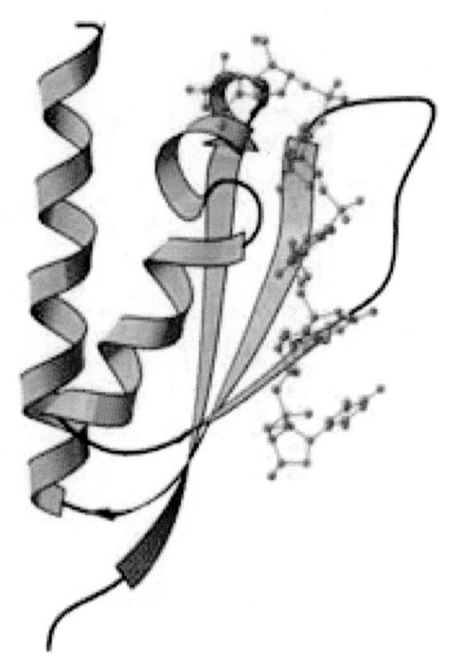

FIGURE 12-3. Structure of a KH3-DNA complex determined with NMR. The DNA (10 mer) is shown in magenta as a ball-and-stick model, and the protein backbone is in cyan. PDB entry 1J5K. Copyright by Professor David C. Richardson. Reprinted with permission. Kinemage graphics, then rendered in Raster3D.

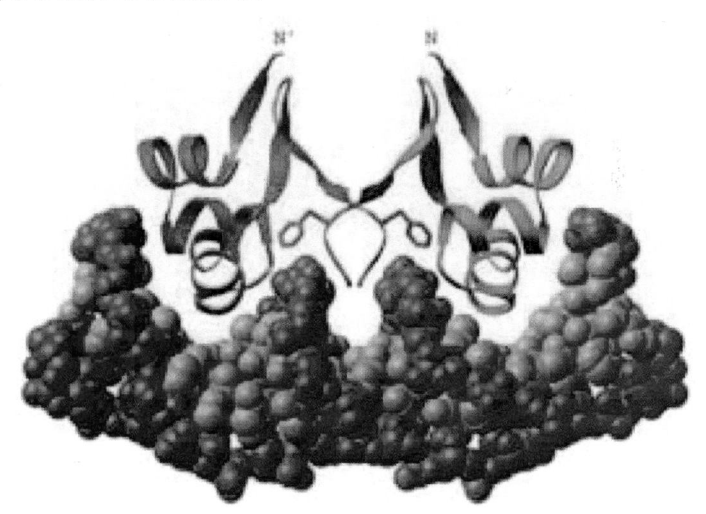

FIGURE 13-5. Structure of the overall complex between Cro protein and λ operator DNA. The direction of the view is parallel with the major grooves of DNA and parallel with the recognition helices. Reproduced from R. A. Albright and B. W. Matthews, "Crystal structure of lambda-Cro bound to a consensus operator" *J. Mol. Biol.* **280**, 137 (1998). Copyright 1998, with permission from Elsevier.

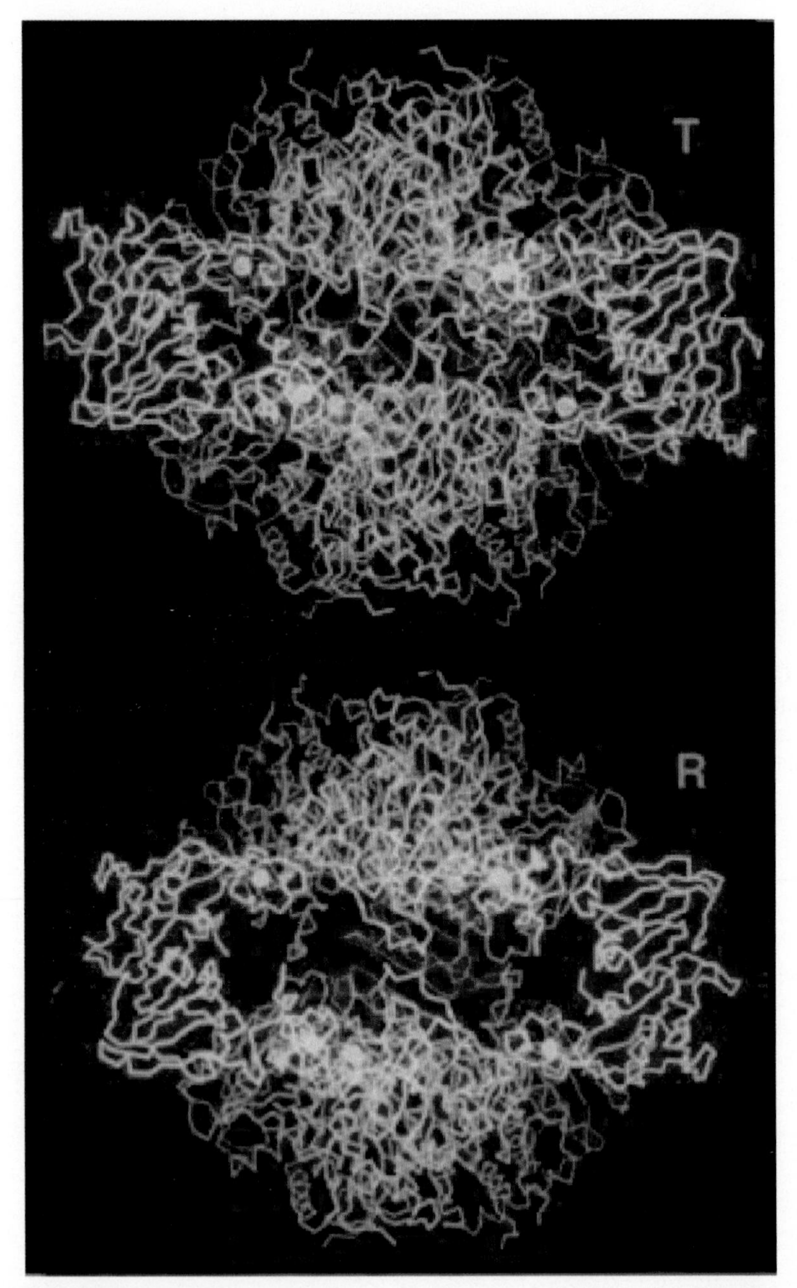

FIGURE 13-16. Cα backbone structures of the R (lower) and T (upper) states of aspartate transcarbamoylase. The two catalytic trimers are in green and are closer together in the T state. The three regulatory dimers (equatorial) are in yellow. One of the regulatory dimers is behind the large central cavity in this view. Adapted from W. N. Lipscomb, *Adv. Enzymol.* **68**, 67 (1994). Copyright © 1994 John Wiley & Sons, Inc. Reprinted by permission of John Wiley & Sons, Inc.

constant is

$$[(\mathrm{DHFR}) + (\mathrm{DHFR'})](\mathrm{NADPH})/(\mathrm{DHFR}-\mathrm{NADPH}) = (k_{-1}/k_1)(1 + k_2/k_{-2}) = 2.8\,\mu\mathrm{M}$$

The dissociation constant determined from kinetics is only in fair agreement with the dissociation constant determined from direct binding studies.

The overall time course of the enzymatic reaction can be easily monitored by following the absorbance change at 340 nm because NADPH has an absorption peak at this wavelength whereas $NADP^+$ does not. In addition, the individual steps in the catalytic process can be characterized by stopped-flow studies that monitor alterations in the tryptophan fluorescence as the reaction proceeds. A detailed description of these experiments is beyond the scope of this discussion. However, it was possible to determine which substrate binds first to the enzyme, the rate constants for the binding steps, the rate constants for conformational changes within the enzyme–substrate complex, and the rate constants for hydride transfer. This permitted the development of a detailed mechanism for the reaction (11).

Another unique use of fluorescence is fluorescence microscopy. With this technique individual molecules are labeled with fluorescent tags and can be observed with a light microscope (cf. Ref.13). Of course, it is not the single molecule that is seen: it is the light from the molecule. This is analogous to seeing the light from a star, rather than the star itself. With this technique, kinetic and equilibrium properties of individual molecules can be studied. For example, the kinetics of binding of substrates and inhibitors to DHFR and catalysis have been investigated by single-molecule fluorescence microscopy (14).

REFERENCES

1. K. Rosenheck and P. Doty, *Proc. Natl. Acad. Sci. USA* **47**, 1775 (1961).
2. R. E. Dickerson and I. Geis, *Hemoglobin: Structure, Function, Evolution and Pathology*, Benjamin/Cummings, Redwood City, CA, 1983.
3. D. G. Anderson, G. G. Hammes, and F. G. Walz, Jr., *Biochemistry* **7**, 1637 (1968).
4. L. J. Roman and S. C. Kowalczykowski, *Biochemistry* **28**, 2863 (1989).
5. L. J. Roman, A. K. Eggleston, and S. C. Kowalczykowski, *J. Biol. Chem.* **267**, 4207 (1992).
6. T. Förster, *Discuss. Faraday Soc.* **27**, 7 (1959).
7. L. Stryer and R. P. Haugland, *Proc. Natl. Acad. Sci. USA* **58**, 719 (1967).
8. R. E. McCarty and G. G. Hammes, *TIBS* **12**, 234 (1987).
9. J. Abrahams, A. Leslie, R. Lutter, and J. Walker, *Nature* **370**, 621 (1994).
10. M. R. Sawaya and J. Kraut, *Biochemistry* **36**, 586 (1997).
11. C. A. Fierke, K. A. Johnson, and S. J. Benkovic, *Biochemistry* **26**, 4085 (1987).
12. P. J. Cayley, S. M. J. Dunn, and R. W. King, *Biochemistry* **20**, 874 (1981).

13. S. Weiss, *Science* **283**, 1676 (1999).
14. N. M. Antikainen, R. D. Smiley, S. J. Benkovic and G. G. Hammes, *Biochemistry* **45**, 7990 (2006).

PROBLEMS

8.1. Colored pH indicators are dyes that have different spectra for different ionization states. Assume the ionization of a pH indicator with p$K = 6.00$ can be written as

$$\text{HIn} \rightleftarrows \text{In}^- + \text{H}^+$$

The measured absorbance in a 1-cm cell at an indicator concentration of 2×10^{-5} M is given in the table below.

λ (nm)	A at pH 3.00	A at pH 9.00
400	0.200	0.000
420	0.300	0.030
440	0.150	0.150
460	0.000	0.200
480	0.000	0.150

For the same concentration of indicator, an absorbance of 0.100 is measured at $\lambda = 400$ nm.

a. What is the pH of the solution?

b. What is the absorbance at $\lambda = 440$ nm?

c. At pH 7.00, for an indicator concentration of 3.00×10^{-5} M and a 1-cm cell, what is the absorbance at $\lambda = 400, 440$, and 480 nm?

8.2. A sample of RNA is hydrolyzed and separated into three fractions by column chromatography. Two of the three fractions are pure nucleotides, but the third contains both adenylic and guanylic acids. At pH 7.0, the absorbance of the mixture is 0.305 at 280 nm and 0.655 at 250 nm in 1-cm cells. The molar extinction coefficients for each pure component at pH 7.0 are

	$\varepsilon_{280}(\text{M}^{-1}\,\text{cm}^{-1})$	$\varepsilon_{250}(\text{M}^{-1}\,\text{cm}^{-1})$
Adenylic acid	2300	12,300
Guanylic acid	9300	15,700

Calculate the mole ratio of adenine to guanine in the RNA.

8.3. The enzyme alcohol dehydrogenase catalyzes the oxidation of alcohol by NAD^+ to give acetaldehyde and NADH. NADH has an absorption maximum at 340 nm with an extinction coefficient of $6.20 \times 10^3\ M^{-1}\ cm^{-1}$, whereas NAD^+ and the other reactants do not absorb significantly at 340 nm. Consequently, the oxidation reaction can be conveniently monitored by following the increase in absorbance at 340 nm.

a. What is the rate of production of NADH if it is observed that the rate of absorbance increase in a 1-cm path length cell is 0.05/min?

b. When an excess of enzyme is added to NADH, the absorbance at 340 nm decreases by 13%. What is the extinction coefficient of NADH at 340 nm when it is bound to the enzyme?

c. The difference in extinction coefficient can be used to determine the binding constant for the association of NADH and the enzyme. The following data were obtained for the difference absorbance when NADH is added to 20 μM enzyme in a 1-cm path length cell.

$(NADH)_{total}$ (μM)	ΔA (340 nm)
8.33	0.004
13.50	0.006
20.0	0.008
45.0	0.012

Calculate the binding constant from these data.

8.4. When complementary strands of deoxyoligonucleotides are mixed, they form a double-stranded DNA at low temperatures but dissociate to single strands as the temperature is raised. This can be observed by a decrease in the absorbance of the solutions at 260 nm. The equilibrium for this process can be written as

$$D \rightleftarrows S_1 + S_2$$

with an equilibrium constant $K = (D)/[(S_1)(S_2)]$. Data are given below for two experiments. In one case, equal length strands and concentrations of oligoA and oligoT were mixed, whereas, in the other case, the same length strands and concentrations of oligoG and oligoC were mixed.

a. Assume that the molecules are duplexes at the lowest temperatures for which data are given and single strands at the highest temperatures. Calculate the fraction present as duplex at each temperature. Determine the melting temperature of each duplex, that is, the temperature at which half of the duplex has been converted to single strands.

b. What do these results tell you about the relative stability of A–T and G–C pairs?

c. Estimate the melting temperature of complementary oligonucleotides that are 50% A–T and 50% G–C, with the same length as in the above experiments. (Your result is not exact, but methods are available for calculating the melting temperatures of short DNA; cf. Chapter 3.)

T (°C)	*A* (oligoA–T)	*T* (°C)	*A* (oligoG–C)
5	0.711	50	0.780
10	0.720	55	0.785
15	0.732	60	0.812
20	0.740	65	0.836
25	0.767	70	0.862
30	0.801	75	0.880
35	0.846	80	0.896
40	0.874	85	0.924
45	0.891	90	0.948
50	0.903	95	0.975
55	1.003		

8.5. Pyrenylmaleimide is a convenient fluorescent probe for labeling sulfhydryl residues on proteins. The fluorescence lifetime of pyrene on a protein sulfhydryl was measured by determining the relative fluorescence intensity after excitation of pyrene with a flash lamp. Typical data are given below.

Relative fluorescence	Time (ns)
0.716	20
0.513	40
0.367	60
0.264	80
0.189	100

a. Determine the fluorescence lifetime.

b. If the quantum yield is 0.7, what is the natural lifetime?

c. When a ligand is bound to the protein, the fluorescence lifetime is 10% shorter. What might be the cause of this change?

8.6. Bovine rhodopsin is a photoreceptor protein that is an integral part of the disc membranes of retinal rod cells and plays a key role in vision. It has a tightly bound 11-*cis*-retinal that has a strong absorbance at about 500 nm. As part of the vision cycle, the retinal is bleached by conversion to the *trans*-isomer. The protein has a molecular weight of 28,000–40,000. Three sites were labeled on the protein with fluorescent probes, sites A, B, and C. Fluorescence resonance energy transfer was measured between these three sites and the retinal and between the three sites themselves [C.-W. Wu and L. Stryer, *Proc. Natl. Acad. Sci. USA* **69**, 1104 (1972)]. Some of the results are summarized in the table below.

Energy donor	Energy acceptor	Transfer efficiency	R_0 (Å)
A	11-*cis*-Retinal	0.09	51
B	11-*cis*-Retinal	0.36	52
C	11-*cis*-Retinal	0.12	33
A	B	0.90	51
A	C	0.92	48
B	C	0.92	47

a. Calculate the distances between these six sites.

b. A protein of molecular weight 28,00–40,000 that is spherical has a radius of 40–45 Å. What can you say about the shape of rhodopsin? Sketch a model for the molecule based on the distances that have been measured.

CHAPTER 9

Circular Dichroism, Optical Rotary Dispersion, and Fluorescence Polarization

9.1 INTRODUCTION

Most biological molecules possess molecular asymmetry, that is, their mirror images are not identical. Such molecules are said to be *chiral*. Probably, the most well-known example is a carbon atom that is tetrahedrally bonded to four different atoms or groups of atoms. This example is described in organic chemistry textbooks and will not be dwelled on here. Less obvious examples of importance in biology are macromolecules that possess chirality. For example, helices can be wound in a left- or right-hand sense. The most common helix in proteins, the α-helix, is wound in a right-hand sense. Although most polynucleotides are wound in a right-hand sense, helices that wind in a left-hand sense also exist.

Chiral molecules can be distinguished by their interactions with polarized light. As we have discussed in Chapter 6, a light wave can be described as an electromagnetic sine wave with a characteristic frequency. In this description, a wave generating an electric field is perpendicular to a wave generating a magnetic field. If light is *unpolarized*, these waves are oriented randomly in space, as shown schematically in Figure 9-1. Looking along the electromagnetic wave, the random orientations look like the spokes of a wheel. For *linearly*, or *plane*, *polarized* light, the wave is oriented in only one direction, as depicted in Figure 9-1. Although the polarized light in the figure looks like a line (linearly polarized), the electric field wave is projected as a plane (plane polarized) so that the two terms are used interchangeably. When a chiral molecule interacts with plane polarized light, the plane is rotated, with the direction and amount of rotation depending on the characteristics of the molecule.

Circularly polarized light sweeps out a circle as the electric wave propagates. The circle can be either right handed or left handed, as shown in Figure 9-1. The convention used is that if the circle moves clockwise, it is right handed, whereas if it

Physical Chemistry for the Biological Sciences by Gordon G. Hammes

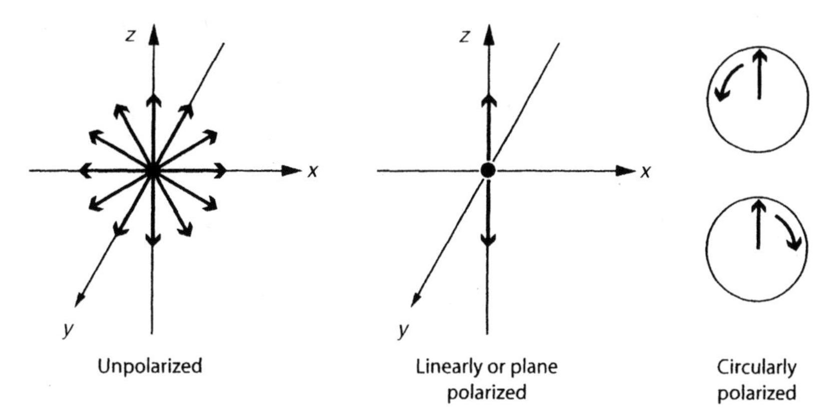

FIGURE 9-1. Diagrams of the electric field components of unpolarized, linerarly or plane polarized, and circularly polarized light. The light is moving along the y axis. The arrows indicate the directions of the electric field. For unpolarized light all directions occur, whereas for linearly or plane polarized light only the z direction is found. For circularly polarized light, the direction of rotation can be clockwise or counterclockwise.

moves counterclockwise, it is left handed. Again, the interaction of circularly polarized light with chiral molecules will alter the circularly polarized light. Elliptically polarized light is also used and is essentially the same as circularly polarized light, but an ellipse is swept out by the electric field wave instead of a circle. We will not dwell on how polarized light is produced, except to note that light can be polarized when it is passed through certain materials. A pair of Polaroid sunglasses is a good example of how light can be polarized.

The interaction of a chiral molecule with polarized light is very specific and has proved to be an important method for characterizing both small molecule and macromolecular structures. Small molecule examples should be already well known to you. Amino acids in most biological systems are *levo* rotary (L), that is they rotate plane-polarized light to the left, and sugars have various optical isomers. Enzymes, in fact, can distinguish between optical isomers and typically will only react with a single or a restricted group of small molecule isomers. Essentially, two types of measurements are commonly made to determine the effects of molecules on polarized light, optical rotation, and circular dichroism (CD). Optical rotation is a measure of the rotation, of linearly polarized light by a molecule, and the wavelength dependence of the optical rotation is called optical rotary dispersion (ORD). CD, on the other hand, is the difference in absorption of left-hand and right-hand circularly polarized light. These effects are relatively small but can be measured readily with modern instrumentation. For a typical protein or nucleic acid with a 100 μM solution of chromophore, the polarized light plane with a wavelength around the electronic absorption maxima is rotated about 0.001–0.1 degrees for a sample 1 cm thick. For CD, the difference in absorption coefficients is about 0.03–0.3% of the total absorption.

9.2 OPTICAL ROTARY DISPERSION

The first studies of the optical properties of biological macromolecules were done with ORD because adequate equipment was not available for CD measurements. At the present time, the opposite is true: CD is the method of choice. Nevertheless, it is instructive to discuss both methods. As we shall see later, they are intimately related.

A typical experimental setup for measuring ORD is shown in Figure 9-2. The principle is that plane polarized light is passed through the sample, and the rotation of the plane of polarization of the light by the sample is measured. The analyzer for the emergent beam is an element that polarizes the beam. It can be rotated until the light intensity after the analyzer disappears. Its polarization axis is then perpendicular to the direction of polarization of the beam emerging from the sample. The amount of rotation required for this disappearance is a direct measure of how much the original beam was rotated by the sample. Clockwise rotation is assigned as positive rotation and counterclockwise as negative rotation. The physical origin of this rotation is the fact that the sample is *circularly birefringent*, that is, the refractive index of the sample is different for left-hand circularly polarized light and right-hand circularly polarized light. A plane polarized beam of light can also be described as two opposite circularly polarized beams. A difference in the refractive index means that the speed at which the light passes through the sample is different

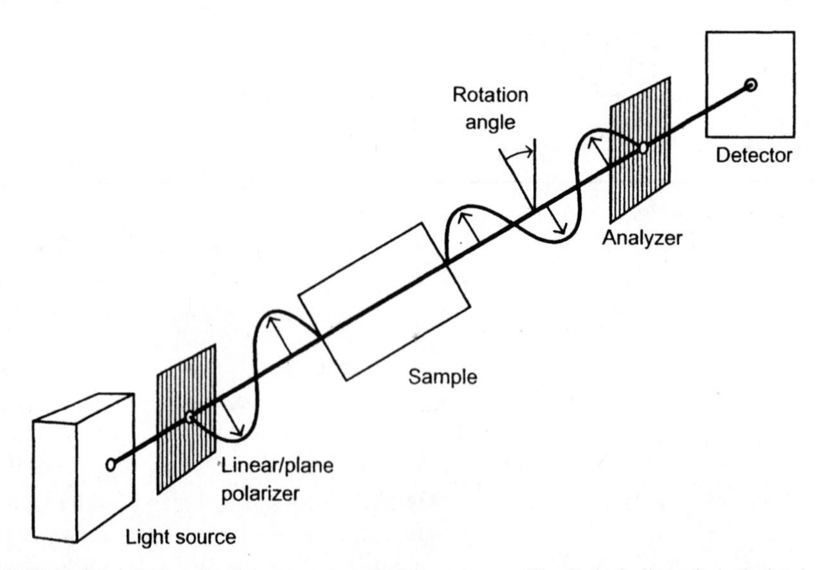

FIGURE 9-2. Schematic diagram of an ORD apparatus. The light is linearly polarized and the polarization plane is rotated by the sample. A second polarizer, the analyzer, can be rotated to obtain the maximum (or minimum) light intensity. The angle of rotation of the analyzer is a direct measure of the angle of rotation by the sample. For CD, circularly polarized light is used, and the difference in absorbance between left and right circularly polarized light is measured.

for right-hand and left-hand circularly polarized light. This means the two polarized beams get out of phase as they pass through the sample. This phenomenon is described in Chapter 6. This phase difference is manifested as a rotation of the light.

If the measured angle of rotation is α, then the specific rotation, $[\alpha]$, is defined as

$$[\alpha] = \alpha/(dc) \tag{9-1}$$

where c is the concentration in g/cc and d is the path length in decimeters. The origin of this definition lies in the history of optical rotation: the first measurements of optical rotation were made in cells that were 10 cm (1 dm) long. Instead of the specific rotation, data are often reported as the molar rotation, $[\phi]$, which is defined as

$$[\phi] = 100\alpha/(lM) \tag{9-2}$$

where l is the path length in centimeters and M is the concentration in mol/l. (For macromolecules, the concentration is often given in terms of the monomers.)

9.3 CIRCULAR DICHROISM

Materials display CD because the absorbance of left and right circularly polarized light is different for molecules with molecular asymmetry. One measure of the CD is simply the difference in absorbance of the material for left and right circularly polarized light, ΔA:

$$\Delta A = A_{\mathrm{L}} - A_{\mathrm{R}} \tag{9-3}$$

Circular dichroism can arise only in the spectral region where absorbance occurs—if the absorbance is essentially zero, there cannot be a measurable difference. For ORD, on the other hand, large rotations are measured in the regions where the sample absorbs light, but measurable rotation also occurs when the absorbance is essentially zero. This is helpful, for example, when looking at the optical activity of substances such as sugars, for which the absorbance is in the far ultraviolet. The large optical rotations in regions of high absorption are called Cotton effects.

As previously discussed, optical rotation arises because of the phase shift of the circularly polarized light that leads to circular birefringence. Circular dichroism arises because not only does the phase shift, but there is a differential decrease in the amplitude for right and left circularly polarized light. This leads to elliptical polarization. As with optical rotation, circular dichroism can be either positive or negative. Circular dichroism is reported either as a differential extinction coefficient,

$$\Delta\varepsilon = \varepsilon_{\mathrm{L}} - \varepsilon_{\mathrm{R}} = \Delta A/(lM) \tag{9-4}$$

or more often as the ellipticity, θ:

$$\theta = 2.303\Delta A 180/(4\pi) \text{ degrees} \tag{9-5}$$

As before, the molar ellipticity is

$$[\theta] = 100\theta/(lM) \tag{9-6}$$

Despite this definition of $[\theta]$, it is usually reported in the units of deg $cm^2\,dmol^{-1}$ because of the historical precedents. The differential extinction coefficient and the specific rotation are related by the relationship

$$[\theta] = 3300\Delta\varepsilon \tag{9-7}$$

The numerical coefficient follows from the Beer–Lambert law for absorption and the various relationships in previous equations.

Since ORD and CD arise from the same physical phenomena, namely the effect of molecular asymmetry on polarized light, one might imagine that the two should be closely related. In fact, these two measurements are not independent: if the optical rotation is known, the circular dichroism can be calculated and vice versa. Although the formal equations for these transformations are well known, in practice the mathematics are sufficiently difficult so that this transformation is rarely done. In fact, CD is the experimental method of choice for a variety of reasons, mostly because circular dichroism is only found at absorption bands so that the interpretation of the spectra is somewhat easier.

9.4 OPTICAL ROTARY DISPERSION AND CIRCULAR DICHROISM OF PROTEINS

The simplest model for the optical activity of a protein is to assume that the optical activity is the sum of the optical activity of the individual amino acids. This model is clearly incorrect: The sum of the optical activity of the amino acids is usually very small relative to the measured optical activity of a protein. This observation is not surprising: We saw previously that the ultraviolet spectra of proteins were primarily due to the peptide bonds. Similarly, the optical activity of proteins is primarily due to the macromolecular structure itself. In fact, specific protein structures have characteristic ORD and CD spectra.

Three of the fundamental structures of proteins are the α-helix, the β-sheet, and the random coil (cf. Chapters 3 and 7). The α-helix is a right-hand helix stabilized by short-range hydrogen bonds between backbone peptide bonds, whereas the β-sheet structure is composed of parallel polypeptide chains, also stabilized by hydrogen bonds between the backbone peptide bonds (Figs. 3-4 and 3-5, see color plates). The random coil is envisaged as a random arrangement of the backbone although it is rarely random—irregular might be a better term. The ORD and CD spectra of many different homopolypeptide chains have been determined under experimental conditions where the structures are known. The results are summarized in Figure 9-3 (1). Within relatively small deviations, the spectra are the same for the common structures of the homopolypeptides. Note that, as expected, the

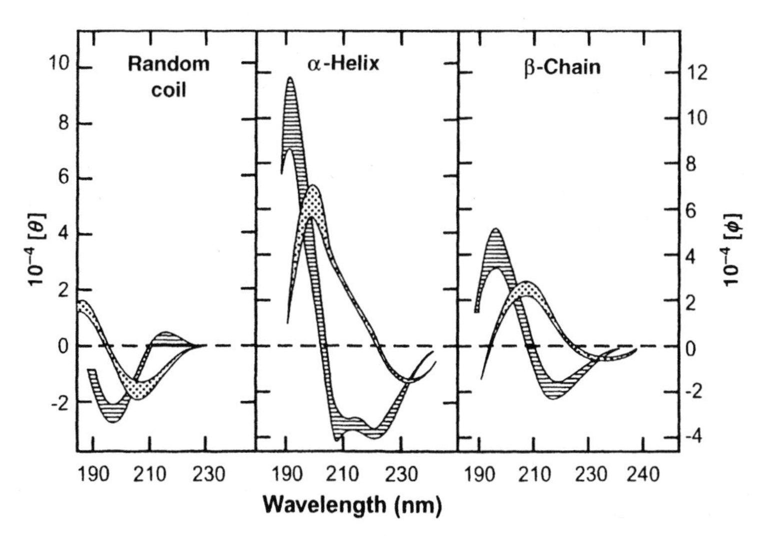

FIGURE 9-3. ORD (*dotted*) and CD (*crosshatched*) of homopolypeptides in the random coil, α-helix, and β-sheet conformations. The range of values found is indicated by the shaded areas. Reproduced with permission from W. B. Gratzer and D. A. Cowburn, *Nature* **222**, 426 (1969).

CD is centered around the ultraviolet absorption spectra of the proteins. Also note that the spectra are reasonably distinct for these three different structures.

If the assumption is made that only α-helix, β-sheet, and random coil structures are present and that these three structures have known optical properties, it should be possible to develop algorithms to deconvolute the ORD and CD spectra of a given protein. The implicit assumption in these algorithms is that ORD and CD spectra of the three different structures are additive, that is, the presence of one structure does not influence the contributions from another structure. Thus, measurement of the ORD or CD spectrum over the wavelength region of 175–250 nm would permit the calculation of the fraction of the protein structure present in each of the three conformations. A number of attempts have been made to do this, and the results have been compared with known crystal structures (cf. Ref. 2). In some cases, crystal structures have been used to derive a set of self-consistent parameters, assuming that the crystal and solution structures are identical. Reasonable agreement between the calculated structures and the actual structures is often found, but the calculations are not quantitative for several reasons. First, the optical properties of homopolymers are not accurate predictors of the optical properties of the protein. The ORD and CD spectra depend on the micro-environment of each residue, so that they will certainly be dependent on the specific sequence and size of the protein. Second, protein structures are not so simple that their domains can be described by the three relatively simple structures assumed for the algorithms. Other structures that can influence CD and ORD spectra include various helices other than the α-helix, disulfide bonds, and various β-turns. Many of the fitting algorithms include structures other than the three discussed here.

Finally, the side chains of amino acids influence the spectra—this is particularly true at wavelengths above about 250 nm where tyrosine, phenylalanine, and tryptophan have strong absorption bands. Prosthetic groups, such as hemes, would have a similar effect at the wavelength where they have strong absorption bands. Nevertheless, ORD and CD spectra are useful semiquantitative predictors of protein structure and also serve as good indicators of structural changes that may occur as experimental conditions are changed (cf. Ref. 3).

9.5 OPTICAL ROTATION AND CIRCULAR DICHROISM OF NUCLEIC ACIDS

The bases of nucleic acids are not intrinsically optically active; however, the sugars are optically active and can induce optical activity in the bases. This optical activity is relatively small, but as might be anticipated from the discussion of proteins, the ORD and CD spectra of nucleic acids are not just the sum of those of the monomers. In fact, CD and ORD are very good indicators of secondary structure such as helices and are the method of choice for following changes in secondary structure. As an example, consider polyadenylic acid (polyA) (4). As shown in Figure 9-4, at neutral

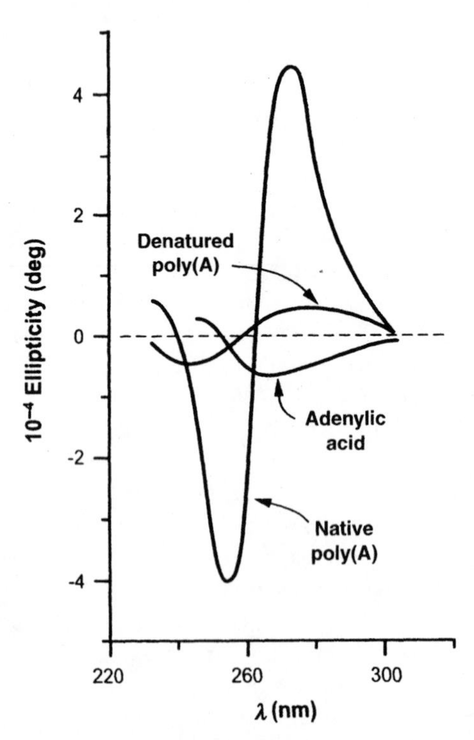

FIGURE 9-4. Circular dichroism of adenylic acid, denatured polyadenylic acid, and native polyadenylic acid. D. Freifelder, *Physical Biochemistry*, 2nd edition, Freeman, New York, 1982, p. 594. © 1976, 1982 by Freeman. Used with permission.

pH and room temperature, native polyA has a very large ellipticity due to the fact that it exists in a helical structure. If the structure is destroyed by raising the temperature or going to extremes of pH, the ellipticity essentially disappears. Also shown in the figure is the ellipticity of adenylic acid, which is quite small. The ORD and CD of single-stranded polynucleotides have been extensively studied. The large ellipticity is primarily due to the interactions of nearest neighbors, that is, adjacent bases. They are stacked on top of each other, and this stacking is primarily responsible for the altered optical properties of the helical structures. To a good approximation, the ORD and CD properties of a base are influenced only by their nearest neighbors. Consequently, the spectra of polymers can be readily simulated by adding together the spectrum of each base, taking into account nearest-neighbor interactions only.

The situation is much more complex for double-stranded nucleic acids such as DNA, and structural predictions are much more difficult, but CD and ORD are still useful tools. As an example, the CD of *E. coli* DNA is shown in Figure 9-5 for both the native and denatured forms (5). Although structural predictions from ORD and CD spectra are difficult, it is true, nevertheless, that specific structures of DNA and

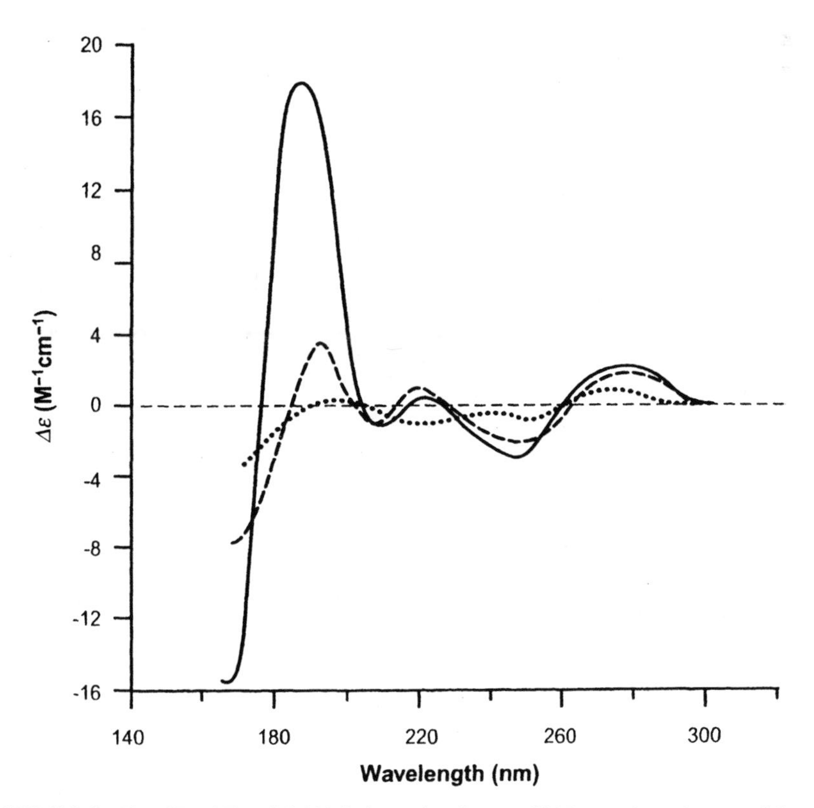

FIGURE 9-5. The CD of *E. coli* DNA in its native form at 20°C (—), heat denatured form at 60°C (—), and the average CD of the four deoxynucleotides (. . .). From C. A. Sprecher and W. C. Johnson, Jr., *Biopolymers* **16**, 2243 (1977). Reprinted with permission of Wiley © 1977.

RNA have very characteristic spectra, and changes in the spectra are good monitors of structural changes (cf. Ref. 3).

9.6 SMALL MOLECULE BINDING TO DNA

DNA is an obvious target for drug intervention since disruption of its structure clearly will have significant biological implications. The negatively charged phosphate groups along the backbone structure of DNA can bind small molecules through electrostatic interactions, but the major sites of drug binding in double-stranded DNA are the major and minor grooves in the double helical structure and intercalation between the bases. These three modes of binding are shown schematically in Figure 9-6 (6). Many of the drugs used contain aromatic rings, and as a specific example of such binding we consider the interaction of DNA with acridine orange, even though acridine orange is not actually a drug (7).

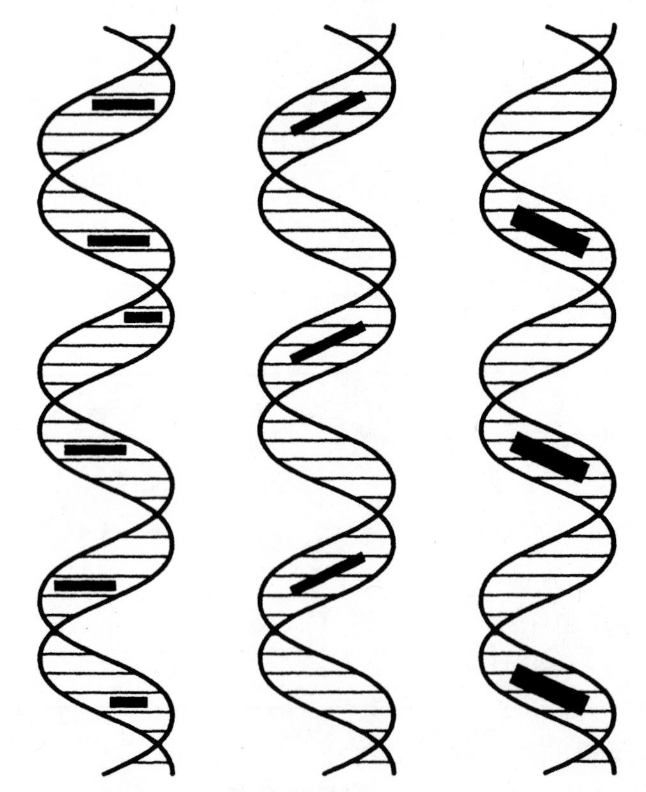

FIGURE 9-6. Schematic diagram of the intercalation of ligands between the bases, and in the minor and major grooves of a DNA double helix. From M. Ardhammar, B. Norden, and T. Kurucsev in *Circular Dichroism: Principles and Applications* (N. Berova, K. Nakanishi, and R. W. Woody, eds.), 2nd edition, Wiley, New York, 2000, p. 746. Reprinted with permission of Wiley © 2000.

The structure of acridine orange is shown in Figure 9-7. It is not optically active by itself but binding to DNA induces optical activity into the molecule. This is because binding causes the electronic energy states of achiral acridine orange to be coupled with the electronic energy states of chiral DNA. Induced optical activity is quite common when achiral small molecules bind to chiral macromolecules. The CD spectra of acridine orange bound to DNA are shown in Figure 9-7 for various ratios of (bound dye)/DNA (7). As expected, the optical activity is centered around the electronic absorption band of acridine orange, which has a maximum at about 430 nm. At very low values of (bound dye)/(acridine orange), a negative value of $\Delta\varepsilon$

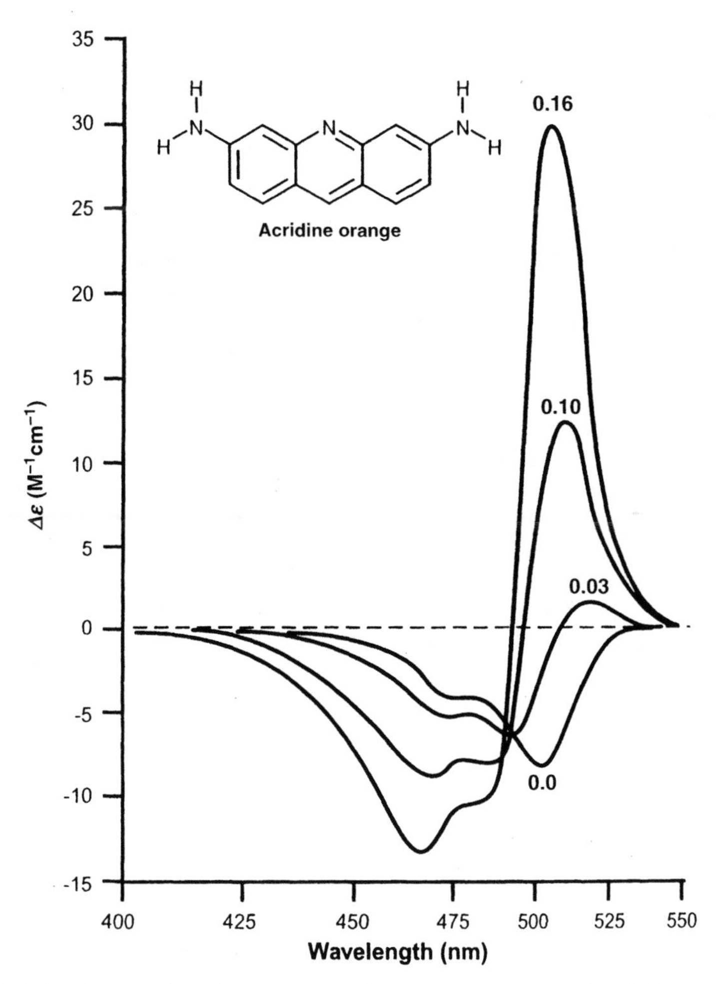

FIGURE 9-7. Circular dichroism per mole of acridine orange bound to DNA at the indicated values of [bound dye]/[DNA]. The total dye concentration is 10^{-5} M. Reprinted in part with permission from D. Fornasiero and T. Kurncsev, *J. Phys. Chem.* **85**, 613 (1981). © 1981 by American Chemical Society.

is observed, but as this ratio increases, a large positive band develops. At the highest ratio shown, both strong negative and positive bands are observed. A molecular interpretation of these results has been developed. At very low concentrations of acridine orange, the binding is through intercalation between the bases, with the acridine orange ring structure parallel to the bases. This produces a negative induced circular dichroism with a maximum $\Delta\varepsilon$ of about $-8\,M^{-1}\,cm^{-1}$. As the concentration of the ligand increases, the dye binds to the groove which induces a positive CD with a maximum value of about $60\,M^{-1}\,cm^{-1}$. These two induced CD signals are enhanced at higher dye concentrations due to interactions between the electronic energy levels of the bound dye molecules.

This example illustrates the exquisite sensitivity of the CD spectrum to the nature of the binding process and to the secondary structure of the DNA. Studies with many different types of ligands have been carried out (cf. Ref. 6).

9.7 PROTEIN FOLDING

Understanding protein structure in molecular terms is a long-standing goal of biochemistry. While great progress has been made toward this goal, we still cannot predict the three-dimensional structure of a protein from a knowledge of the one-dimensional sequence of amino acids that make up a protein. One approach to this end is to study the folding and unfolding of proteins with the rationale that understanding these processes will provide a better understanding of protein structure. In addition, understanding the folding of proteins is of great physiological significance since proteins are synthesized unfolded *in vivo* and must be folded into specific structures in order to perform their biological activities. Protein folding already has been briefly discussed in Chapter 3. Many reviews of protein folding are available (cf. Ref. 8), and only one particular example is presented here. CD and ORD are particularly useful tools for the study of protein folding and unfolding since they are extremely sensitive to secondary and tertiary structure of protein so that large changes are generally seen as a result of the folding process.

Traditionally, proteins were thought to be synthesized and then immediately folded into their biologically active form. However, in recent years a number of examples have emerged in which the unfolded form exists *in vivo* until it is folded into its biologically active form by specific conditions in the external environment such as the presence of a specific ligand. The biological significance is not understood, but the evidence for such structures is quite convincing. An example of this phenomenon is the protein associated with the ribozyme ribonuclease P (9). Ribonuclease P is a ribonucleic acid (RNA) catalyst, previously described in Chapter 5. Its activity and/or specificity is enhanced by the binding of the catalytic RNA to a protein, and the protein is essential for *in vivo* catalysis (10). We will not discuss the eatalytic reaction itself which is the processing of the 5′-leader sequences from precursor tRNA. This ribozyme is quite ubiquitous in nature: in bacteria, the complete catalytic unit consists of a single RNA of about 400 nucleotides and a single protein of about 120 amino acids.

When the protein from *B. subtilis*, is isolated, it is unfolded at physiological temperatures in the absence of ligands (11). This can be ascertained from the CD spectrum in sodium cacodylate (pH 7), which does not bind to the protein. The spectrum of the protein in the far and near ultraviolet is shown in Figure 9-8. The far ultraviolet

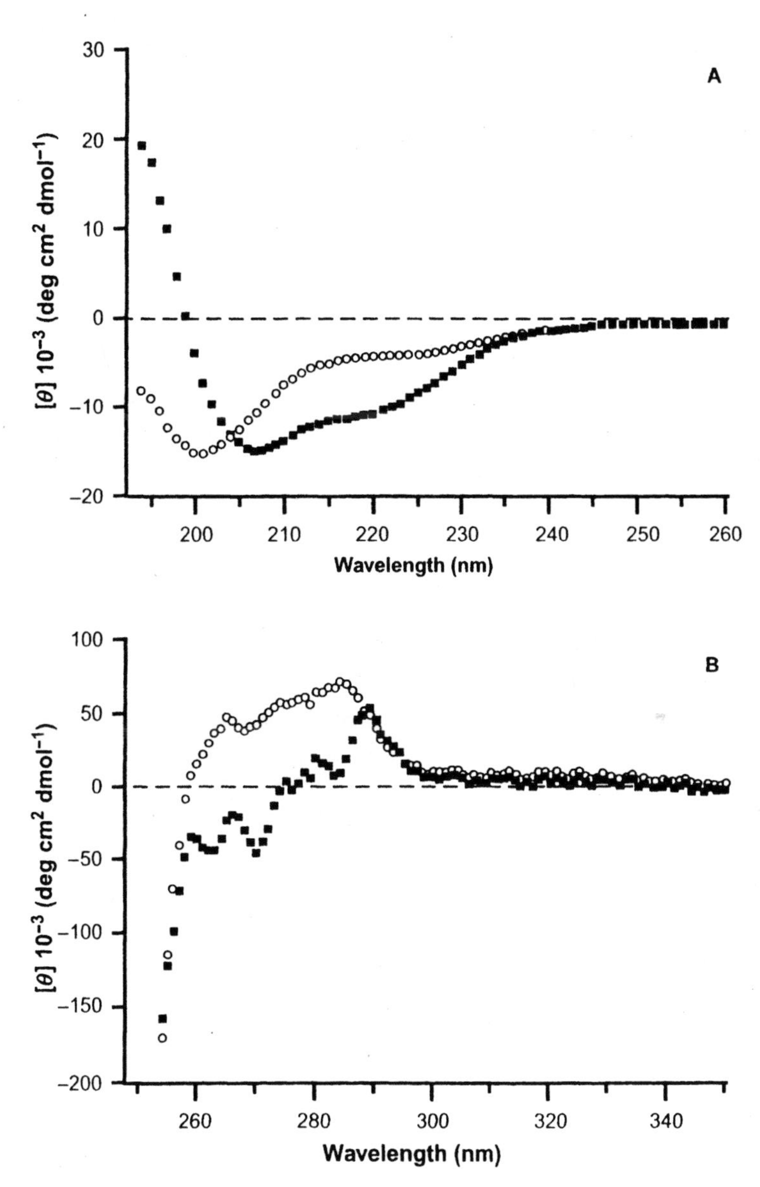

FIGURE 9-8. Circular dichroism of P protein in the presence (○) and absence (■) of 10 (B) or 20 (A) mM sulfate at pH 7.0 in 10 mM sodium cacodylate. Reprinted in part with permission from C. J. Henkels, J. C. Kurz, C. A. Fierke, and T. G. Oas, *Biochemistry* **40**, 2777 (2001). © 2001 by American Chemical Society.

spectrum is due to the protein backbone whereas the near ultraviolet spectrum is primarily due to tyrosine residues in the protein. Included in Figure 9-8 are the spectra in the presence of 10 or 20 mM sulfate. An obvious change in the spectrum occurs. The far-ultraviolet spectrum is consistent with the known crystal structure of the protein which contains α-helices and β-sheets, and the near-ultraviolet spectrum changes because the electronic environments of the tyrosines are altered.

The change in structure can be quantified by measuring the change in the CD at 222 nm as a function of the ligand concentration. This is shown for a variety of ligands in Figure 9-9. A variety of anions cause the protein to go from a disordered structure to a more folded structure, with highly charged anions such as polyphosphates being more effective than monovalent ions such as chloride. The stoichiometry of the ligand binding was determined for pyrophosphate and was found to be 2 pyrophosphates/protein. A simple model that explains the observations is that the ligand binds to the folded protein but not the unfolded protein. The apparent equilibrium constant, K_{app}, for the ratio of the denatured (unfolded) protein, D, to the native state (folded) protein, N, can be obtained directly from the data in Figure 9-9 if the assumption is made that only two forms of the protein exist, folded and unfolded. For example, the apparent equilibrium constant is unity when the change in the CD is half of the total change. The apparent equilibrium constant at any point on the curve is given by

$$K_{app} = (\theta_L - \theta_D)/(\theta_N - \theta_L) = \Sigma N/\Sigma D \qquad (9\text{-}7)$$

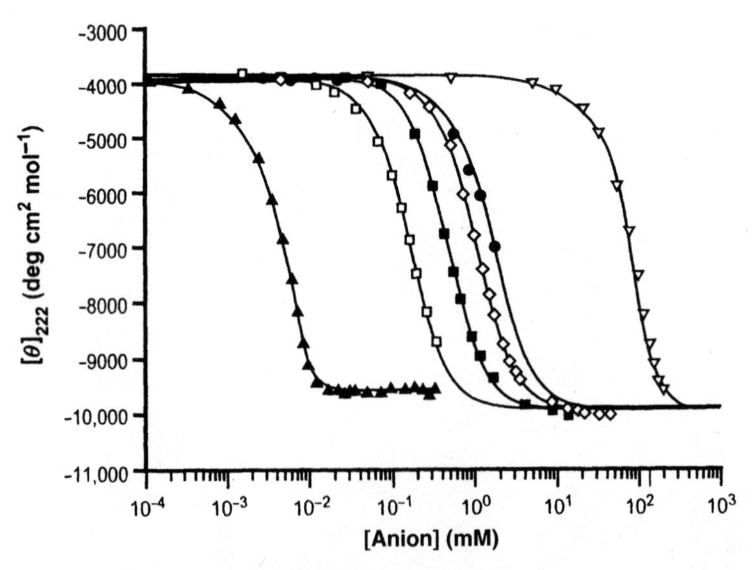

FIGURE 9.9. Anion-induced folding transitions of the P protein followed by the change in the CD signal at 222 nm at 37°C. The anions are dCTP (▲), CMP (□), sulfate (■), phosphate (◇), dCMP (•), and formate (▽). Selected data from C. J. Henkels, J. C. Kurz, C. A. Fierke, and T. G. Oas, *Biochemistry* **40**, 2777 (2001). Figure courtesy of Dr. C. J. Henkels, Duke University. Reproduced with permission.

where θ's are the absolute values of the circular dichroism in the figure for a given ligand concentration, L, the denatured form, D, and the native form, N. The summation is over all liganded forms of the denatured and native proteins. If the ligand is assumed to bind to two independent and equivalent sites, the folding mechanism can be written as

$$\mathrm{D} \underset{}{\overset{K_{\mathrm{fold}}}{\rightleftharpoons}} \mathrm{N} + 2\mathrm{L} \overset{K_{\mathrm{a}}}{\rightleftharpoons} \mathrm{NL} + \mathrm{L} \overset{K_{\mathrm{a}}}{\rightleftharpoons} \mathrm{NL_2} \tag{9-8}$$

This mechanism gives

$$\begin{aligned} K_{\mathrm{app}} &= \frac{(\mathrm{N}) + (\mathrm{NL}) + (\mathrm{NL_2})}{(\mathrm{D})} = \frac{(\mathrm{N})[1 + (\mathrm{NL})/(\mathrm{N}) + (\mathrm{NL_2})/(\mathrm{N})]}{(\mathrm{D})} \\ &= (\mathrm{N})/(\mathrm{D})\left[1 + 2K_{\mathrm{a}}(\mathrm{L}) + \mathrm{K}_{\mathrm{a}}^2(\mathrm{L})^2\right] = K_{\mathrm{fold}}[1 + K_{\mathrm{a}}(\mathrm{L})]^2 \end{aligned} \tag{4-9}$$

where $K_{\mathrm{fold}} = (\mathrm{N})/(\mathrm{D})$ and K_{a} is the is the microscopic association constant for the binding of ligand to the native protein $[K_{\mathrm{a}} = (\mathrm{NL})/[2(\mathrm{N})(\mathrm{L})] = 2(\mathrm{NL_2})/[(\mathrm{NL})(\mathrm{L})]$. The concept of microscopic association constants is considered more extensively in Chapter 13. The two constants K_{fold} and K_{a} cannot be determined independently from the data in Figure 9-9. However, an independent method, which will not be described here, gives an estimated value of 0.0071 for K_{fold} at 37°C. With this value in hand, it is possible to calculate the microscopic association constants for the ligands.

This example illustrates two important concepts. First, it shows that measurement of the CD provides a means of following the transition between the denatured and native state of proteins. Second, it shows how the binding of a ligand to the native state can convert the denatured protein to its native structure. (Coupled equilibria are common in biological systems.) These concepts are of both practical and theoretical value for studying and understanding biochemical processes.

9.8 INTERACTION OF DNA WITH ZINC FINGER PROTEINS

Zinc finger motifs are found in proteins that regulate transcription. As the name implies, the structure is a finger consisting of a loop of β-structure with two cysteines and an α-helix with two histidines. The four side chains of these amino acids tightly bind a zinc ion, as shown schematically in Figure 9-10. Transcription factors contain multiple "fingers," and the α-helix of each finger binds in the major groove of double-stranded DNA, thereby regulating transcription. The interaction of zinc finger proteins with DNA has been extensively studied. For example, the interaction of a transcription factor that regulates the expression of the protein metallothionein in mouse and human cells with DNA has been studied using CD (12,13). The transcription factor contains six zinc fingers, which were isolated from the rest of the transcription factor. Preparations contained three to six fingers. The six-finger protein binds three zinc ions very tightly, and far-ultraviolet CD

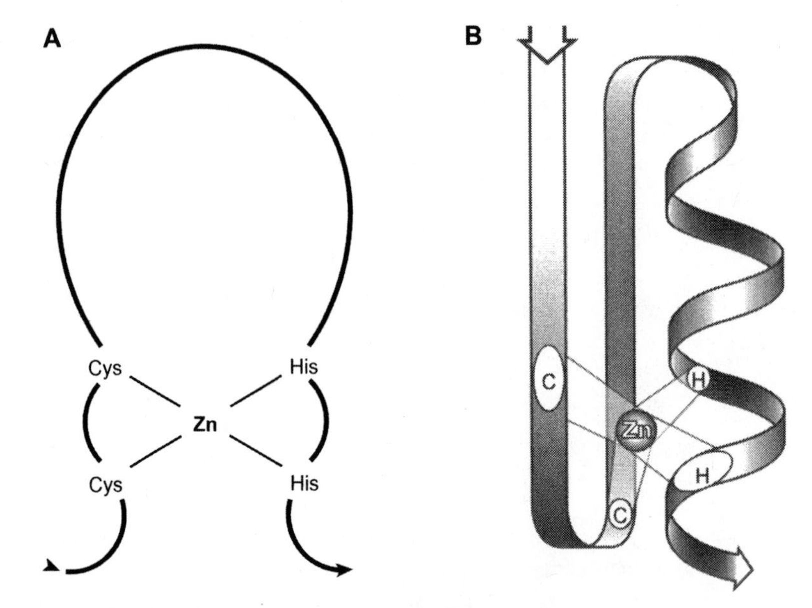

FIGURE 9-10. Schematic drawing of the Zn finger structure. (A) Coordination of Zn to cysteine and histidine ligands. (B) Secondary structure. Reprinted from R. M. Evans and S. M. Hollenberg, Zinc fingers: gilt by association, *Cell* **52**, 1(1988). © 1988, with permission of Elsevier.

spectra showed that the conformation of the protein was unchanged once the three tight binding sites were occupied by zinc, although three additional zinc ions were bound more weakly.

The binding of the isolated zinc fingers to a 23-mer DNA that contains the sequence involved in the regulatory process was studied with CD. As the number of zinc fingers was increased from 3 to 6, a negative ellipticity at about 210 nm decreased in magnitude (approached zero) with a concomitant increase in a positive ellipticity at 280 nm. This change is not due to the protein, whose CD spectrum was subtracted from the experimentally determined spectrum. It is due to a change in the DNA structure from the B to A form. The B form of DNA is the structure stable at high humidity and is the form most prevalent in cells, whereas the A form is stable at lower humidity. The A form is characterized by a tilting of the base pairs relative to the axis of the helix and a change in the position of the sugars. This change in DNA structure appears to be a general consequence of binding zinc fingers and is thought to be involved in the regulation of transcription. The A and B forms of DNA have somewhat different CD spectra, which make it possible to interpret the CD changes associated with zinc finger binding in structural terms.

The interaction of many different proteins and nucleic interactions have been probed with CD (12). These interactions are of great physiological importance, and CD has proved to be a valuable tool for probing the structural changes that occur.

9.9 FLUORESCENCE POLARIZATION

With the introduction of the concept of polarized light, we can introduce an interesting phenomenon observed with fluorescence. Normal fluorescence experiments are not carried out with polarized light so the fluorescence observed also is not polarized. However, if the light used for excitation of fluorescence is polarized, then the fluorescence emitted is also often polarized. The basic idea is not hard to understand: If the electric field wave is oriented in space, it will interact with the electric dipole of the molecule, and this interaction can ultimately lead to the absorption of energy and fluorescence. The polarization of the light emitted will depend on the relative orientation of the electric field and the electric dipole and how much it changes during the fluorescence process. (The polarization of the emitted light is determined by the "transition dipole," which is generally oriented differently than the electric dipole.) Now let us consider what happens with a population of molecules interacting with polarized light.

The experiment itself is straightforward. A polarizer is put in front of the excitation beam so that molecules are excited with polarized light. A second polarizer is put in front of the detector for the emitted light. Two measurements are made: the intensity of the emitted light with the axis of the detection polarizer parallel to the polarization axis of the excitation polarizer, $I_{\|}$, and the intensity when the two axes are perpendicular, $I_{\perp}$. The polarization, P, is defined as

$$P = (I_{\|} - I_{\perp})/(I_{\|} + I_{\perp}) \quad (9\text{-}10)$$

If the emitted light is completely unpolarized, the parallel and perpendicular intensities would be identical and the polarization would be zero. This would happen if the molecules rotate very rapidly during the time that the fluorescence occurs. Since typical fluorescence lifetimes are in the range of 1–100 ns, this would mean that the molecules would have to rotate many times during this relatively short lifetime. This is, in fact, what happens for small molecules where polarization is usually not observed. The opposite extreme is the case when the molecules do not rotate at all during the lifetime of the fluorescence but are randomly oriented in space. This would be the situation, for example, in extremely viscous solvents. The calculation for this situation is somewhat complex, but can be done. For this limit, the polarization is ½. For some macromolecules and ligands bound to macromolecules, the rate of rotation of the fluorescent species is comparable to the fluorescent lifetime. For such cases, the polarization lies between 0 and ½, and measurement of the polarization gives information about the rate of rotation of the fluorophore.

The time dependence of the polarization will provide quantitative information about the rate of rotation of fluorescent molecules. Usually, the data are analyzed in terms of the anisotropy, A, rather than the polarization, although the two are conceptually equivalent. It turns out the anisotropy is more easily related to rotational motion.

$$A = (I_{\|} - I_{\perp})/(I_{\|} + 2I_{\perp}) \quad (9\text{-}11)$$

If the time dependence of the fluorescence is measured for the parallel and perpendicular components, the time dependence of the polarization can be calculated. These are not easy measurements since fluorescence decays in nanoseconds. Theoretical considerations show that for a simple rotation

$$A(t) = A(0)\exp(-t/\tau_c) \tag{9-12}$$

where t is the time, $A(0)$ is a constant, and τ_c is the rotational correlation time and is related to the rotational diffusion constant, D_{rot}, by $\tau_c = 1/(6D_{rot})$. For example, the rotational correlation time for a fluorescent derivative of chymotrypsin is 15 ns, which is typical for small proteins. The rotational correlation time becomes longer as the molecular volume of the macromolecule increases, so that it is a direct measure of the size of the macromolecule. The exact relationship is complex, as both molecular weight and shape are important. If more than one mode of rotation is possible, multiple exponential decays may be observed.

Polarization and/or anisotropy provide information about the orientation and rotational freedom of fluorescent molecules. This, in turn, can often provide useful information about biological structures and mechanisms.

9.10 INTEGRATION OF HIV GENOME INTO HOST GENOME

HIV is a retrovirus that is unfortunately well known to everyone. In order for the virus to infect its host, the HIV genome must be integrated into the host genome. This integration is quite complex: it involves processing the HIV DNA, strand transfer, and DNA repair. Integrase is the enzyme responsible for the 3′ -processing and strand transfer. These reactions can be monitored *in vitro* with short oligonucleotides that are models for the end of the viral DNA. The catalytic core of the enzyme recognizes a specific sequence of DNA. The self-association of integrase appears to be important for its function. The oligomeric state of integrase bound to viral DNA can be monitored through time-resolved fluorescence anisotropy measurements (14). It is a useful tool for this purpose because very low protein and DNA concentrations can be monitored, and the rotational correlation time is directly related to the size of the macromolecule.

The rotational correlation time for integrase was determined by monitoring the time decay of tryptophan fluorescence. The rotational correlation time decreased from about 90 ns to 20 ns at 25°C as viral-specific DNA, a 21-base sequence, was added. This was attributed to association of integrase with the DNA. The concentration of integrase was only 100 nM in these experiments. Independent studies of integrase established that the rotational correlation time in the absence of DNA was associated with a tetrameric structure of the protein. The actual situation is a bit more complex than this, as some other aggregates are present, but the dominant species is a tetramer, which becomes even more predominant at 37°C. The rotational correlation time of the 21-mer DNA duplex was determined by modifying the DNA with fluorescein. The dominant rotational correlation time is about 2 ns. A much faster rotational correlation time is also observed due to rotation of fluorescein within the DNA. Multiple rotational correlation times are commonly observed so that it is necessary to establish what is being measured on a molecular basis. As the fluorescein-labeled DNA was titrated with integrase, an additional rotational correlation time was observed of about 15 ns. This rotational correlation time is attributed to the complex formed and is consistent with that determined from measurement of the correlation time associated with the tryptophan fluorescence. The rotational correlation time for monomer integrase, molecular weight 32,000, is 16 ns.

These elegant experiments with the integrase–DNA complex, and the many control experiments, demonstrate that the integrase depolymerizes when it binds to DNA. Furthermore, the predominant binding state at 25°C is monomeric, whereas at 37°C a mixture of monomers and dimers is present. This raises the interesting question as to what is the active oligomeric state of integrase? Prior to this study, the prevailing mechanism was thought to involve a multimeric structure of integrase. These results suggest that *in vitro* monomers and dimers may be enzymatically active, although the *in vivo* activity, which is considerably more complex, probably involves structures larger than dimers.

9.11 α-KETOGLUTARATE DEHYDROGENASE

The α-ketogluarate dehydrogenase complex from *E. coli* contains three enzymes that catalyze the overall reaction

$$\alpha\text{-ketogluarate} + \text{CoA} + \text{NAD}^+ \rightarrow \text{succinyl-CoA} + \text{CO}_2 + \text{NADH} + \text{H}^+ \qquad (9\text{-}13)$$

The first enzyme decarboxylates the α-ketoglutarate. The intermediate formed with thiamine pyrophosphate transfers the succinyl moiety to lipoamide, and this intermediate is oxidized to form a succinyl–lipoic acid intermediate. The succinyl group is transferred to CoA, and finally the dihydrolipoamide is oxidized. This last reaction involves an enzyme-bound flavin and reduces NAD^+ to NADH. The multienzyme complex has a molecular weight of about 2.5×10^6 and contains 12 copies of the first enzyme, 24 copies of the second, and 12 copies of the third (15). The intermediates in the reaction sequence are bound to lipoamide, that is, lipoic acid covalently attached to a lysine through an amide linkage. The structure of lipoic acid is shown in Figure 9-11. It can exist in an oxidized form with a disulfide at the end of the chain, or in a reduced form with two sulfhydryl groups. The proposed mechanism involves the reduced lipoic acid-bound intermediates moving between the active sites of the three enzymes. Two questions arise: Are the active sites close enough to permit lipoic acid to span the distances between the active sites, and does the lipoic acid rotate between the active sites fast enough for the reaction to proceed

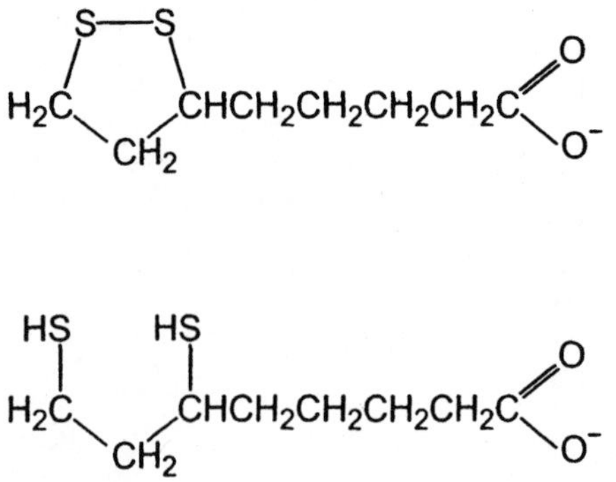

Oxidized lipoic acid

Reduced lipoic acid

FIGURE 9-11. Structures of oxidized and reduced lipoic acid.

efficiently? Answers to both of these questions can be obtained with fluorescence methodology.

Fluorescence resonance energy transfer measurements between the active sites of the three enzymes show they are about 3 nm apart, approximately the maximum span of a lipoic acid (16). Other experiments showed that succinyl and electron transfer between lipoic acids can occur, so that the potential distance an intermediate can be transferred is longer than a single lipoic acid. Thus, the energy transfer measurements demonstrated that the transfer of intermediates occurs over a relatively long distance and that lipoic acid is a viable intermediate.

Does the lipoic acid rotate fast enough to serve the postulated transfer function? This is not an easy question to answer because no easy method exists for measuring the rate of rotation. However, fluorescence anisotropy measurements provide some insight into this rate (17). The lipoic acid was labeled with a fluorescent probe on the sulfhydryl that normally carries a reaction intermediate. The fluorescence lifetime of the probe and the dynamic fluorescence anisotropy were then measured. A typical time course for the fluorescence and anisotropy is shown in Figure 9-12. As might be expected, the dynamic fluorescence anisotropy is quite complex. Three components are found. One has a very long rotational correlation time and is due to the overall rotation of the very large multienzyme complex. The second component has a rotational correlation time of about 25 ns and is due to local rotation of the fluorescent probe in a hydrophobic environment. Rotational correlation times of this magnitude have been observed often for local motion of ligands bound to macromolecules. The third component has a rotational correlation time of about 350 ns. This can be attributed to rotation of lipoic acid between the catalytic sites. This component of the anisotropy decay is not seen if the enzyme to which the lipoic acid is bound is separated from the other two enzymes. It cannot be ascertained whether the rate-limiting process is the actual rotation or dissociation of the probe from one of the catalytic sites, but in any event these results establish that the rate of rotation is sufficiently fast to support the observed catalytic rate, namely a turnover number of about 130 s^{-1}.

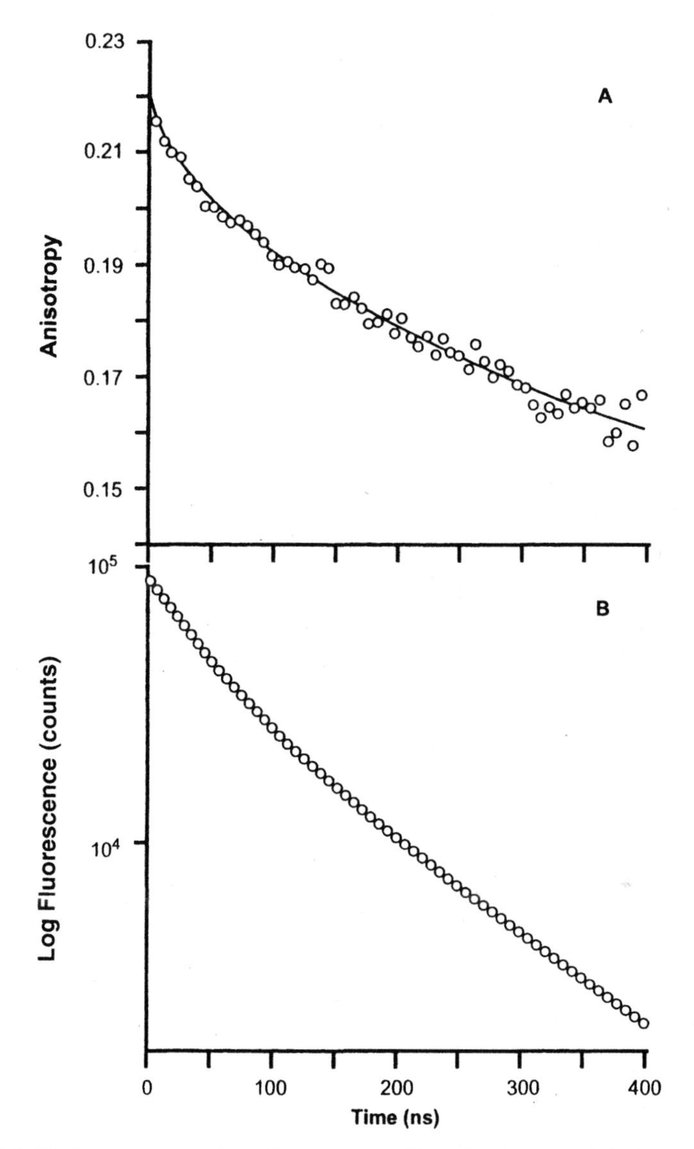

FIGURE 9-12. Time course of the fluorescence (B) and anisotropy (A) decay of pyrene covalently attached to the lipoic acid of *E coli* α-ketoglutarate dehydrogenase. Reprinted in part with permission from D. E. Waskeiwicz and G. G. Hammes, *Biochemistry* **22**, 6489 (1982). © 1982 by American Chemical Society.

The uses of fluorescence anisotropy to probe biological mechanisms discussed are intended as illustrative examples of the unique information that can be obtained. They also illustrate the difficulty of interpreting the results that are obtained. As with any physical method, great care must be taken to explore all possible explanations and careful control experiments are required.

REFERENCES

1. W. B. Gratzer and D. A. Cowburn, *Nature* **222**, 426 (1969).
2. N. Sreeramama and R. W. Woody, *J. Mol. Biol.* **242**, 497 (1994).
3. N. Berova, K. Nakanishi and R. W. Woody (eds.), *Circular Dichroism: Principles and Applications*, 2nd edition, Wiley, New York, 2000.
4. D. Freifelder, *Physical Biochemistry*, 2nd edition, Freeman, New York, 1982, p. 594.
5. C. A. Sprecher and W. C. Johnson, Jr., *Biopolymers* **16**, 2243 (1997).
6. M. Ardhammar, B. Norden, and T. Kurucsev in Ref. 4, pp. 741–768.
7. D. Fornasiero and T. Kurucsev, *J. Phys. Chem.* **85**, 613 (1981).
8. J. K. Myers and T. G. Oas, *Annu. Rev. Biochem.* **71**, 783 (2002).
9. S. Altman and L. Kirsebom, *Ribonuclease P in the RNA World* (R. F. Gesteland, T. R. Cech, and J. F. Atkins, eds.), Cold Spring Harbor Laboratory Press, Plainview, NY, 1999, pp. 351–380.
10. R. Kole, M. F. Baer, B. C. Stark, and S. Altman, *Cell* **19**, 881 (1980).
11. C. H. Henkels, J. C. Kurz, C. A. Fierke, and T. G. Oas, *Biochemistry* **40**, 2777 (2001).
12. D. M. Gray in Ref. 3, pp. 769–796.
13. X. Chen, A. Agarwal, and D. P. Giedroc, *Biochemistry* **37**, 11152 (1998).
14. E. Duprez, P. Tanc, H. Leh, J.-F. Mouscadet, C. Auclair, M. E. Hawkins, and J.-C. Brochon, *Proc. Natl. Acad. Sci. USA* **98**, 10090 (2001).
15. L. J. Reed, *Acc. Chem. Res.* **7**, 40 (1974).
16. K. J. Angelides and G. G. Hammes, *Biochemistry* **18**, 5531 (1979).
17. D. E. Waskiewicz and G. G. Hammes, *Biochemistry* **22**, 6489 (1982).

PROBLEMS

9.1 The sugar D-mannose can exist as two enantiomers that rotate light in opposite directions. The (+) configuration has a specific rotation of 29.3° d^{-1} g/cc at 589.3 nm, 20°C, whereas the (−) configuration has a specific rotation of −17.0° d^{-1} g/cc at the same wavelength. When either pure enantiomer is put into water, the optical rotation changes until a specific rotation of 14.2° d^{-1} g/cc is reached. Calculate the ratio of (+) and (−) enantiomers at equilibrium.

9.2. Solutions of RNA and DNA at a concentration of 2.00×10^{-5} M nucleotides (monomers) have the following differential absorption characteristics in a 1-cm cell:

λ (nm)	$A_L - A_R$, DNA	$A_L - A_R$, RNA
240	-4.40×10^{-4}	0.00
260	0.00	12.0×10^{-4}
280	6.00×10^{-4}	3.20×10^{-4}
300	0.20×10^{-4}	-1.00×10^{-4}

a. Calculate $\varepsilon_L - \varepsilon_R$ for the DNA and RNA at each wavelength.

b. Calculate the molar ellipticity for DNA and RNA at each wavelength.

c. A mixture of the DNA and RNA has the following differential absorption characteristics.

λ (nm)	$A_L - A_R$
240	-0.53×10^{-4}
260	4.00×10^{-4}
280	1.05×10^{-4}
300	-0.31×10^{-4}

What are the concentrations of DNA and RNA in the mixture?

d. If the DNA and RNA are hydrolyzed to give the individual nucleotides, will the molar ellipticity increase, decrease, or remain the same?

9.3. A common feature of many DNA binding proteins is the "leucine zipper." It is two similar sequences of 30–35 amino acids containing multiple leucines. Each sequence forms a right-handed α-helix, and the two helices wrap around each other to form a left-handed super helix. A number of studies have been carried out of this "coil–coil" structure with model peptides. In one such study, the circular dichroism spectum at 0°C had a large positive peak at 195 nm and negative minima at 208 and 222 nm. This was interpreted as being the spectrum of a completely helical structure.

a. At 0°C, the molar ellipticity at 222 nm is $-33{,}000$ deg cm^2 dmol^{-1}. At 80°C, the molar ellipticity at 222 nm is essentially zero. Interpret this result.

b. At 55°C, the molar ellipticity at 222 nm is $-16{,}000$ deg cm^2 dmol^{-1}. What percentage of the peptide is α-helical at this temperature?

c. For some model systems, it is found that the temperature at which the ellipticity approaches zero increases as the concentration of model peptide increases. How would you explain this result?

9.4. Estrogen receptors are ligand-activated transcription factors that mediate the effects of female sex hormone on DNA transcription. The interaction of estrogen receptors with a specific DNA fragment that binds the receptor and estradiol has been studied with circular dichroism [N. Greenfield, V. Vijayanathan, T. J. Thomas, M. A. Gallo, and T. Thomas, *Biochemistry* **40**, 6646 (2001)]. The circular dichroism spectrum was analyzed to show that the receptor is approximately 75% α-helical, 3% β-sheet, 10% turns, and 12% random coil. The following data were obtained for the ellipticity at 222 nm in the presence and absence of the DNA and estradiol.

Ligand	$[\theta]_{222}$, deg cm^2 dmol^{-1}
None	$-25{,}000$
5 μM DNA	$-32{,}000$
5 μM Estradiol	$-22{,}000$

a. Assume that the specific rotation at 222 nm is due only to the α-helix and that the receptor is saturated with the ligand. Indicate what is happening to the structure of the receptor and calculate the percentage of helix when the ligand is bound.

b. When the temperature is raised, the ellipticity at 222 nm approaches zero. Explain what is happening to the structure of the receptor.

c. The temperature at which the difference between the initial and final ellipticities has reached half of its value, T_m, is 38.0°C in the absence of ligands. It is 43.6°C and 46.1°C in the presence of 5 μM estradiol and 5 μM of the specific binding DNA, respectively. Explain these results.

d. In the presence of 750 nM estradiol, T_m is 40.8°C, Explain this result and estimate the dissociation constant for the binding of estradiol to its receptor protein with the assumption that the concentration of estradiol is much greater than the receptor concentration. At 40.8°C, the ratio of denatured protein to native protein was determined to be 1.20 in the absence of ligands. (This is a hypothetical result as this result was not reported in the publication.)

9.5. Adenosine monophosphate has an extinction coefficient of about 15,000 cm^{-1} M^{-1} at 260 nm, 0°C. The value of $\varepsilon_L - \varepsilon_R$ at 260 nm is about 2.00 cm^{-1} M^{-1}. The corresponding value of $\varepsilon_L - \varepsilon_R$ for a polymer of polyrA is −17.0 cm^{-1} M^{-1} at 0°C, and its magnitude decreases greatly as the temperature is raised. (The monomer concentration was used in calculating this number.)

a. Explain the reason for the large difference in the differential extinction coefficient between AMP and polyrA.

b. Why does the magnitude of the differential extinction coefficient decrease as the temperature is increased?

c. Calculate the observed ellipticity in degrees for a 10^{-4} M solution of AMP and for a 10^{-4} M solution of polyrAMP (monomer concentration) at 260 nm, 0°C.

d. Calculate the absorbance of a 10^{-4} M solution of AMP. Would the absorbance of the polyrA at the same concentration (0°C) be greater than, less than, or the same as the ATP solution?

9.6. The system for transport of mannitol across the *Escherichia coli* membrane involves a membrane-bound enzyme that couples phosphorylation of the sugar to its translocation through the membrane. Fluorescence lifetime and anisotropy measurements have been carried out of the tryptophans in this enzyme dissolved in detergent micelles. [D. Dijkstra, J. Broos, A. J. W. G. Visser, A. van Hoek, and G. T. Robillard, *Biochemistry* **36**, 4860 (1997)]. The fluorescence anisotropy decay of one of these tryptophans has two components. The faster component, with a rotational correlation time of about 1 ns, is due to local motion of the tryptophan. The decay of anisotropy for the slower component can be approximated by the following data.

Anisotropy	Time (ns)
0.290	0.00
0.245	5.00
0.208	10.00
0.149	20.00
0.126	25.00
0.106	30.00
0.090	35.00

a. Calculate the rotational correlation time.

b. The expected correlation time for the known molecular weight of the macromolecule is 120–140 ns. How do you explain the difference between the calculated and observed correlation times?

c. When the protein is phosphorylated, the rotational correlation time increases to 51 ns. What might be the cause of this change?

CHAPTER 10

Vibrations in Macromolecules

10.1 INTRODUCTION

Thus far, we have considered only spectral phenomena in the visible and ultraviolet regions of the spectrum. In this spectral region, the interaction between light and molecules causes transitions between energy levels of the electrons and alters the populations of electronic energy levels. Molecules also have characteristic vibrational motions that are influenced by much longer wavelengths, typically in the far red, or infrared. Since the energy associated with a given wavelength, λ, is hc/λ, this means that the difference in energy between levels is much smaller than for the electronic energy scaffold. (Recall that h is Planck's constant and c is the speed of light.) For macromolecules, the number of different vibrations is approximately $3N$, where N is the number of atoms in the molecule. For small molecules, the number of translational and rotational degrees of freedom must be subtracted from this number, but this is a small correction for macromolecules.

In principle, some type of coupling might be expected when transitions occur between electronic and vibration energy levels. However, this is not the case because electronic transitions occur much more rapidly than the time scale of nuclear motions—nuclei are quite sluggish compared to electrons (10^{-16} s versus 10^{-13} s). Thus, the nuclei can be assumed to be stationary during an electronic transition. This is called the Franck–Condon principle. The Franck–Condon principle is why electronic transitions between vibrational energy levels of different electronic energy states can be drawn as straight lines (Fig. 8-11).

A detailed analysis of all of the vibrational modes of freedom of a macromolecule is not possible, but a few general conclusions bear mention. First, the characteristic vibration of each degree of freedom is a combination of the motions of many different bonds. These characteristic motions are called *normal modes* of vibration. The nature of these normal mode vibrations can be calculated very precisely for small molecules, and the characteristics of some of the normal modes can be associated with similar normal modes in macromolecules. Second, in some limiting cases these normal modes are dominated by the movement of a single or a restricted number of chemical bonds. For example, the vibrations of C—H bonds, double bonds between C, and double bonds between C and O are each associated with similar spectral characteristics in many molecules. Third, the energy level distribution for each normal mode can be approximated by the relationship

$$E = \left(\mathrm{v} + \frac{1}{2}\right)h\nu \tag{10-1}$$

where E is the energy and v is the vibrational quantum number which, is 0 or a positive integer. This relationship is derived directly from quantum mechanics assuming that the vibration is characterized as a simple harmonic oscillator, essentially a spring moving back and forth. Note that in the lowest energy state, $\mathrm{v} = 0$, the normal mode still has an intrinsic energy, $(\frac{1}{2})h\nu$. This is called the zero-point energy and is possessed by all molecules, even at the hypothetical temperature of absolute zero. The zero-point energy is a manifestation of the uncertainty principle (Eq. 6-4): the energy cannot be zero because this implies that the positions would be known exactly. The potential energy, U, for a harmonic oscillator is shown in Figure 10-1, together with the energy levels inside the potential well. For a simple harmonic oscillator,

$$U = \left(\frac{1}{2}\right)kx^2 \tag{10-2}$$

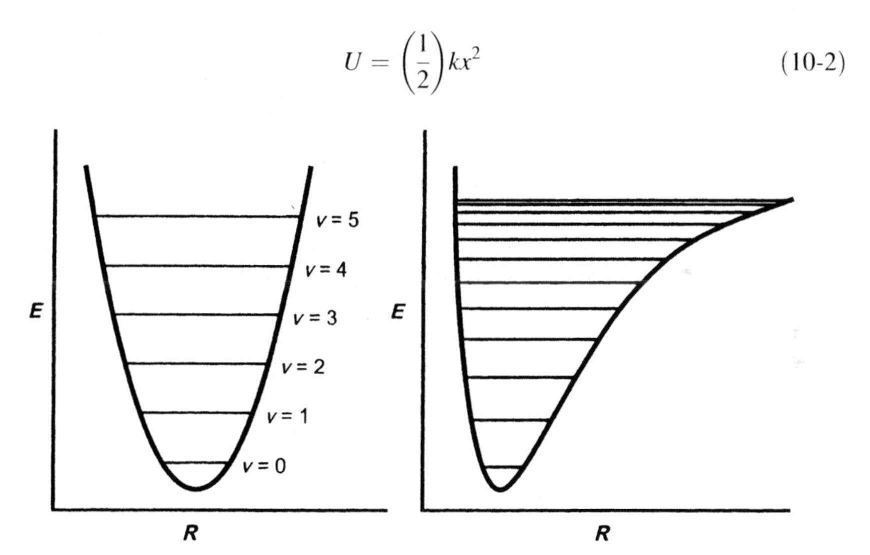

FIGURE 10-1. Schematic diagrams of the potential energy function for the interaction of two atoms in a diatomic molecule. A harmonic oscillator potential is shown on the left and a more realistic intermolecular potential on the right. R is the distance between the atoms, and the energy levels are shown within the wells. For a potential energy function obeying Eq. 10-2, the abscissa for the potential on the left is the x coordinate with $x = 0$ at the minimum in the curve. The actual number of energy levels is much greater than shown, but the even spacing of the energy levels in the harmonic oscillator potential is apparent, as is the decreasing space between energy levels as v increases for the more realistic potential energy function. When the energy reaches the top of the well, the diatomic molecule dissociates into atoms.

In this equation, k is a constant and x is the vibrational coordinate. If the oscillator is thought of as a one-dimensional spring, the x coordinate is the coordinate that the spring moves back and forth along, and k is a measure of how stiff the spring is. For a single bond, the x coordinate is the distance along the bond, and the origin of the coordinate system is the equilibrium bond distance. The characteristic frequency of the bond motion is

$$\nu = (1/2\pi)(k/\mu)^{1/2} \tag{10-3}$$

where μ is the reduced mass of the system. For a complex normal mode, the reduced mass is a weighted average of the masses: for our purposes, an exact calculation of the reduced mass is not necessary. For a C—H bond, the reduced mass can be approximated as the mass of the hydrogen atom. A useful tool for identifying the nature of a given vibrational mode is to substitute deuterium for hydrogen. According to Eq. 10-3, the ratio of the characteristic frequency of the vibration is

$$\nu_{\mathrm{H}}/\nu_{\mathrm{D}} = (m_{\mathrm{D}}/m_{\mathrm{H}})^{1/2} = \sqrt{2} \tag{10-4}$$

This substitution of deuterium for hydrogen is very useful for determining if the motion of a hydrogen atom is the dominant factor in a given vibrational mode.

For real molecules, the vibrational motion deviates from this simple harmonic model at very high energies, and the energy levels become more closely spaced at the top of the potential well. The approximate potential energy function for a vibrational coordinate of a real molecule (anharmonic oscillator) is included in Figure 10-1. For a diatomic molecule, the abscissa is the internuclear distance, whereas for larger molecules it is a combination of internuclear distances associated with the normal mode vibration. Quantum mechanical calculations can be carried out with anharmonic potential energy functions to provide direct correlation of theory with experimental findings. Note that near the bottom of the potential well, harmonic and anharmonic oscillators behave quite similarily so that the harmonic oscillator is a good model at room temperature.

10.2 INFRARED SPECTROSCOPY

The difference in energy between different vibration energy levels is about 100 J/mol. This corresponds to light in the infrared region, $\lambda \sim 1$ mm. Wave numbers, $\bar{\upsilon}$, are usually used when discussing vibration spectra, rather than wavelength. The wave number is simply the reciprocal of the wavelength. Quantum mechanical calculations indicate that there are selection rules that govern the transitions between vibrational energy levels. If absorption of light is to occur, the vibration must cause a change in the electric dipole moment. Recall that a dipole moment is simply a measure of the balance of charges within a molecule. For example, HCl has a permanent dipole moment because H has a partial positive charge and Cl has a partial negative charge. On the other hand, H_2 does not have a permanent dipole moment because no net imbalance in charge is present. From a practical standpoint, it is important to note that water has a dipole moment and absorbs light in the infrared very strongly. Consequently, it is very difficult to measure the infrared absorption of macromolecules in water as the absorption–emission of the molecule is significantly obscured by that of water. Sometimes, a combination of H_2O and D_2O can be used to partially circumvent this problem, but it severely restricts the use of infrared spectroscopy for biological systems. In some cases, dried and/or hydrated films are used, although this is far from the conditions in the milieu of biology.

Two experimental methods are available for measuring infrared spectra. One is essentially the same as the visible ultraviolet spectrometer previously discussed. Infrared light is passed through the sample, and the absorption is measured. The only difference is the light source, typically a glowing wire, and the detector, typically a thermocouple. This method has been largely supplanted by Fourier transfer methods. These methods were briefly described in Chapter 6. Basically with Fourier transform infrared measurements, a beam of light is split in two, with only half of the light going through the sample. The difference in phase of the two waves creates constructive and/or destructive interference and is a measure of the sample absorption. The waves are rapidly scanned over a specific wavelength region of the spectra, and multiple scans are averaged to create the final spectrum. This method is more sensitive than the conventional dispersion spectrometer.

10.3 RAMAN SPECTROSCOPY

Another method exists for studying transitions between vibration energy levels that uses a concept not yet discussed. In addition to absorbing light, samples also scatter light. The amount of scattered light is a maximum 90° to the direction of the incident beam. Most of the scattered light is at the same frequency as that of the incident light. This is called Rayleigh scattering. At the molecular level, the electric field of the light perturbs the electron distribution, but no transitions between energy levels occur so that the molecule immediately returns to its unperturbed state. This scattering is inversely proportional to the fourth power of the wavelength so the scattering is much greater at shorter wavelengths. This is essentially why the sky is blue, the shorter wavelengths of the visible spectrum (blue) are scattered more than the longer wavelengths (red). Rayleigh scattering is observed at all wavelengths. The intensity of the scattered light is related to the *polarizability* of the molecule, that is, to how easily electrical charges can be shifted within the molecule to make it more polar. A small number of the molecules return to a different vibrational energy level after scattering. The vibrational energy level can be either higher or lower than the initial state. As a result of this change in energy level, some of the scattered light will be at a slightly lower or higher frequency than the incident light. This is called *Raman* scattering, after the Indian scientist who discovered the phenomena.

A typical setup for measuring Raman scattering is shown in Figure 10-2. A very intense light source is needed to observe Raman scattering because only a very small amount of the scattered light displays a change in frequency. The advent of lasers has permitted this to be done routinely with visible light and in the

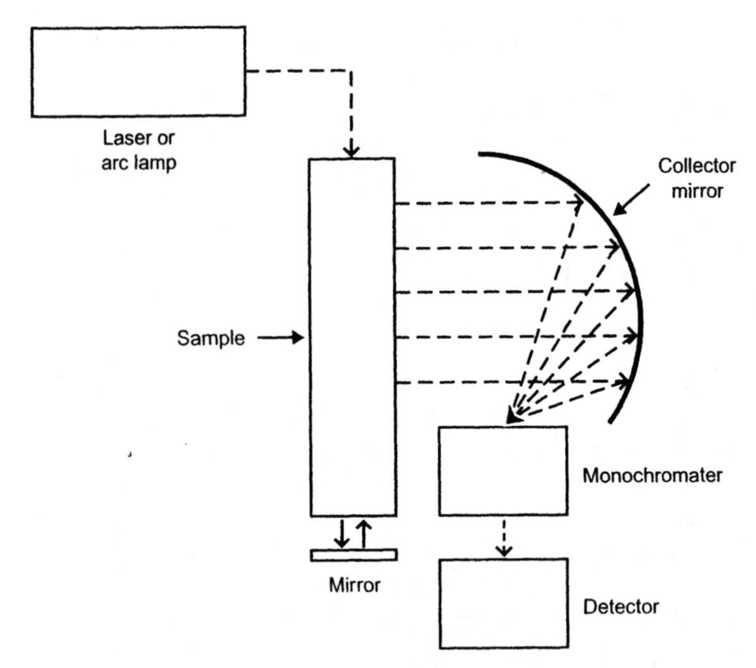

FIGURE 10-2. Schematic diagram of an apparatus used for Raman spectroscopy. The mirror at the end of the sample cell is to put the light through the sample a second time, effectively increasing the path length of the cell, and the collector mirror is designed to collect as much of the scattered light 90° to the source light beam as possible.

ultraviolet with rather expensive lasers. Prior to lasers, very large arc lamps were utilized. In addition, very high concentrations are required. However, Raman scattering does not require the light to be at a wavelength comparable to the energy of vibrational transitions since it is a *scattering* phenomenon, as contrasted to absorption spectroscopy. The Rayleigh line is very intense, but much less intense scattering can be detected at a lower frequency than the incident light. This is because the incident light was used to promote the molecule to a higher vibrational energy level. These are called Stokes lines. An even smaller fraction of the scattered light occurs at a higher frequency than the incident light because energy is added to the incident light by the movement of the molecule to a lower vibrational energy level. These are called anti-Stokes lines. Most molecules are found in their ground vibrational energy level at room temperature so that the observation of anti-Stokes lines is rare.

Raman spectroscopy has two major advantages over infrared spectroscopy for studying transitions between vibrational energy levels. First, a permanent dipole moment is not required. It is only necessary for the polarizability of the molecule to change between different vibrational energy levels. Second, visible light can be used rather than infrared light so that Raman spectra can be readily obtained in water. They can also be obtained in crystals and films. As previously indicated, the primary disadvantages are that because the intensity of the Raman lines is very weak, intense light sources and high concentrations of the molecule of interest are needed. Raman and infrared spectroscopy should be regarded as complimentary. Since infrared spectroscopy is dependent on the permanent dipole moment and Raman spectroscopy on the polarizability, usually (but not always) a vibrational transition is observed either in the infrared or in Raman scattering, but not in both.

Thus far in our discussion, the implicit assumption is that the wavelength of the exciting light for Raman spectroscopy is not near an absorption band. If the wavelength coincides with the absorption band for an electronic transition, a large increase occurs in the intensity of the Raman spectrum. Basically, this is related to the fact that an electronic absorption band is the superposition of many different vibrational modes. This can sometimes be seen directly in the electronic absorption spectrum by fine structure in the peaks. Essentially, the alteration of the population of vibrational energy levels within the electronic absorption band is responsible for the enhanced intensity of the Raman spectrum. Determining the spectrum within the electronic absorption band is called *resonance Raman spectroscopy*. The twofold advantage of resonance Raman spectroscopy is that lower concentrations can be used and the spectrum is simplified because only the intensified lines are observed. At the present time, resonance Raman measurements are made primarily in the visible region of the spectrum. In biological systems, this means chromophores such as hemes and retinal must be present.

10.4 STRUCTURE DETERMINATION WITH VIBRATIONAL SPECTROSCOPY

Infrared spectroscopy has been a useful tool for the determination of the structure of organic molecules for many years. This is because specific types of bonds and/or chemical groups have characteristic vibrational frequencies. Some of these group frequencies are given in Table 10.1. Infrared spectra are often referred to as "fingerprints" for the molecule and large compilations of data are available. In the case of biological molecules, changes in the group frequencies can be used to derive information about the secondary structure of the molecules (1).

The carbonyl of the amide bond in proteins is particularly useful for the determination of secondary structure (2,3). The stretching normal mode, amide I mode, of the carbonyl has been shown to have a specific frequency associated with α-helices, β-sheets, and other characteristic structures. (Strictly speaking, this normal mode is not just the stretching of the carbonyl. It also involves some bending of the C–N–H angle.) This was ascertained by the study of model peptides for which precise measurements and theoretical calculation of normal modes could be carried out. The approximate wave numbers corresponding to the three common structures found in proteins are as follows: α-helix, $1650\,\mathrm{cm}^{-1}$; β-sheet, 1632 and $1685\,\mathrm{cm}^{-1}$; and random coil, $1658\,\mathrm{cm}^{-1}$.

For proteins, a more empirical approach has been adopted. The most successful approach uses known structures to calibrate the vibrational frequency measurements. The vibrational spectrum of the amide bond for a protein is complex because

TABLE 10-1. Group Frequencies in the Infrared Region[a]

Chemical group	Frequency (cm^{-1})
$-CH_3$	1460
$-CH_2-$	2930
	2860
	1470
C–H	3300
–C–C–	1165
–C=O	1730
–C–H (in CH_3)	2960
	2870
–C–H (in CHO)	2870
	2720
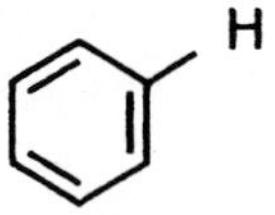 H	3060
–CN	2250
–O–O–	1200–1100
–OH	3600
$-NH_2$	3400
$=CH_2$	3030
–SH	2580
–C=N–	1600
C–Cl	725
C=S	1100

[a]Reproduced with permission from D. Sheehan, *Physical Biochemistry*, John Wiley and Sons, 2000, p. 98.

of the many amide bonds present in multiple environments. Nevertheless, the spectrum can be deconvoluted to provide information about the amount and types of secondary structures present. The assumption usually made is that the observed vibrational frequency is a linear combination of the frequencies associated with the various secondary structures that are present, with each specific frequency weighted by the percent of a given structure. Some typical results are shown in Table 10.2 and compared with the known structure of the protein and CD estimates of secondary structure (4). Raman spectroscopy has proven particularly useful for studying secondary structure of proteins since water solutions can be used, and the amide I mode has a spectral band that is well isolated from other protein bands (1630–1700 cm^{-1}). The amide I band is the most common vibrational frequency used as an indicator of secondary structure (cf. Ref. 5), but other bands and their relation to structure have been identified (3).

The structure of nucleic acids can also be investigated with vibrational spectroscopy. The base vibrations and the groups involved in hydrogen bonding are

TABLE 10-2. Protein Secondary Structure Determined by Infrared (IR) and Circular Dichroism (CD) Spectra and X-Ray Crystallography[a]

	Secondary structure (%)				
Protein	α-Helix	β-Sheet	Turn	Random	Method
Hemoglobin	78	12	10	[b]	IR
	87	0	7	6	X-ray
	68–75	1–4	15–20	9–16	CD
Myoglobin	85	7	8	[b]	IR
	85	0	8	7	X-ray
	67–86	0–13	0–6	11–30	CD
Lysozyme	40	19	27	14	IR
	45	19	23	13	X-ray
	29–45	11–39	8–26	8–60	CD
Cytochrome *c* (oxidized)	42	21	25	12	IR
	48	10	17	25	X-ray
	27–46	0–9	15–28	28–41	CD
α-Chymotrypsin	9	47	30	14	IR
	8	50	27	15	X-ray
	8–15	10–53	2–22	38–70	CD
Trypsin	9	44	38	9	IR
	9	56	24	11	X-ray
Ribonuclease A	15	40	36	9	IR
	23	46	21	10	X-ray
	12–30	21–44	11–22	19–50	CD
Alcohol dehydrogenase	18	45	23	14	IR
	29	40	19	12	X-ray
Concanavalin A	8	58	26	8	IR
	3	60	22	15	X-ray
	3–25	41–49	15–27	9–36	CD
Immunoglobin G	3	64	28	5	IR
	3	67	18	12	X-ray
Major histocompatability complex antigen A2	17	41	28	14	IR
	20	42			X-ray
	8–13	74–77			CD
β_2-Macroglobulin	6	52	33	9	IR
	0	48			X-ray
	0	59			CD

[a]Reproduced with permission from A. Dong, P. Huang, and W. S. Caughey, *Biochemistry* **29**, 3303 (1990). © 1990 American Chemical Society.

[b]The band due to random structure appears as a shoulder on the α-helix band and is too small to be separated from α-helix structure. The random structure is estimated at <5% and is included in the α-helix value.

particularly sensitive to the secondary structure of nucleic acids. Most of the work with nucleic acids has involved Raman spectroscopy (3). For example, the melting of DNA and RNA structures can be readily followed, and the specific vibrational frequencies provide molecular details about the melting process. The different types of helices formed by DNA can be distinguished. Again, standards are established with known structures and then used to determine the structures of unknown samples, that is, the amount of various types of structures present.

10.5 RESONANCE RAMAN SPECTROSCOPY

Heme proteins have been extensively studied with resonance Raman spectroscopy (6). They are very prevalent in nature and have a very intense absorbance in the visible region of the spectrum due to the porphyrin ring structure. Excellent Raman spectra can be obtained at very low concentrations, often in the micromolar range. The vibrational spectra obtained are characteristic of the porphyrin ring structure. However, the highest frequency normal modes, 1350–1650 cm^{-1}, are very sensitive to the state of the Fe atom that is bound to the porphyrin. Thus, these frequencies can establish the spin state and coordination states of the Fe in both of its oxidation states. The binding of ligands to the Fe and the distortion of the porphyrin skeleton can also be detected. As an example, the high-frequency region of the spectrum for deoxy- and CO-myoglobin is shown in Figure 10-3 (6). Myoglobin is a protein that is used for oxygen transport in some organisms, a function carried out by hemoglobin in humans. The frequency

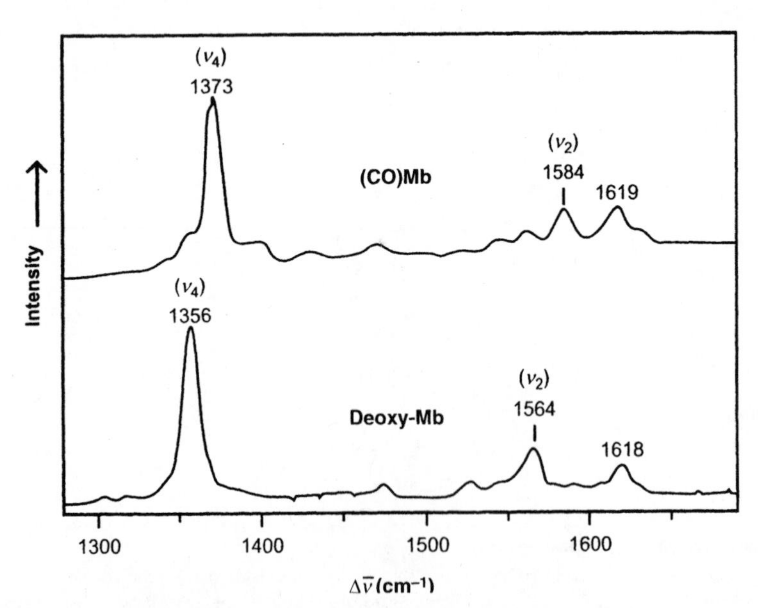

FIGURE 10-3. Comparison between resonance Raman spectra for carbonmonoxymyglobin [(CO)Mb] and deoxymyoglobin at room temperature. The excitation was in the wavelength region corresponding to the heme absorption. Note that the two species can be readily distinguished by their Raman spectra. Reprinted from T. Spiro and R. S. Czernuszewicz, Resonance Raman spectroscopy of metalloproteins, *Methods Enzymol.* **246**, 416 (1995). © 1995, with permission from Elsevier.

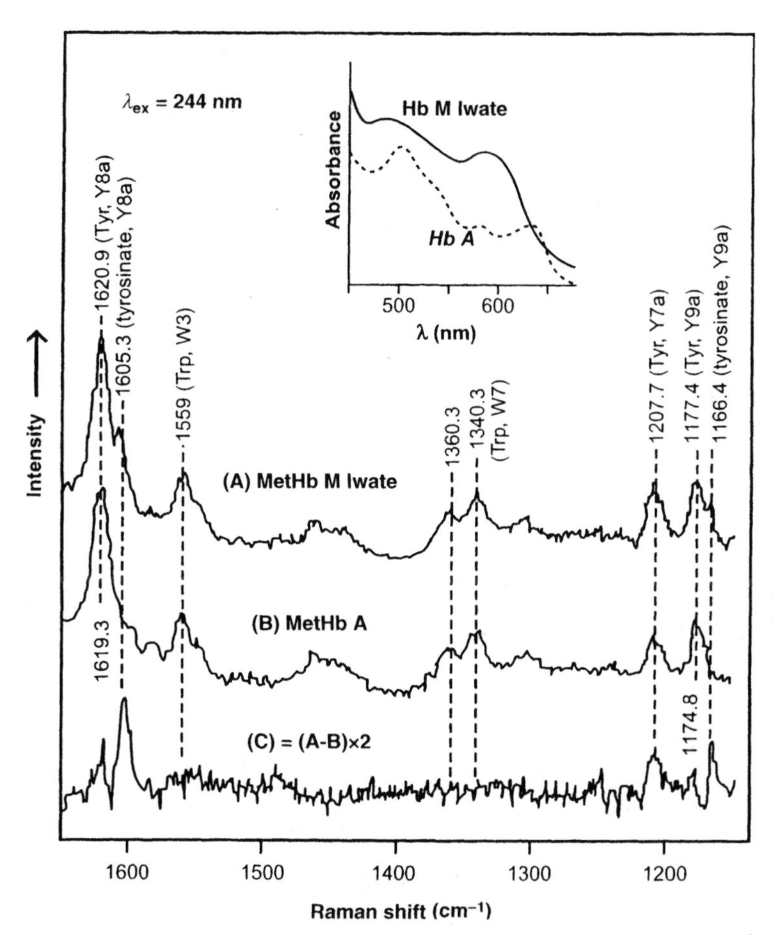

FIGURE 10.4. Resonance Raman spectra of two different hemoglobin (Hb) variants with excitation at 244 nm. The spectra of metHb M Iwate (A), metHb (B), and their difference, expanded two-fold (C), are shown. The inset shows the absorption spectra of the two hemoglobins in the visible region. Reprinted in part with permission from M. Nagai, M. Aki, R. Li., Y. Jin, H. Sakai, S. Nagatomo, and T. Kitagawa, *Biochemistry* **39**, 13093 (2000). © 2000 by American Chemical Society.

shifts in the vibration spectra between these two forms of myoglobin can be readily discerned.

The details of ligand binding in hemoglobin have been frequent targets of Raman spectroscopy (cf. Ref. 7). As an example, consider the study of a naturally occurring mutant labeled hemoglobin M Iwate (8). This mutant has a tyrosine substituted for a histidine in the α chain. The resonance Raman spectrum of the tyrosine was determined by the scattering of 244 nm radiation, and the resonance Raman of the heme was determined using scattering at 406 and 441 nm. The ultraviolet Raman spectra for normal hemoglobin and hemoglobin M Iwate are shown in Figure 10-4, along with the difference spectrum. The vibrations seen in the difference spectrum are attributed to the extra tyrosine in the mutant. By comparison with known compounds, it was shown that the tyrosine is deprotonated in the mutant and is bound to the Fe(III) heme in the α-subunit.

The effect of this binding on the overall heme structure was determined from the resonance Raman spectra in the visible range. Furthermore, reduction of the Fe(III) to Fe(II) resulted in the elimination of the tyrosinate vibrations, and it could be shown that a histidine residue was coordinated to the Fe(II) heme. This study is a good example of how vibrational spectroscopy can be used to elucidate the detailed structure of ligand–protein interactions.

10.6 STRUCTURE OF ENZYME–SUBSTRATE COMPLEXES

One of the advantages of vibrational spectroscopy is that subtle changes in the electronic environment can be related to structural changes that may be difficult to observe by more direct methods, such as X-ray crystallography. This feature has been used to elucidate the catalytic process for enzymatic reactions (cf. Ref. 9). As an example, consider the enzyme dihydrofolate reductase (DHFR), which has been discussed earlier (Chapter 8). The reaction catalyzed converts 7,8-dihydrofolate (H_2folate) to 5,6,7,8-tetrahydrofolate by catalyzing the transfer of a hydride ion from NADPH to C6 and addition of a solvent proton to N5 (Fig. 10-5).

An important mechanistic question centers on the protonation state of N5 of dihydrofolate. Is the nitrogen protonated prior to the hydride transfer or is the proton added after the hydride transfer occurs? Raman spectroscopy showed that a N5–C6 stretching vibration changed upon protonation of N5 from $1650\,cm^{-1}$ to $1675\,cm^{-1}$ (10). This was established by the use of model compounds, isotope effects on the vibration spectrum, and theoretical calculations. On the basis of these extensive studies of model systems, N5 was found to be unprotonated in the DHFR–H_2folate complex at neutral pH, but it was protonated int the DHFR–H_2folate–$NADP^+$ complex. A pH titration indicated that the pK of N5 is 6.5 in this complex, as compared to 2.6 when the substrate is not bound to the enzyme. In the actual reaction, NADPH is present rather than $NADP^+$ so that

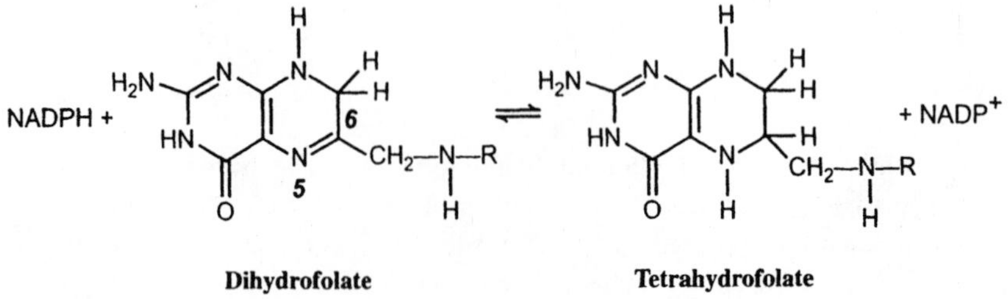

Dihydrofolate **Tetrahydrofolate**

FIGURE 10-5. The reaction catalyzed by the enzyme dihydrofolate reductase. The NADPH reduces dihydrofolate to tetrahydrofolate with production of $NADP^+$. Note that a proton, obtained from the solvent, is needed on the left-hand side of the equation to give a balanced chemical reaction.

the implicit assumption is that the complex studied is a good model for the catalytic reaction. Thus, N5 can be protonated much more readily in the enzyme–$NADP^+$ complex.

From the point of view of understanding enzyme catalysis, this result means that the population of the N5 protonated substrate is four orders of magnitude larger in the environment of the enzyme. The positive charge on the N5 would also make C6 more positive and would presumably make the hydride transfer reaction much faster than with unprotonated N5. Much of the catalytic effect of the enzyme, therefore, apparently is due to providing an environment that stabilizes the protonated substrate prior to the hydride transfer. A structural explanation for this stabilization is not obvious. The N5 is in a hydrophobic region of the protein, and no negative charges are conveniently close that might stabilize the protonated species. However, conformational changes of the protein have been established as part of the overall mechanism of action of the enzyme, and these changes may be the source of the stabilization. In any event, Raman spectroscopy has provided unique insight into the catalytic process.

Although vibration spectroscopy has not been used as extensively in biological systems as ultraviolet–visible absorption and fluorescence spectroscopy, it can sometimes provide unique and important information.

REFERENCES

1. K. Sauer, (ed.), *Methods Enzymol.* **246** (1995).
2. F. Siebert, *Methods Enzymol.* **246**, 501 (1995).
3. W. L. Peticolas, *Methods Enzymol.* **246**, 389 (1995).
4. A. Dong, P. Huang, and W. S. Caughey, *Biochemistry* **29**, 3303 (1990).
5. W. K. Surewicz, H. H. Mantsch, and D. Chapman, *Biochemistry* **32**, 389 (1993).
6. T. Spiro and R. S. Czernuszewicz, *Methods Enzymol.* **246**, 416 (1995).
7. D. L. Rousseau and J. M. Friedman, in *Biological Application of Raman Spectroscopy*, Vol. 3 (T. G. Spiro, ed.), Wiley, New York, 1988 p. 133.
8. M. Nagai, M. Aki, R. Li, Y. Jin, H. Sakai, S. Nagatomo, and T. Kitagawa, *Biochemistry* **39**, 13093 (2000).
9. H. Deng and R. Callender, in *Infrared and Raman Spectroscopy of Biological Materials* (H.-U. Gremlich and B. Yan, eds.), Marcel Dekker, Inc., New York, 2001, pp. 477–514.
10. Y.-Q. Chen, J. Kraut, R. E. Blakley, and R. Callender, *Biochemistry* **33**, 7021 (1994).

PROBLEMS

10.1. The zero-point energies for two different vibrational manifolds are 9.55×10^3 and 1.19×10^4 J/mol.

a. Calculate the frequency of light emitted for a transition between the first energy level and the zero-point energy for each of these manifolds.

b. In terms of the harmonic oscillator model, which of these manifolds has the "stiffer" spring?

c. For both cases, the radiation associated with a transition between energy levels 20 and 19 occurs at a longer wavelength (smaller wave number) than the radiation associated with a transition between the energy levels 2 and 1. How do you explain this?

10.2. From infrared studies of model proteins, the wave numbers for the amide II band (due to N-H deformation in the peptide bond) are 1540–1550 cm^{-1} for the α-helix, 1520–1525 cm^{-1} for β-sheet, and $<1520\ cm^{-1}$ for "random coils."

a. Calculate the zero-point energy for these three vibration energy levels.

b. Explain why the three fundamental wave numbers differ in terms of the protein structure.

c. For polyglutamic acid, the amide II band is at a wave number of about 1545 cm^{-1} at low pH (pH 4). As the pH increases (>pH 9), the wave number decreases to below 1520 cm^{-1}. Explain this result.

CHAPTER 11

Principles of Nuclear Magnetic Resonance and Electron Spin Resonance

11.1 INTRODUCTION

In this chapter, we consider the interaction of molecules with radiation when molecules are placed into a strong magnetic field. The fundamental properties of atoms that are important for this discussion are the nuclear spin, for nuclear magnetic resonance (NMR), and the electron spin, for electron spin resonance (ESR) or, equivalently, electron paramagnetic resonance (EPR). Strictly speaking, the concept of spin can be rigorously defined only by the use of quantum mechanics. However, we will use a semiclassical approach in which spin in the nucleus or in an electron can be represented as a charge moving in a circular path. This movement creates a magnetic dipole that can be thought of as a bar magnet. In the absence of a magnetic field, the magnetic dipole is oriented randomly, and only one energy level is associated with the electron or nucleus. But in the presence of a magnetic field, the magnetic dipoles (or bar magnets) tend to be oriented either in the direction of the field or opposed to it, thus creating multiple energy states. Application of quantum mechanics to this situation indicates that the orientation of the magnetic dipole and the energy states are quantized with characteristic quantum numbers.

For electrons, the spin quantum number, S, is 1/2, and the two spin states are $+1/2$ and $-1/2$, represented as the familiar arrows pointed up or down. Neutrons and protons also have a spin quantum number of 1/2 so that the nucleus of an atom has a characteristic spin quantum number, I. Simple rules exist for determining the nuclear spin quantum number. Nuclei with an even mass number and even charge number have no nuclear spin ($I = 0$). Nuclei with an odd mass number have a half integral spin ($I = 1/2, 3/2, 5/2$, etc.). Finally, nuclei with an even mass number and odd charge number have integral spin ($I = 1, 2$, etc.).

The number of energy levels associated with spin is determined by the spin quantum number. For an electron, the number of energy levels is $2S + 1$, or 2. In

Physical Chemistry for the Biological Sciences by Gordon G. Hammes

the absence of a magnetic field, no distinction can be made between these two quantum states: They have the same energy, that is, they are degenerate. In the presence of a magnetic field, however, the alignment of the magnetic moments with and against the magnetic field creates two distinct energy levels. The hydrogen nucleus also has a spin quantum number, I, of 1/2, with two possible orientations of its magnetic dipole (or magnetic moment) in a magnetic field and two energy levels ($2I + 1$). On the other hand, ^{23}Na has $I = 3/2$ and four energy levels in the presence of a magnetic field. In general, the quantum states in the presence of a magnetic field are characterized by quantum numbers ranging from I to $-I$ in integral steps. Thus, for sodium, these quantum numbers are 3/2, 1/2, −1/2, and −3/2. Although the average orientation of the nuclear spin state of a proton is either aligned with the magnetic field or against it, the conservation of angular momentum requires the magnetic moment associated with the orientation to rotate about the direction of the field. This is analogous to a top spinning on its axis and rotating around a vertical line due to gravity.

A quantum mechanical treatment of nuclear spins in a magnetic field of a strength H provides an explicit equation for the energy levels, E:

$$E = -g_N\beta_N HM_I = -(h/2\pi)\gamma Hm_I \tag{11-1}$$

In this equation, β_N is the nuclear magneton and is a universal constant calculated from the properties of nuclei: $\beta_N = 5.051 \times 10^{-27}$ Joules/Tesla. The nuclear g factor, g_N, is a constant, but it is different for each atom, and m_I is the spin quantum number characterizing the orientation of the magnetic moment in the magnetic field (I to $-I$ in integral steps). This equation also defines the gyromagnetic ratio, γ, which is a frequently used constant. Values of the nuclear spin quantum mumber, I, and γ for some nuclei of biological interest are presented in Table 11.1.

The dependence of the energy on magnetic field for atoms with a total nuclear spin quantum number of 1/2 is shown in Figure 11-1. The difference in energy between the quantized levels increases as the magnetic field increases. Also shown is the precession of the magnetic moment about the field direction for the two possible orientations.

TABLE 11-1. Magnetic Properties of Selected Nuclei

Isotope	Spin	$10^7\gamma$ (T^{-1} s^{-1})	Natural abundance
^1H	1/2	26.75	99.98
^2H(D)	1	4.11	0.0156
^{13}C	1/2	6.73	1.108
^{14}N	1	1.93	99.63
^{15}N	1/2	−2.75	0.37
^{19}F	1/2	25.18	100.0
^{31}P	1/2	10.84	100.0
^{17}O	5/2	−3.63	0.037

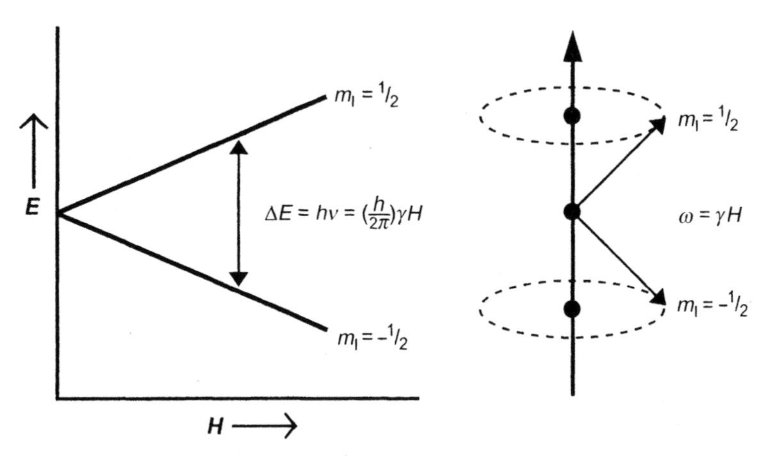

FIGURE 11-1. On the left, a schematic plot of the energy versus the magnetic field is shown for a nuclear or electron spin with a quantum number of 1/2. The frequency of the radiation emitted/absorbed for a transition between the two energy states created by the magnetic field is dependent on the strength of the magnetic field (Eq. 11-2). On the right, the magnetic dipole associated with the nuclear spin is shown precessing around the direction of the magnetic field at an angular frequency ω. The two orientations, up and down, correspond to the two energy levels in the diagram on the left.

The energy difference between the energy levels can be calculated, again with the assistance of quantum mechanics, by the requirement that the change in quantum number for allowed transitions between energy levels is 1. In this case,

$$\Delta E = g_N \beta_N H = (h/2\pi)\gamma H \qquad (11\text{-}2)$$

With this equation we can calculate the frequency, ν, associated with transitions between energy levels since $\Delta E = h\nu$. The magnetic field of modern instruments varies from about 7 to 19 Tesla. For a magnetic field of 11.75 Tesla, the frequency associated with transitions between the energy levels of protons is 500 MHz. This is in the radio frequency range. The actual energy difference between levels is quite small, only 3.4×10^{-25} J/proton. Because the energy difference is small, the actual population difference between energy levels is also very small at room temperature: the ratio of populations in the two states is $\exp(-\Delta E/kT) = 0.99993$ for an 11.75 Tesla magnetic field at 37°C. For this same magnetic field, the resonant frequency for ^{13}C is 130 MHz.

The quantum mechanical calculation of the energy levels for an unpaired electron in a magnetic field is quite similar to that for nuclear spin, except that the electron always has spin 1/2. The energy difference between levels is

$$\Delta E = g_s \beta_s H \qquad (11\text{-}3)$$

where $g_S = 2.0023$ and $\beta_S = 9.274 \times 10^{-24}$ Joules/Tesla for electrons. Because the Bohr magneton is much larger for an electron than a proton, the energy levels are further apart, and the resonance energy frequency is much larger. (β_S and β_N differ by the ratio of the mass of the proton to the mass of the electron, 1836.) For a 1 Tesla field, the resonant frequency is 28,000 MHz = 28 GHz. This frequency is in the microwave region, and quite different experimental techniques are required for ESR and NMR.

Finally, we return to the precession of the magnetic moment about the direction of the magnetic field, as depicted in Figure 11-1. The angular precession frequency, that is, how fast the magnetic moment is rotating about the vertical line, is $\omega = 2\pi\nu = \gamma H$. This is called the *Larmor* frequency. The visualization of rotating magnetic moments is useful when considering the effects of changing magnetic fields on the nuclear spins.

We first discuss NMR in some detail because it is extensively used in biology. Although ESR has been used to obtain important information about biological systems, it is less extensively used and will receive relatively brief consideration.

11.2 NMR SPECTROMETERS

The first NMR spectrometers placed a sample in a fixed magnetic field and applied a fixed radiofrequency by means of a coil perpendicular to the field direction. The magnetic field was varied by a coil until resonance was achieved, with the absorption being detected by a third coil. This is analogous to the methods used in visible and ultraviolet spectroscopy. Although the magnetic field applied to the sample is uniform, the actual magnetic field at the nucleus is dependent on a number of environmental factors so that the field must be scanned to find the energy absorption (resonance) condition. We will return to this matter a bit later. A typical spectrum would display the absorption of energy versus frequency, as shown schematically in Figure 11-2.

The absorption of radiation is difficult to detect because the populations of the energy states are very similar, as discussed above. In practice, this means that relatively high concentrations of the species being observed must be present. The sensitivity of detection depends on the characteristics of the nucleus being observed and its natural abundance. The sensitivity is also enhanced by increasing the energy difference between the two states, which is the reason that instruments with larger magnetic fields are being developed continuously. The proton, ^1H, provides the best sensitivity, and its natural abundance is 99.98%. Consequently, the most extensive measurements have been carried out with proton NMR. Although the natural abundance of ^{31}P is essentially 100%, the sensitivity of detection relative to protons is only about 6%. Probably the second most studied atom with NMR is ^{13}C. Its natural abundance is only about 1%, but its prevalence in biological compounds is very extensive. Furthermore, its abundance in molecules of interest can be enhanced through synthesis or bacterial growth with ^{13}C-enriched compounds. Regrettably, the most abundant isotope of carbon, ^{12}C, does not have a nuclear spin. A variety of other isotopes have been studied with NMR. Most notable for biologists are ^{14}N, ^{15}N, and ^{19}F, which can often be substituted for H in substances of biological interest.

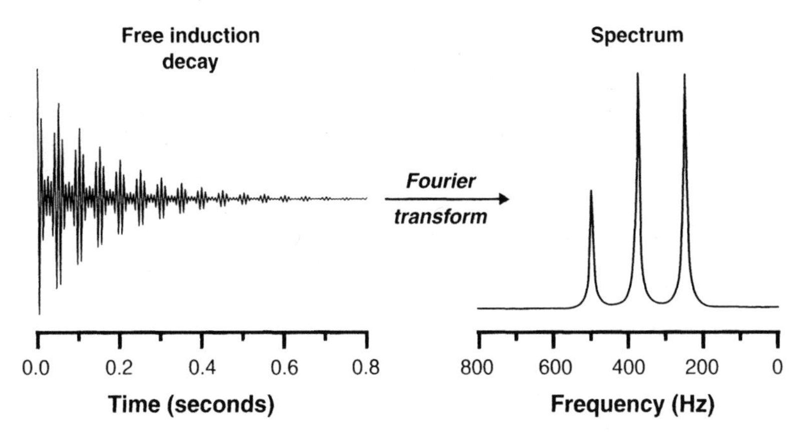

FIGURE 11-2. The free induction decay of the signal from nuclei in a magnetic field is shown (*left*) after a short pulse of radiofreqency radiation is applied to the sample. A Fourier transform of the free induction decay gives the familiar NMR spectrum on the right where the absorption is plotted versus the frequency. Copyright by Professor T. G. Oas, Duke University. Reproduced with permission.

As discussed in Chapter 6, the frequency dependence of the absorption can be transformed into a time dependence and vice versa through Fourier transforms. The most common method of obtaining an NMR spectrum today is to apply a timed radio frequency pulse and then watch the nuclei return to their equilibrium configurations. The time dependence of the return to equilibrium or *free induction decay* (FID) can be transformed from a time-dependent signal into a frequency spectrum, as shown schematically in Figure 11-2. Multiple FIDs can be combined to produce an average FID that has less noise than a single FID. Consequently, Fourier transform instruments are considerably more sensitive than continuous wave instruments so that lower concentrations can be used. The timing and nature of the pulses can be quite complex, but this does not alter the basic concept underlying Fourier transform methods.

11.3 CHEMICAL SHIFTS

Thus far, we have discussed NMR as though the nuclei were isolated in the magnetic field, but the utility of NMR derives from the interaction of the nuclei with surrounding electrons and other nuclei in the molecule. The external magnetic field interacts with the electrons to induce magnetic moments in the electrons that usually oppose the external field. Consequently, the magnetic field at the nucleus is usually lower than the external field. This shielding effect of the electrons can be incorporated into the standard equations by noting that the field at the nucleus, H, in a static magnetic field, H_0, is

$$H = H_0(1 - \sigma) \tag{11-4}$$

where σ is the shielding constant, typically about 10^{-5}. The value of σ can be positive or negative, depending on whether the magnetic field from the electrons aligns against or with the external magnetic field.

The shielding effect is directly proportional to the strength of the external field, but the same relative change in resonance frequency is observed, regardless of the external field strength. This shielding effect, therefore, can be expressed as a relative change in frequency with respect to a standard, thereby rendering it independent of the external field. These frequency changes are called *chemical shifts*, δ, and are given in units of parts per million (ppm). The chemical shift can be written as

$$\delta = \frac{\nu - \nu_{\text{ref}}}{\nu_{\text{ref}}} 10^6 = (\sigma - \sigma_{\text{ref}}) 10^6 \qquad (11\text{-}5)$$

where ν is the frequency of the nucleus, ν_{ref} is the frequency of a standard compound, and σ_{ref} is the shielding constant of the standard compound. The most common reference compound for protons is tetramethylsilane, but it is insoluble in water so trimethylsilylpropionate-d_4 is usually used in aqueous media (cf. Ref. 1 for a discussion of chemical shift references). In principle, the chemical shift depends on the orientation of the sample with respect to the magnetic field. In liquids, this is generally not a problem because molecules are rapidly tumbling and sampling all possible orientations. For large molecules that tumble relatively slowly, however, the resonances can become so broad that the spectrum is obscured, and in solids special conditions are required to obtain high-resolution spectra.

The electron density at the nucleus is often the dominant factor in determining the chemical shift. A high electron density creates a large shielding, and the applied magnetic field must be increased to get resonance: this results in an upfield shift and a decrease in the magnitude of δ because reference compounds are generally highly shielded. Conversely, a low electron density at the nucleus causes a downfield shift and increase in δ. The range of chemical shifts for ^{1}H and ^{13}C in various compounds is shown in Figure 11-3.

In principle, the chemical shifts of nuclei in proteins should provide information about the protein structure. However, chemical shifts alone are not sufficient to determine protein structure. The average chemical shifts of various nuclei for amino acids in denatured proteins are given in Table 11.2. Note that they are quite similar for all of the amino acids. When secondary structures such as α-helices or β-sheets are present, changes in chemical shifts occur so that the amounts of various secondary structures present can be inferred from the NMR spectra.

Some of the largest changes in chemical shifts in proteins and nucleic acids are observed for aromatic rings of nucleotides, tyrosine, phenylalanine, and tryptophan. These shifts are due to the interaction of the external magnetic field with the delocalized electrons of the aromatic ring. Nuclei above or below the ring usually have decreased chemical shifts, whereas those near the edges have increased chemical shifts. These effects are called *ring currents* and can cause unusually large chemical shifts. These chemical shifts can provide information about the structure of

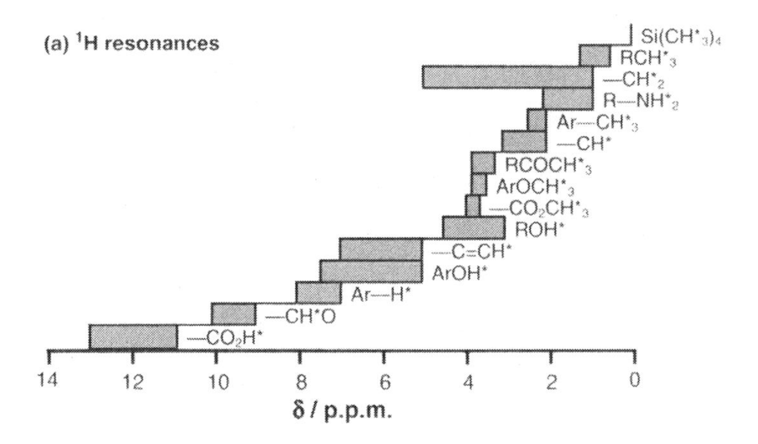

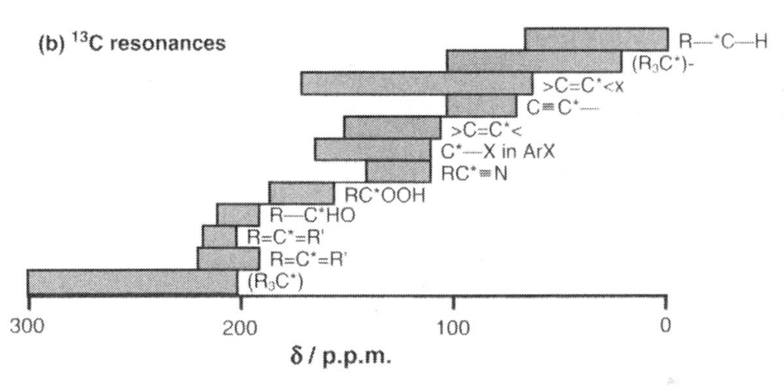

FIGURE 11-3. Range of typical chemical shifts for ^{1}H and ^{13}C resonances. P. W. Atkins, *Physical Chemistry*, 3rd edition, W. H. Freeman, New York, NY, 1986, p. 489. © 1978, 1982, 1986 by Peter W. Atkins. Used with permission of Freeman.

macromolecules and about alterations in structure due to changes in the environment such as temperature and the addition of various chemical agents.

11.4 SPIN–SPIN SPLITTING

The chemical shift is caused by interactions between the nucleus and nearby electrons. A conceptually different interaction is transmitted between nearby nuclei by intervening electrons participating in chemical bonds. Basically, the spin state of a neighboring nucleus alters the shielding a nucleus experiences. This effect is smaller than typical chemical shifts and is called *spin–spin splitting* and is often referred to as scalar coupling. Unlike chemical shifts, the magnitude of spin–spin coupling is independent of the magnitude of the applied magnetic field.

TABLE 11-2. Average Chemical Shifts of Random Coil Amino Acids (ppm)

Amino acid	α-^{1}H	Amide-^{1}H	α-^{13}C	Carbonyl-^{13}C	Amide-^{15}N
Ala	4.33	8.15	52.2	177.6	122.5
Cys	4.54	8.23	56.8	174.6	118.0
Asp	4.71	8.37	53.9	176.8	120.6
Glu	4.33	8.36	56.3	176.6	121.3
Phe	4.63	8.30	57.9	175.9	120.9
Gly	3.96	8.29	45.0	173.6	108.9
His	4.60	8.28	55.5	174.9	119.1
Ile	4.17	8.21	61.2	176.5	123.2
Lys	4.33	8.25	56.4	176.5	121.5
Leu	4.32	8.23	55.0	176.9	121.8
Met	4.48	8.29	55.2	176.3	120.5
Asn	4.74	8.38	52.7	175.6	119.5
Pro	4.42	—	63.0	176.0	128.1
Gln	4.33	8.27	56.0	175.6	120.3
Arg	4.35	8.27	56.0	176.6	120.8
Ser	4.47	8.31	58.1	174.4	116.7
Thr	4.35	8.24	62.0	174.8	114.2
Val	4.12	6.19	62.2	176.0	121.1
Trp	4.66	8.18	57.6	173.6	120.5
Tyr	4.55	8.28	58.0	175.9	122.0

Reproduced with permission from D. S. Wishart and B. D. Sykes, Chemical shifts as a tool for structure determination, *Methods Enyzmol.* **239**, 363 (1994). © 1994, with permission of Elsevier.

This effect is most easily understood by considering the proton NMR spectra of ethanol shown in Figure 11-4. the low-resolution spectrum (left) shows the three peaks that might be expected on the basis of our discussion of chemical shifts, namely, CH_3 protons, CH_2 protons, and the OH proton. The ratio of the areas under the resonances is proportional to the number of protons in each chemical environment, 3:2:1. The high-resolution structure (right) shows that these three resonances are multiplets of peaks due to spin–spin splitting. The number of peaks within a multiplet is determined by the number of spin orientations of neighboring nuclei. In this case, the spins of the two hydrogens on the methylene (CH_2) carbon can have four possible arrangements of orientations

$$\uparrow\uparrow \quad \underline{\uparrow\downarrow \quad \downarrow\uparrow} \quad \downarrow\downarrow$$

The middle two arrangements are equivalent so that the methyl proton resonances are split by the neighboring CH_2 into three resonances with relative areas under the peak of 1:2:1. The OH proton, also next to the methylene carbon, is split into three peaks. Similarly, the methyl proton can have three different orientations, with the two underlined groups being equivalent:

$$\uparrow\uparrow\uparrow \quad \underline{\uparrow\uparrow\downarrow \quad \uparrow\downarrow\uparrow \quad \downarrow\uparrow\uparrow} \quad \underline{\uparrow\downarrow\downarrow \quad \downarrow\uparrow\downarrow \quad \downarrow\downarrow\uparrow} \quad \downarrow\downarrow\downarrow$$

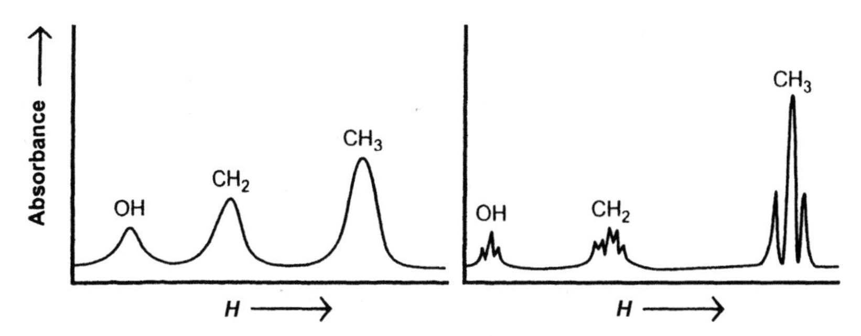

FIGURE 11-4. Schematic representation of the NMR spectrum of dry ethanol at low (*left*) and high (*right*) resolution. The ratio of the areas under the peaks is 3:2:1. The octet expected for the methylene protons (*right*) is not shown as the splitting of the four peaks is very small.

The methylene resonance, therefore, is split into four peaks, with area ratios of 1:3:3:1. The hydroxyl proton will split these four peaks further into an octet, but this detail is difficult to see in the spectrum. In order for scalar coupling to occur, the two nuclei must have distinct resonance peaks. Thus, the methyl protons will not split each other's resonances.

Spin–spin splitting is of special importance in considering the NMR spectra of proteins because coupling between the α-C proton and the N proton of the amide bond occurs. Scalar coupling can also occur between different nuclei, for example, a proton and ^{15}N. The magnitude of the spin–spin splitting is called the coupling constant and is designated by J. The number of covalent bonds separating the nuclei in question is often appended as a prior superscript and a post subscript designates the atoms involved. For example, the coupling constant for the amide proton and the α-C proton would be written as $^{3}J_{\text{HN}-\text{H}\alpha}$.

The spin–spin coupling constant for protons on adjacent atoms varies considerably, from about 0 to 10 Hz. This variation is due to different torsional angles between the protons as defined below:

$\theta = 0°$ $\theta = 180°$

If the torsional angle is 90°, the coupling is close to zero, whereas when it is 0° or 180°, it is about 10 Hz. If free rotation about the bond occurs, an average coupling constant is obtained. The dependence of the coupling constant on the torsional angle is given by the Karplus equation:

$$J = A + B\cos\theta + C\cos^2\theta \tag{11-6}$$

The constants A, B, and C can be calculated or established from measurements with molecules having known dihedral angles. If free rotation about the bond occurs, the coupling constant will be some average of the possible dihedral angles. However, if the bond is constrained, such as in a peptide linkage that has double bond character or in a folded macromolecular structure, the calculated dihedral angel can provide useful information about the structure. This is especially true when used in conjunction with other information.

This simple picture of spin–spin splitting is not rigorous. It provides a useful and adequate explanation for relatively simple situations but breaks down when considering multidimensional NMR, which is discussed later.

11.5 RELAXATION TIMES

Thus far, we have not dealt explicitly with the time scale for nuclear spin transitions. If we think of a pulsed NMR experiment, we can envisage nuclear spins being oriented in a specific direction by the pulse, with the precession of spins about the direction of the field. When the pulse is turned off, the nuclei will come to equilibrium with regard to their environment. The rate at which a particular nuclear spin returns to equilibrium depends on interactions with other nuclear spins, and this in turn will depend on the fluctuating fields experienced as the molecules tumble. This mode of relaxation involves a change of energy between the spin systems and their environment. The return of a spin population in a magnetic field to its equilibrium population follows first-order kinetics, and the reciprocal of the first-order rate constant is called the *spin–lattice relaxation time*, T_1. Typically, it is in the range of tenths of a second to seconds for protons.

Simply put, the value of T_1 depends on the interactions of a nuclear spin with its neighbors, including the solvent, and how fast the molecule rotates. The molecular rotation is characterized by a rotational correlation time that is basically a measure of the rotational diffusion constant of a molecule. Measurements of T_1 can, in fact, be used to determine rotational diffusion constants of rigid molecules. In some cases, chemical exchange of nuclei in different environments can contribute to T_1, although this is somewhat unusual. In general, anything that gives rise to magnetic fluctuations in the environment can contribute to this relaxation, for example, unpaired electrons or dissolved oxygen.

A second mode of nuclear spin relaxation is possible that does not involve the exchange of energy of the magnetic moment with its environment. In terms of the picture of a nuclear spin being oriented in a field and precessing about the field direction, the spin–lattice relaxation time can be thought of as characterizing the return of the spin orientations to their equilibrium positions. The *spin–spin relaxation time*, T_2, on the other hand, is associated with the rate of precession about the field direction. Basically, T_2 is a measure of alterations in the precession frequency during nuclear spin relaxation. This alteration is different for different nuclear spins so that the rates of precession change with respect to each other. This is called a loss of phase coherence. This can be viewed as an exchange of energy within the spin system. It does not change the net population of the excited states.

The dominant factor determining T_2 is the rate of molecular tumbling. The molecular tumbling effect is quantitatively different for T_2 and T_1. Because T_2 is generally much shorter than T_1 in liquids, the line widths of spectra are determined by T_2, $1/T_2 = \pi\nu_{1/2}$ where $\nu_{1/2}$ is the peak width at one-half of its maximum value. As previously stated, rapidly tumbling molecules produce relatively sharp lines whereas slowly tumbling molecules have relatively broad lines.

Chemical reactions provide one of the most interesting examples of how T_2 can be altered. Consider the simple chemical reaction

$$A \rightleftharpoons B \tag{11-7}$$

Assume that a specific proton has a different chemical shift in A and B. (This means they have a different Larmor frequency.) If the chemical exchange rate is very slow compared to the difference in chemical shifts (i.e., smaller than the difference in Larmor frequencies), two distinct peaks will be seen in the NMR spectrum, as shown in Figure 11-5 (top). At the other extreme, if the chemical exchange rate is very fast relative to the difference in the chemical shift frequencies, states A

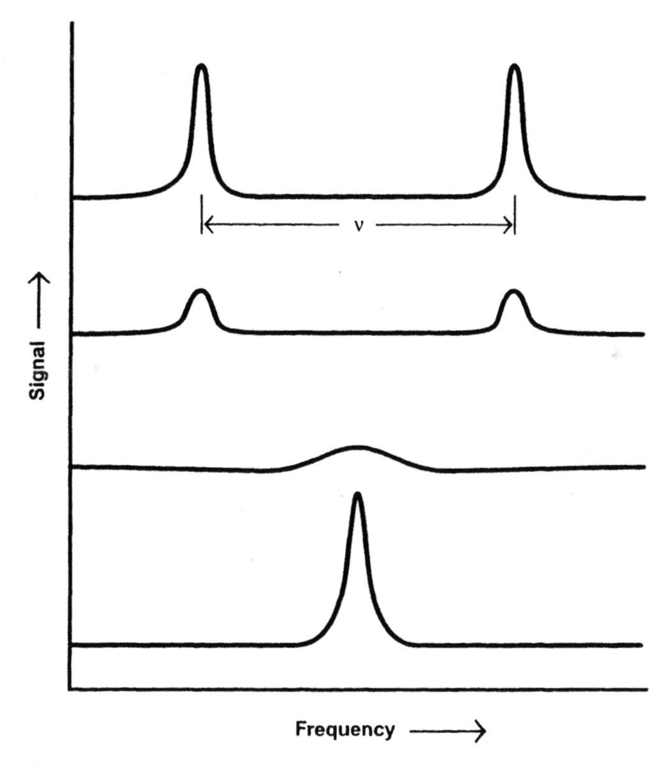

FIGURE 11-5. Schematic drawing of the effect of chemical exchange on NMR spectra. In the top spectrum, resonances from two protons are shown. In this case, the rate of exchange between the two environments is much less than ν, or $1/\tau_A \ll \nu$, whereas in the lowest spectrum, $1/\tau_A \gg \nu$, so that only a single resonance is seen that is the average of the positions in the upper spectrum. The two spectra in the middle represent the cases where $1/\tau_A \sim \nu$, with $1/\tau_A$ becoming progressively larger from the top spectrum to the bottom spectrum.

and B will interconvert many times during the NMR experiment, and the two frequencies are effectively averaged. Consequently, only a sharp single line will be seen, located at the average of the two lines weighted by the relative populations of A and B (Fig. 11-5, bottom). In the intermediate cases, where the rate of the chemical reaction is comparable to the difference in chemical shift frequencies, the two lines will broaden, coalesce into a single broad peak, and finally sharpen, as shown in Figure 11-5. For a single line, the spin–spin relaxation time can be written as

$$1/T_2 = 1/T_{2A} + 1/\tau_A \tag{11-8}$$

where T_{2A} is the relaxation time of the A state without significant chemical exchange and τ_A is the relaxation time for the chemical reaction. We will not delve into the details, but the line shape can be quantitatively analyzed to determine the rate constants for the chemical reaction. The time scale of the reactions that can be studied is determined by the chemical shift difference between the two states. For example, if the difference is 100 Hz, then rates of chemical reactions in the range of $1/100 = 10^{-2}$ s can be studied. Reactions such as the rate of exchange of hydrogens between proteins or amino acids and water have been investigated with this method.

11.6 MULTIDIMENSIONAL NMR

Thus far, we have considered what is now called one-dimensional NMR, namely determining the spectrum for a specific nucleus by scanning the magnetic field or by analysis of the frequencies associated with the free induction decay following a frequency pulse. One-dimensional analysis has been invaluable in determining the structures of small molecules and can also be used to obtain information about macromolecules. However, in order to get definitive structural information about macromolecules, multidimensional NMR is necessary. The genesis of this field was in the early 1970s and its vigorous evolution continues to this day. The underlying principle of multidimensional NMR is to find "cross-peaks" that link two resonances. This linkage can be either through space or through a small number of chemical bonds. Finding these cross-peaks allows the spatial relationships to be determined between the nuclei responsible for the two resonances. These connections between resonances are sometimes called coherence pathways.

A detailed presentation of multinuclear NMR is beyond the scope of this text. However, the concepts can be understood by considering two-dimensional NMR in a qualitative manner. Two-dimensional NMR can be viewed as the assembly of one-dimensional spectra in an array. The experiments consist of four stages, illustrated in Figure 11-6. In the preparation phase, a frequency pulse is applied to the system. An evolution phase of length t_1 then occurs and is followed by a mixing period ending in a second pulse. (A pulse may also be applied between the evolution and mixing phases, depending on the specific two-dimensional experiment being carried out.) Finally, a free induction decay occurs for a time t_2 in the data acquisition

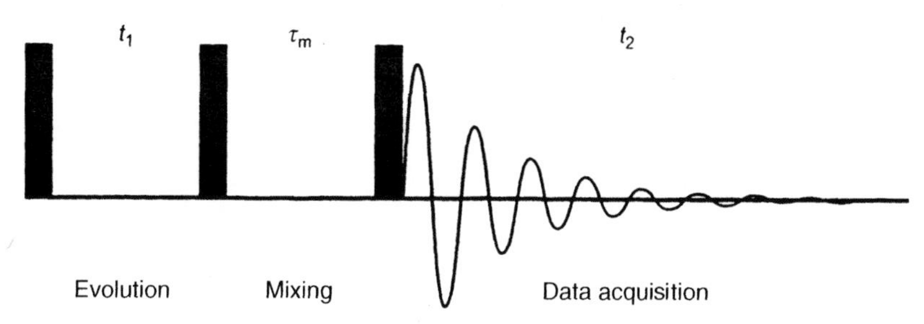

FIGURE 11-6. Schematic representation of a two-dimensional NMR experiment in which a radiofrequency pulse (black bar) is applied to the sample initially. After a period t_1, a second pulse is applied, followed by a mixing time, τ_m. The data are acquired after a final pulse. The number of pulses and the various times depend on the type of experiment being carried out.

phase. In the actual experiment, t_1 is varied incrementally and the signal is collected for the period t_2, ultimately giving a large number of time points. This experiment is repeated many times and signal averaged. The nature of the pulses will depend on the specific experiment. Qualitatively, the pulses flip the spins (magnetic moments) in the field that is created, and the end of the mixing phase flips them again. In the evolution phase, the magnetic moments associated with the nuclear spin will partially return to equilibrium at a rate that depends on T_1 and T_2. The return to equilibrium also occurs in the collection phase. In contrast, one-dimensional NMR uses a single pulse, followed by free induction decay and data collection.

The design of these experiments can be quite tricky, as it depends on what connections between spin states are being probed and what the relaxation times are. The representation of the result is done by carrying out a Fourier transform on both the t_1 and t_2 data sets and converting them into a two-dimensional plot of ν_1 versus ν_2 with the third dimension being the amplitude of any resonance peaks that are observed. A very simple illustration is given in Figure 11-7 for two interacting spin systems. Peaks that occur on the diagonal have the same frequency in both dimensions and correspond to the one-dimensional spectrum, whereas those occurring off the diagonal represent cases where different spin systems interact during the mixing period (coherence transfer). Contours are usually used to indicate the amplitudes of the resonance peaks, rather than a third dimension. The interactions (coherence transfer) between nuclear spins can be either homonuclear (same nuclei) or heteronuclear (different nuclei).

One of the first two-dimensional experiments carried out was COSY (COrrelated SpectroscopY). This experiment identifies pairs of nuclei that are linked by scalar coupling (spin–spin splitting connectivities). The interaction between these nuclei occurs during the relaxation taking place in the mixing phase and results in cross-peaks in the spectrum. Scalar coupling (off diagonal peaks) between nuclei means they are within three covalent bonds. A typical COSY spectrum is shown in Figure 11-8 for λ cro repressor, a DNA binding protein that regulates phage development.

One of the most important two-dimensional NMR experiments is Nuclear Overhauser Effect SpectroscopY (NOESY). This is a through space interaction that

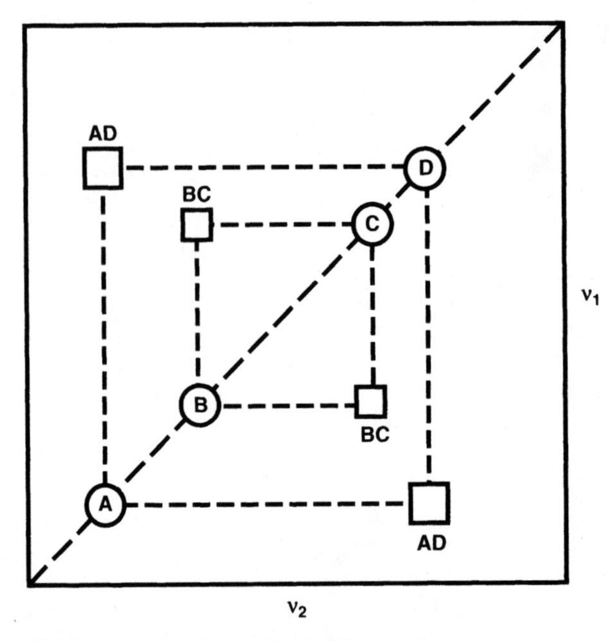

FIGURE 11-7. Schematic representation of a two-dimensional NMR spectrum for two spins. The frequency axes labeled ν_1 and ν_2 are the positions of the resonances for the two different spins. The diagonal (circles) corresponds to the one-dimensional spectrum. If coherence transfer occurs during the mixing period, off-diagonal resonances (squares) will be seen. In this example, coherence transfer occurs between nuclei with resonances at A and D and between nuclei with resonances at B and C.

takes place because of the interactions between the magnetic dipoles of two nuclear spins. In terms of our previous discussion, this primarily involves the T_1 mode of relaxation and is coupled with rotational motion. The NOE is the NMR equivalent of fluorescence resonance energy transfer discussed in Chapter 8. In fact, an NOE can be observed in one dimension. If a sufficiently large magnetic field is applied at the resonance condition of a given nucleus, the spin system becomes saturated, that is, the ground and excited states become equally populated so that no more energy can be absorbed. If the magnetic moment of this nucleus is sufficiently close in space to another magnetic moment, energy can be transferred. This perturbs the intensity of the resonance of the nucleus to which energy is transferred and decreases the resonance intensity of the nucleus that was initially irradiated. The interaction between magnetic dipoles varies as the inverse sixth power of the distance between the two dipoles so that only nuclear spin systems that are very close to each other give rise to NOEs. In practice, this means distances of 5 Å or less.

The NOESY experiment is quite similar to the two-dimensional experiment described previously, except that a pulse is applied at both the beginning and the end of the mixing period, with the time of the mixing period being constant. An example of a NOESY spectrum is given in Figure 11-9 where proton–proton

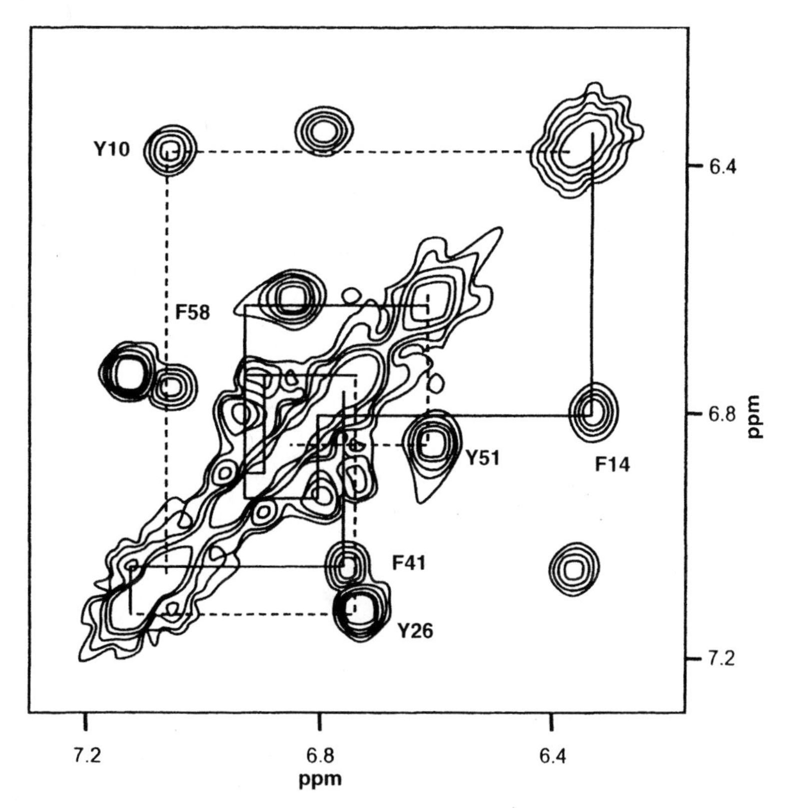

FIGURE 11-8. Example of a COSY spectrum for the protons of the aromatic spin systems in λ cro repressor protein. Tyrosine rings are connected with a dashed line and phenylalanine rings by a solid line. The diagonal and cross-peaks can be easily seen. Reprinted in part with permission from P. L. Weber, D. F. Wemmer, and B. R. Reid, *Biochemistry* **24**, 4553 (1985). © 1985 by American Chemical Society.

NOEs for a complex of DNA and an antibiotic, distamycin A, are presented. Because of the distance dependence of this effect, the distance between spins can be estimated from the cross-peaks. However, because only the distance, not the direction, is derived from these measurements, many distances must be determined to arrive at a unique structure.

One of the most useful multidimensional spectra is HSQC (Heteronuclear Single Quantum Correlation), an example of which is shown in Figure 11-10 for a protein involved in proton transport across a membrane (subunit c of ATP synthase from *E. coli*). This spectrum uses both ^{1}H and ^{15}N and selectively detects only pairs of covalently attached nuclei. Each spot in the contour map represents such a pair, with the position on the horizontal axis representing the proton resonance frequency and the position on the vertical axis representing the resonance frequency of the nitrogen nucleus. Every amino acid, except for proline, has a backbone amide so essentially every residue is represented in this spectrum. Side chains containing

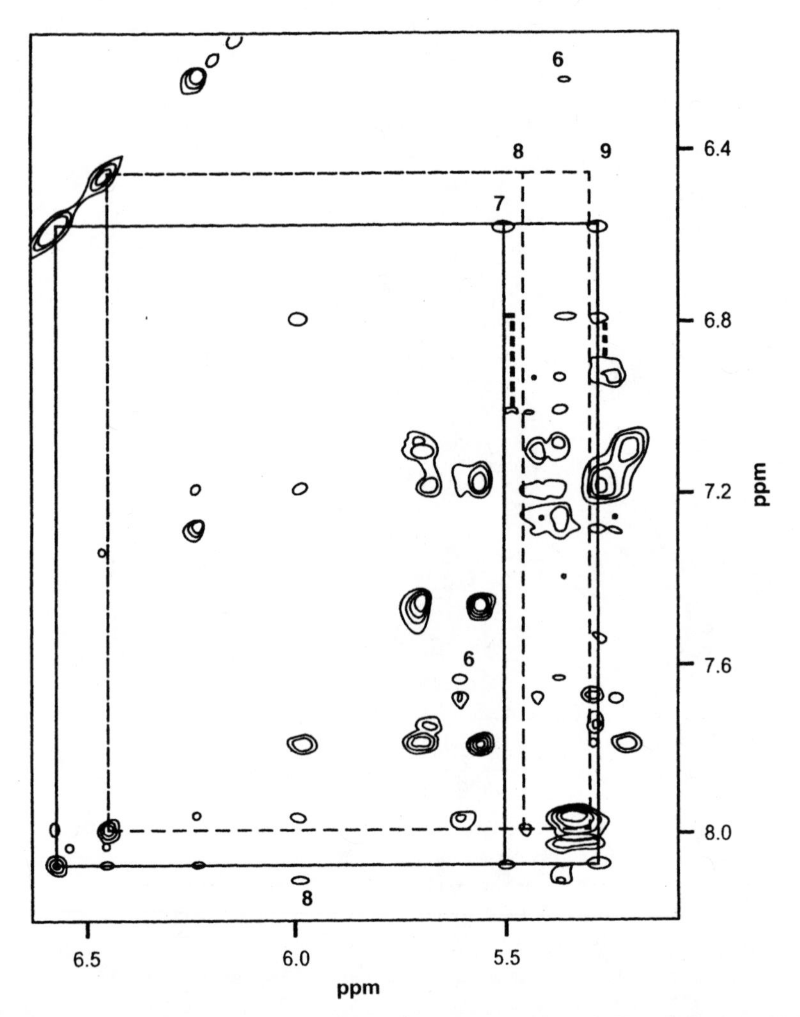

FIGURE 11-9. Example of a NOESY spectrum for a complex of distamycin A and DNA. Aromatic C6H resonances of adenine and guanine and C2H resonances of adenine are shown along the vertical axis. The Cl′H resonances are shown along the horizontal axis. Sets of sequential connectivities are denoted by dotted, dashed, and solid lines. Reprinted in part with permission from J. G. Pelton and D. F. Wemmer, *Biochemistry* **27**, 8088 (1988). © 1988 by American Chemical Society.

amides will give rise to additional resonances. This gives direct information about which nitrogen is coupled to which hydrogen. Furthermore, the resonances are usually quite well resolved. The HSQC spectrum provides information that is useful for assigning the observed resonances to specific amino acid residues.

Extending NMR to dimensions greater than two involves similar concepts. Multiple pulses and even field gradients are used. These dimensions can be in terms of different nuclei and/or combining two-dimensional experiments. The

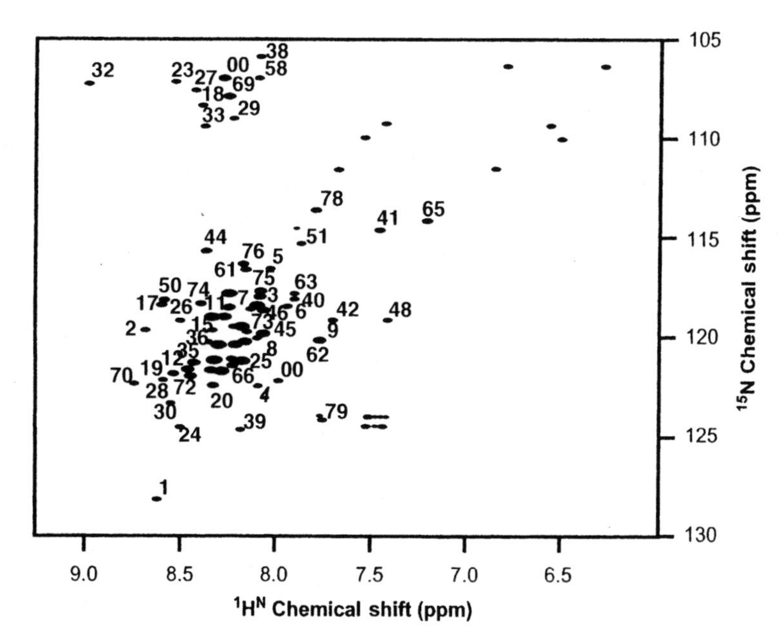

FIGURE 11-10. Example of an HSQC spectrum. All 76 backbone amide cross-peaks are shown for the c subunit protein of ATP synthase from *E. coli*. This membrane-bound protein is involved in proton transport across the membrane. Reprinted in part from M. F. Girvin, V. K. Rastogi, F. Abildgaard, J. L. Markley, and R H. Fillingame, *Biochemistry* **37**, 8817 (1998).

most common nuclei studied in addition to ^{1}H are ^{13}C and ^{15}N. In these cases, proteins and nucleic acids enriched in the nucleus of interest can be used to enhance the sensitivity.

Transforming the NMR results for a macromolecule into a structure is straightforward but not easy. First, the resonances have to be assigned to specific nuclei within the structure. This can be done by analysis of scalar coupling and sequential NOEs. The three-dimensional structure is derived from the distances determined from NOE experiments and dihedral angle information determined from spin–spin splitting and the Karplus equation (Eq. 11-6). In practice, the information gleaned from the NMR spectra provides hundreds of constraints on the structure. A number of computer programs have been written that take the constraints and convert them into a family of structures, usually very similar, consistent with the constraints. These programs involve sophisticated theory as well as data analysis to arrive at final structures. NMR spectroscopy is a very powerful tool for determining protein structures in their biologically active conformations. As proteins increase in size, their rate of rotation slows down, and NMR spectra broaden. Although the size of macromolecules whose structure can be determined with NMR increases year by year, most of the structures to date are for molecules with a molecular weight less than 20,000.

11.7 MAGNETIC RESONANCE IMAGING

One of the most remarkable advances in diagnostic medicine has been the evolution of magnetic resonance imaging (MRI). Although X-rays readily distinguish hard objects such as bones, they do not distinguish soft tissue structure. MRI, on the other hand, provides excellent images of tissues and is able to distinguish between various types of tissues. Protons in water are the primary nucleus used for detection, although applications with ^{13}C, ^{31}P, and ^{19}F have been developed. The principle underlying MRI is the use of a magnetic field gradient. Since the resonance frequency is directly proportional to the magnetic field, the frequency of the resonance will depend on its location in the magnetic field. The intensity of the absorption is dependent on how many protons are present. If a linear magnetic field gradient is applied, the position of the resonance will change, also linearly, as the field is varied. Thus, a plot of the amplitude of the resonance versus frequency is equivalent to a plot of the integrated number of protons versus distance. A series of cross-sections can be obtained by rotating the sample in the field, or by moving the field around the sample. These cross-sections can then be reconstructed to give a three-dimensional image.

In soft tissue, the amount of water varies for different tissues, so the density of protons varies. In addition, the various tissues are characterized by significantly different T_1 values. The difference in proton density can be shown in reconstructions by varying the darkness of the shading. This can be seen in Figure 11-11 where the MRI image of an adult human brain is shown. In addition to being able to distinguish various soft tissues, MRI is noninvasive, as contrasted to X-rays or injections of foreign substances, including radioactive isotopes, required for other imaging techniques.

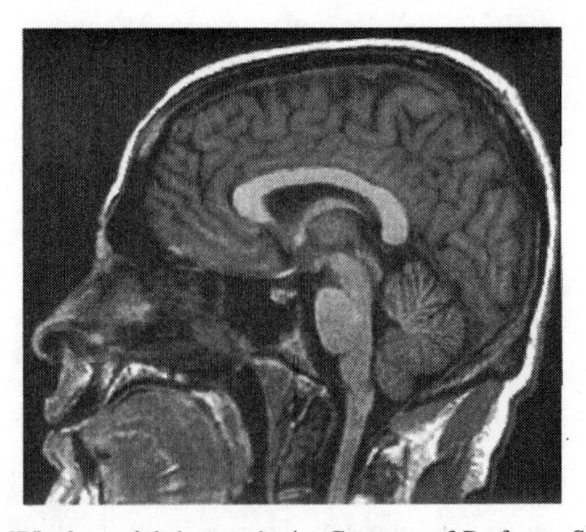

FIGURE 11-11. MRI of an adult human brain. Courtesy of Professor Scott Huettel, Duke University Brain Imaging and Analysis Center. Reproduced with permission.

11.8 ELECTRON SPIN RESONANCE

We now return briefly to ESR. Unlike NMR, ESR is not observed for most materials. This is because electrons are usually paired and consequently have no net magnetic moment. However, free radicals and other paramagnetic substances have unpaired electrons that give rise to ESR spectra. Because the frequencies associated with transitions between the energy levels of unpaired electrons in a magnetic field are in the microwave region, special techniques are required for placement of the sample in the magnetic field. Conceptually, the experiment is the same as for NMR. The magnetic field is varied until resonance is found.

The usefulness of ESR in biological systems arises from the interactions between nuclear spins and the electron spin. This gives rise to "hyperfine structure" in the spectrum. For example, if a neighboring nucleus has a nuclear spin of 1/2, it will have two orientations in the field and will split each of the two energy states of the electron into two, as shown in Figure 11-12. Because of quantum mechanical selection rules, only two transitions between these four energy states are allowed. If the nuclear spin is 1, six energy levels are created and three transitions between the levels are allowed. The hyperfine structure can provide information about the environment of the paramagnetic species.

As with NMR, the sharpness of the spectrum provides information about the rotational mobility of the paramagnetic species. However, the time scale for

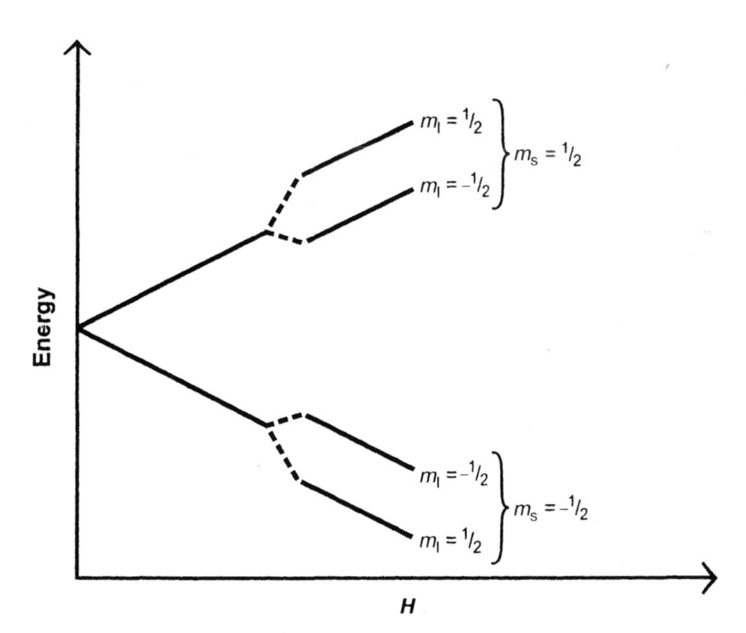

FIGURE 11-12. Hyperfine splitting of the energy levels of an electron in a magnetic field, H, by a nuclear spin. The electron has a spin quantum number, m_S, of $\pm 1/2$ and the nucleus has a spin quantum number, m_I, of $\pm 1/2$.

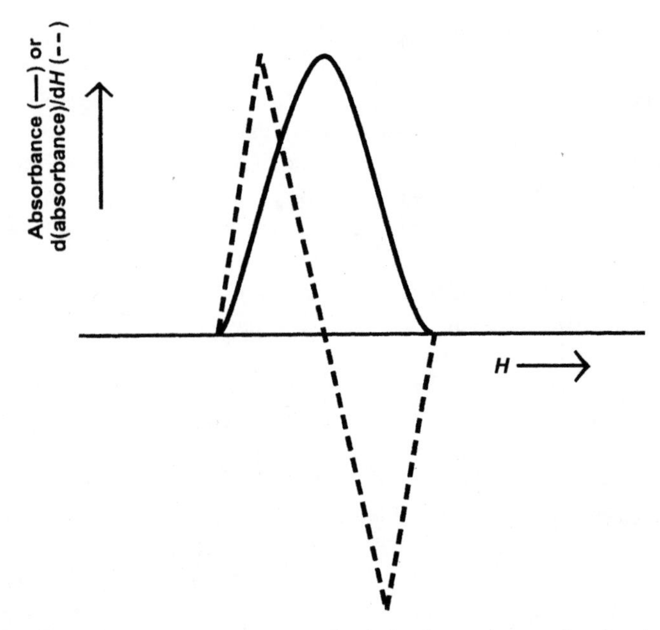

FIGURE 11-13. Schematic representation of a derivative spectrum. An absorbance peak is shown as a solid line and its derivative is shown as a dashed line. ESR spectra are usually presented as derivatives of the absorption.

rotation is much different than NMR because the frequencies are much higher. For ESR, rotational correlation times of approximately 10^{-9} s influence the line width whereas for NMR the time scale is about 10^{-6} s. For reasons we will not dwell on here, the spectra are usually presented as the derivative of the amplitude versus the field (Fig. 11-13).

One of the important developments in the application of ESR to biological systems was the synthesis of stable free radicals that could be reacted with macromolecules and membranes, both noncovalently and covalently (2). The most common element of these spin labels is the nitroxide free radical (Fig. 11-14). Because the ^{14}N nucleus, with a nuclear spin of 1, is next to the free radical, three bands are seen in the spectrum. The derivative of the resonances is three up and down peaks. This is illustrated in Figure 11-14 where ESR spectra are shown for nitroxides under various conditions of rotational mobility. The effect of molecular motion on the spectra of the free radical is clearly illustrated. As these spectra illustrate, spin labels can provide information about the rotational mobility of the macromolecule to which they are bound. However, it should be noted that the probe mobility can be due to macromolecule rotation, segmental motion of the macromolecule, or simply rotation of the probe in the site to which it is bound. The interpretation of what motion is being observed is not always straightforward.

ESR spectra can also provide information about the polarity of the spin label environment. The extent of splitting of the hyperfine structure depends on the

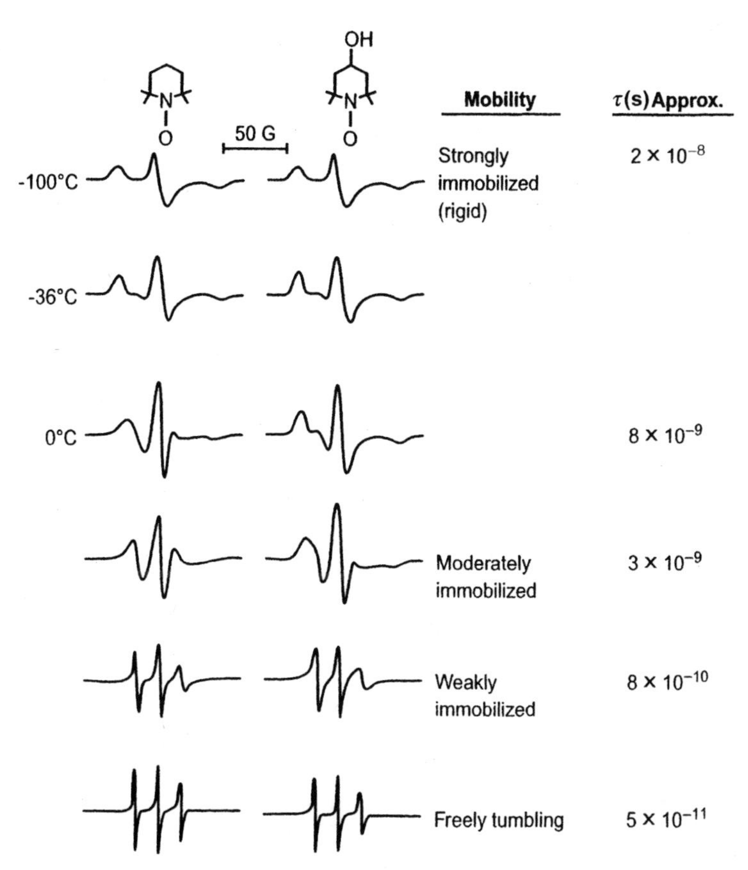

FIGURE 11-14. ESR spectra of spin labels whose structures are shown at the top of the figure. The effect of viscosity on the line shapes and rotational correlation times, τ, is shown. Reproduced with permission from R. A. Dwek, *Nuclear Magnetic Resonance in Biochemistry*, Clarendon Press, Oxford, England, 1973, p. 289. Adapted from data in P. Jost, A. S. Waggoner, and O. H. Griffith, Spin Labeling and Membrane structure, in *Structure and Function of Biological Membranes* (L. Rothfield, ed.), Academic Press, New York, 1971, p. 83. 1971, with permission of Elsevier.

effective dielectric constant of the environment. Thus, for example, the splitting is quite different in a biological membrane or in solution. When bound to a membrane, the splitting will be different for a spin label close to a polar headgroup and for a spin label buried in the hydrocarbon chains.

This chapter is a relatively brief introduction to NMR and ESR. Many texts are available that provide more detailed descriptions of NMR(cf. Refs. 3–7) and ESR (cf. Ref. 8). Unfortunately, most of these are not easy reading and require a working knowledge of quantum mechanics. Applications of these techniques to biological systems are presented in the next chapter.

REFERENCES

1. D. S. Wishart and B. D. Sykes, *Methods Enzymol.* **239**, 363 (1994).
2. S. Ohnishi and H. M. McConnell, *J. Am. Chem. Soc.* **87**, 2293 (1965).
3. I. Tinoco, Jr., K, Sauer, J. C. Wang, and J. D. Puglisi, *Physical Chemistry: Principles and Applications to the Biological Sciences*, 4th edition, Prentice-Hall, Englewood Cliffs, NJ, 2001.
4. K. Wuthrich, *Acc. Chem. Res.* **22**, 36 (1989).
5. T. L. James and N. J. Oppenheimer (eds.), *Methods Enzymol.* **239** (1994).
6. J. Cavenaugh, W. J. Fairbother, A. G. Palmer, III, and N. J. Skelton, *Protein NMR Spectroscopy*, Academic Press, San Diego, CA, 1996.
7. T. L. James, V. Dötsch, and U. Schmitz eds., *Methods Enzymol*, **238–239** (2001).
8. J. A. Weil, J. R. Bolton, and J. E. Wertz, *Electron Paramagnetic Resonance: Elementary Theory and Practical Applications*, Wiley–Interscience, New York, 1994.

PROBLEMS

11.1. Deduce the structure of the compounds below from their schematic NMR spectra. Indicate which protons are assigned to each resonance.

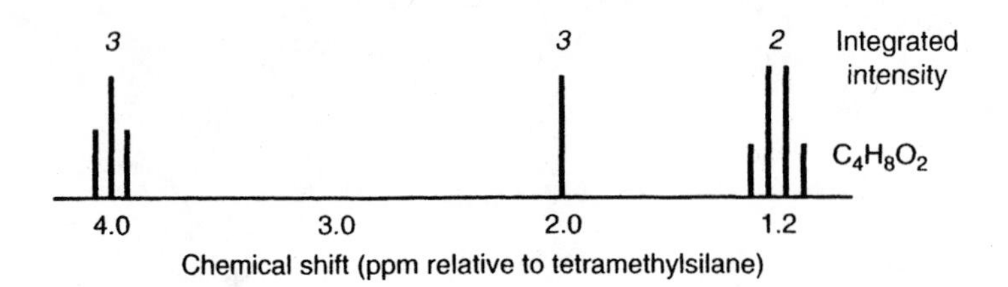

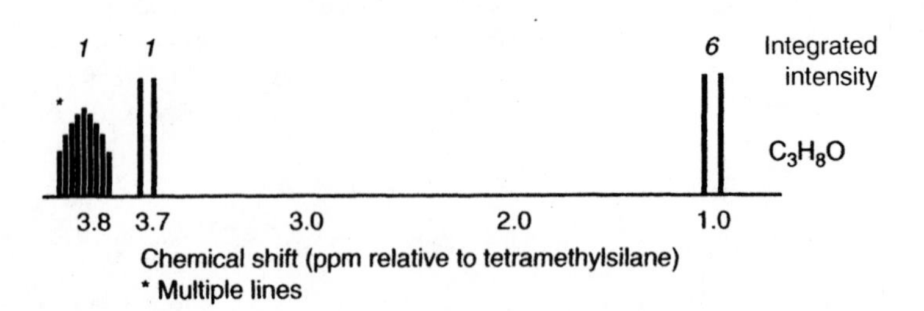

11.2. Nucleotide phosphates are important biological molecules. Sketch the ^{31}P (spin 1/2) NMR spectra of a nucleotide monophosphate, diphosphate, and triphosphate. Order the resonances in terms of chemical shift, with zero on the right-hand side. The phosphates are designated as α, β, and γ with α being closest to the sugar and γ furthest away. For the triphosphate, the γ position has the lowest electron density and the β position has the highest.

11.3. The structure of L-leucine in D_2O is

$$
\begin{array}{ccccccc}
 & & CH_3 & & & H & \\
 & & | & & & | & \\
H & - & C_\gamma & - & C_\beta H_2 & - & C_\alpha - COO^- \\
 & & | & & & | & \\
 & & CH_3 & & & ND_3^+ &
\end{array}
$$

a. Sketch the one-dimensional proton NMR spectrum. The approximate chemical shifts are 1.0 for the methyl groups, 1.4 for the γ hydrogen, 1.5 for the β hydrogens, and 3.3 for the α hydrogen. Assume that the methyl hydrogens and β hydrogens are not resolved, that is, all of the hydrogens in each of the two classes have a single resonance peak. Indicate the integrated intensity of the resonances for each resonance.

b. Sketch the proton COESY spectrum of L-leucine, indicating the coupling that should give rise to off-diagonal peaks.

11.4. Two common structures found in proteins are the α-helix and β-sheet (parallel and antiparallel), as discussed in Chapters 3 and 7. The approximate distances in Å between protons for the two structures are given below:

	α-Helix	β-Sheet
Amide H–amide H	2.8	4.2
α-Carbon H–amide H	3.5	2.2
α-Carbon H–(amide H)$_{i+3}$	3.4	>5

The subscript $i + 3$ means the α-carbon is $i + 3$ residues away from the amide proton. Indicate how the measurement of NOEs could be used to distinguish these structures by considering the expected relative intensities of the NOEs for the three classes of protons in the table.

11.5. a. Use Eq. 11-3 to construct a plot of the resonant frequency for an electron as a function of the field strength from 0 to 4 Tesla.

b. Show schematically the splitting expected in the energy levels if the unpaired electron is located next to a nuclear spin of 3/2.

11.6. An RNA structure containing uracil and adenine was studied by NMR. The structures of these two bases are given below.

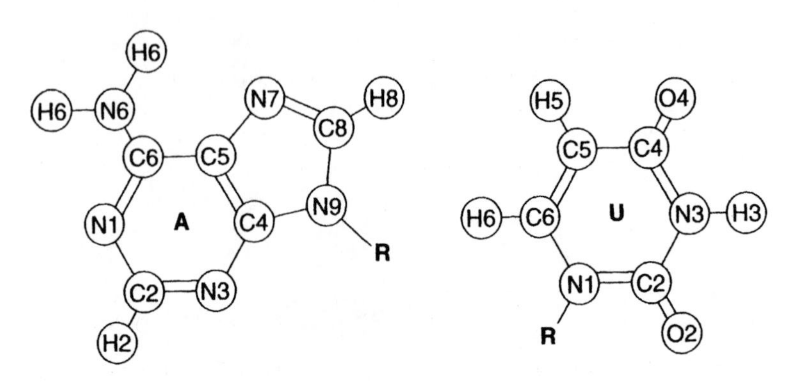

a. In a $^{2}J_{NN}$ COSY-type experiment with ^{15}N-enriched RNA, scalar couplings were observed between N3 of uridine and N1 of adenine [A. J. Dingley and S. Grzesiek, *J. Am. Chem. Soc.* **120**, 8293 (1998)]. In addition, in a NOESY-type experiment, an NOE was observed between H3 of uridine and H2 of adenine. Postulate a structure for the hydrogen-bonded dimer formed between uridine and adenine.

b. As the pH of the solution is lowered, the chemical shift of adenine N1 shifts downfield, and at sufficiently low pH (<4), the scalar coupling and NOE disappeared. Explain this result. (This experiment was not actually done, but this is the expected result.)

c. With free adenosine, this same chemical shift is observed, but it occurs at much lower pH values. Explain this result.

11.7. The spectrum of pure methanol is shown in (a) below. When HCl is added, the spectrum is that shown in (b) [Z. Luz, D. Gill, and S. Meiboom, *J. Chem. Phys.* **30**, 1540 (1959)].

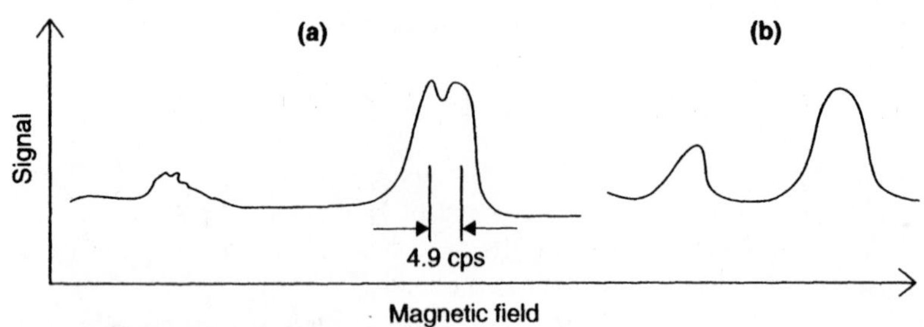

a. Explain this result.

b. The rate of the proton exchange reaction

$$CH_3OH_2{}^+ + CH_3OH \rightarrow CH_3OH + CH_3OH_2{}^+$$

can be written as $R = k(CH_3OH_2^{\ +})(CH_3OH)$ where the rate constant k has a value of about $10^8\ M^{-1}\ s^{-1}$. The chemical relaxation time, τ, for this reaction is $\tau = 1/[k(CH_3OH_2^{\ +})]$. If the above reaction is responsible for the doublet in the spectrum becoming a single peak, estimate the concentration of HCl at which the doublet disappears. Is this consistent with the autoprotolysis constant for methanol: $(CH_3O^-)(CH_3OH_2^{\ +}) = 2 \times 10^{-17}\ M^2$?

CHAPTER 12

Applications of Magnetic Resonance to Biology

12.1 INTRODUCTION

Magnetic resonance has been widely applied to biological systems so that the selection of examples to illustrate the information that can be obtained is necessarily quite arbitrary. We first consider some examples of structural determinations utilizing NMR. As previously indicated, the advantage of NMR is that structures in solution can be obtained and variation of these structures in different environments can be readily assessed. The primary drawback of NMR is that only relatively small structures can be easily determined. NMR can provide important information about biological systems other than macromolecule structures, as will be illustrated. In the case of ESR, spin labels and paramagnetic ions can provide unique information, especially in unusual environments such as membranes.

12.2 REGULATION OF DNA TRANSCRIPTION

The regulation of transcription is of central importance in biology. We have previously discussed the use of CD to study zinc fingers, a structure of special importance for the interaction of transcription factors with DNA (Chapter 9). In this section, we consider the structure of cyclic AMP response element binding protein (CBP). It is a large transcriptional adapter protein that mediates transcription responses to intra- and extracellular signals (cf. Refs. 1,2). This class of proteins has been implicated in the regulation of cell growth, transformation, and differentiation. Defects in this regulator are involved in a multitude of human diseases. The CBP interacts with a variety of transcription factors and other components of transcription regulation to mediate cellular activities.

CBP itself is a very large protein, more than 2000 amino acids, but contains a number of distinct structural and functional domains. Three putative zinc-binding domains (zinc fingers) were identified by sequence homology. The CH3 domain was selected for study because not only is it a representative zinc finger structure, but it also binds known transcription factors (3). Constructs of varying length were examined, but the final studies were done on an amino acid sequence that was 88 residues long. It contains 13 cysteine and 5 histidine residues. Nine of the cysteine residues were coordinated to Zn^{2+}. This was ascertained from the chemical shifts of $^{13}C^{\beta}$ of cysteine. The cysteines that were not coordinated to Zn^{2+} had chemical shifts of 26.5 – 27.1 ppm, whereas the cysteines bound to Zn^{2+} had chemical shifts of 29.2 – 30.7 ppm. The three Zn^{2+}-coordinated histidines also displayed a downfield shift of 8 – 9 ppm for a ^{13}C ring resonance of the imidazole and an upfield shift for an ^{15}N resonance. Thus, the sites of metal coordination could be determined from one-dimensional NMR. Three Zn^{2+} ions are bound/CH3 domain.

If Zn^{2+} is absent, the NMR spectra indicated a disordered structure. However, in the presence of the metal, a well-defined structure formed. This can be seen in Figure 12-1, where the HSQC spectrum is shown for the protein with three bound

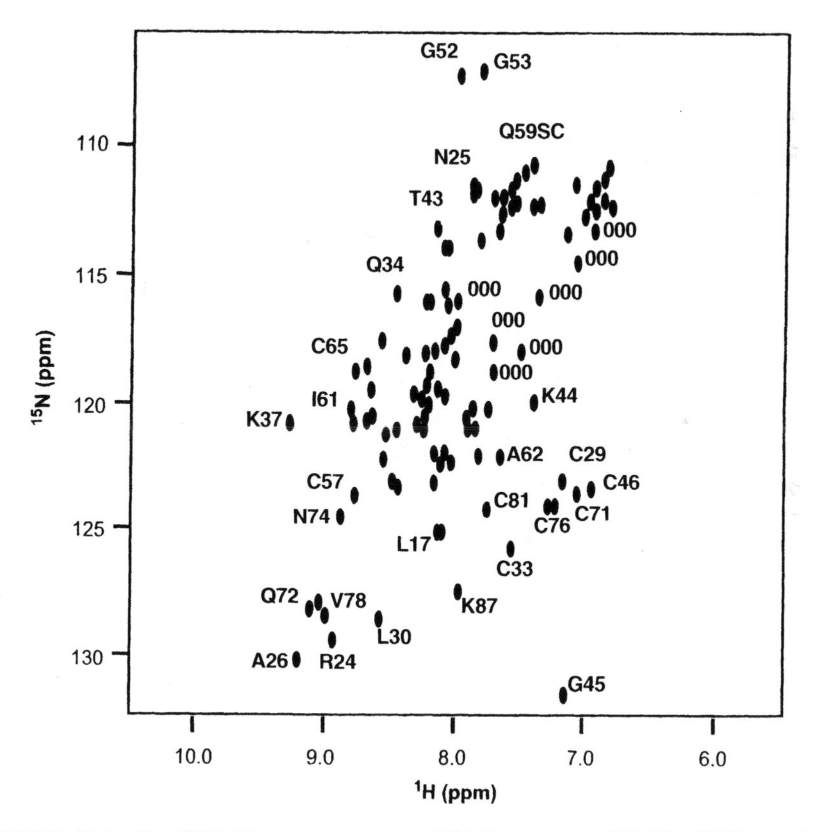

FIGURE 12-1. The 600 MHz proton–nitrogen HSQC spectrum of TAZ2 (CH3) domain, a portion of the c-AMP response element binding protein, with three bound Zn^{2+}. Many cross-peaks can be seen. Some of the amino acid assignments for the resonances are indicated. Reprinted from R. N. De Guzman, H. Y. Liu, M. Martinez-Yamout, H. J. Dyson, and P. E. Wright, Solution Structure of the TAZ2 (CH3) Domain of the Transcriptional Adaptor Protein CBP, *J. Mol. Biol.* **303**, 243 (2000). © 2000, with permission from Elsevier.

Zn^{2+} atoms. Many additional cross-peaks can be seen in the presence of Zn^{2+}. The structure was solved using three-dimensional heteronuclear spectra of uniformly labeled ^{15}N and ^{15}N, ^{13}C proteins. The structure determination involved the fitting of NOESY and dihedral angle constraints. In addition, an energy minimization was carried out to arrive at the final structure(s). Energy minimization is a theoretical calculation of the minimum free energy for the structure, making use of empirical equations for the intramolecular interactions. The final structure is shown in Figure 12-2 (see color plates) in the usual format, namely, the best fit superposition of multiple structures, in this case 20. In essence, this representation provides a measure of the uncertainty in the final structure. The cysteines and imidazoles are also shown as ball-and-stick representations to indicate how the Zn^{2+} is bound to the protein. Table 12-1 is a brief summary of some of the NMR constraints used to determine the structure and the deviations from ideal covalent geometry.

The structure contains four α-helices and three histidine(cysteine)$_3$ Zn^{2+} binding motifs. The helices are tightly packed to form a hydrophobic core. Two of the metal binding regions are quite similar, but the structure of the third is noticeably different. Although detailed data were not presented, preliminary results suggest that the structure of the CH1 domain is quite similar to that of CH3. Surprisingly, the structures of these zinc fingers are different than others that have been determined, presumably because of unique steric constraints within the structure.

As mentioned at the outset, the structure of the CH3 domain is of special interest because it mediates protein–protein interactions that are crucial for regulation of transcription. The interaction with a small peptide (eight residues) from p53, a known activator, with CH3 was investigated with NMR. A specific interaction was found by following shifts in both the backbone and side chains in HSQC spec-

TABLE 12-1. NMR Restraints and Statistics for the Structure of the TAZ2 (CH3) Domain

A. NMR restraints	
Total NMR restraints	1030
Distance restraints	846
Intraresidue	145
Sequential	286
Medium range	252
Long range	163
Total dihedral angle restraints	184
B. NOE violations	
Average violation (Å)	0.11 ± 0.04
Maximum violation (Å)	0.25
C. Deviations from ideal covalent geometry	
Bond lengths (Å)	0.0057 ± 0.0001
Bond angles (deg)	2.41 ± 0.02

Reference 3.

tra. The largest chemical shifts were observed for a small number of residues in three of the helices so that the portion of CH3 interacting with the ligand could be identified. The dissociation constant for the interaction of the peptide and CH3 is about 300 μM.

This study indicates the variety of information that can be obtained with NMR: the structure of an important biological molecule was determined; the specific residues binding the Zn^{2+} were determined; and the binding region for the interaction of CH3 with a regulatory molecule was identified.

12.3 PROTEIN–DNA INTERACTIONS

Proto-oncogenes are segments of DNA that code for proteins that have a normal function, but can be mutated or altered to become cancer-causing oncogenes. One of these, *c-myc*, encodes a nuclear protein involved in the regulation of transcription. The expression of *c-myc* is in turn regulated by FUSE binding protein (Far-UpStream Element). This protein binds to single-stranded DNA (ssDNA) that is about 1500 base pairs upstream from the *c-myc* promoter. This protein, FBP, contains four homologous repeats (KH) that are separated by linkers of varying lengths (4,5). The minimal ssDNA binding domain is designated as KH3–KH4. The structure of the KH3–KH4 domain bound to ssDNA (20 – 29 nucleotides long) has been solved using multidimensional NMR (6). The total molecular weight of the complex is about 30,000, and 3153 NMR restraints were used to determine the structure. A representation of the best-fit structure for the KH3–ssDNA is shown in Figure 12-3 (see color plates). The ssDNA is shown as a ball-and-stick model, whereas the protein chain is shown to delineate the overall fold and the α-helices and β-sheets.

The protein fold has three α-helices packed onto a three-stranded antiparallel β-sheet. The ssDNA binds in a groove formed by helices 1 and 2. The center of the groove is hydrophobic and the edges are hydrophilic so that the DNA bases point toward the center and the sugar phosphates toward the left-hand side of the protein. As expected, recognition of the DNA involves a number of intermolecular hydrogen bonds. This protein binding scaffold for nucleic acid binding can be fine-tuned for either ssDNA or RNA. The KH3 and KH4 domains do not interact with each other. The flexible linker between the domains is 30 amino acids long. NMR relaxation time measurements indicate that the two domains wobble with respect to each other, with time constants of nanoseconds. The flexibility of the domains appears to be of functional significance. If the flexibility of the motion is restricted by deleting four of the DNA bases in the intervening ssDNA between domains, the *c-myc* expression is reduced by a factor of four whereas adding four bases has no effect.

What is the significance of this structure? The functional concept is that DNA transcription can be controlled at a significant distance along the DNA by recognition of the ssDNA that is formed during transcription. The structure determined shows that the regulatory element (FUSE) upstream from the *c-myc* promoter is recognized specifically and forms a very stable complex with the FBA. This work, when combined with functional studies, can be used to construct a mechanism for transcriptional control. Furthermore, the over-expression of *c-myc* has been linked to cancer: the loss of FBP prevents *c-myc* expression and halts cellular proliferation. The disruption of the FBP–ssDNA interaction, therefore, is a potential target for cancer therapy.

12.4 DYNAMICS OF PROTEIN FOLDING

Understanding how proteins fold into their native conformation is of central importance in biology. First, protein folding is a necessary part of cellular metabolism and development. Second, understanding how proteins fold provides information about the intramolecular forces in proteins. Many proteins fold and unfold very fast, in times less than a second, so that study of the dynamics of protein folding requires special methods, including the use of NMR (cf. Ref. 7). Protein folding has been previously discussed in Chapters 3 and 9.

A prototypical fast folding protein is the N-terminal domain of bacteriophage λ repressor, a crucial molecule in gene regulation (cf. Ref. 8). The structure of the N-terminal domain has been determined with both crystallography and NMR, and it is essentially identical to the full-length version of λ repressor. The protein unfolds in a simple two-state process, both thermally and in the presence of urea. The folding reaction can be written as

$$\mathrm{N} \underset{k_{\mathrm{f}}}{\overset{k_{\mathrm{u}}}{\rightleftarrows}} \mathrm{D} \tag{12-1}$$

where N and D are the native and denatured states, and k_{f} and k_{r} are the first-order rate constants for folding and unfolding, respectively.

The aromatic region of the NMR spectrum is quite different for the native and denatured states as shown in Figure 12-4, where the NMR spectra are shown at various urea concentrations (9 – 11). Both the native and denatured states show well-resolved spectra. Some of the central peaks are in fast exchange at all urea concentrations so that only sharp resonances are observed, and the resonance most downfield is in slow exchange. However, some of the peaks show broadening at intermediate urea concentrations because the rate of interconversion of the native and denatured states is comparable to the chemical shift between the native and denatured resonances, about 100 cps.

The resonances chosen for analysis are associated with two specific tyrosine residues. The analysis of the data calculated the line shapes from theoretical considerations and compared the calculated and experimental line shapes with varying rates of reaction until the two matched. Basically, this measured the effect of the reaction rate on the spin–spin relaxation time, T_2, and yielded the relaxation time for the reaction in Eq. 12-1, τ (Eq. 11-8). The reciprocal of the relaxation time for the chemical reaction is the sum of the two rate constants (Chapter 4). The propor-

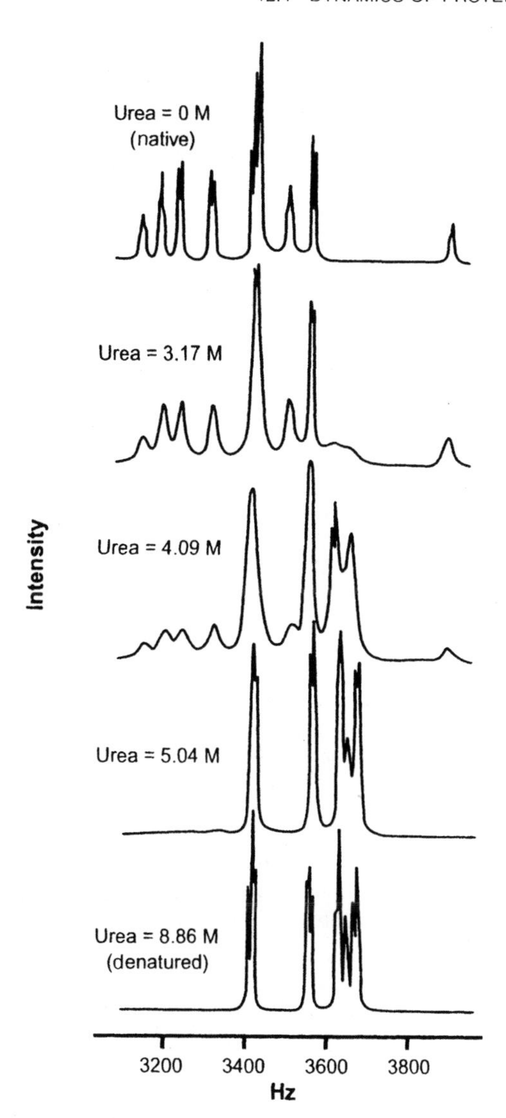

FIGURE 12-4. NMR spectra of the aromatic region of λ repressor at various concentrations of urea. Differences between the native and denatured (top and bottom) can be seen. The sharp peaks in the middle are in fast exchange. At intermediate urea concentrations, line broadening of several of the peaks can be seen because of chemical exchange. Analysis of the line broadening permitted determination of the rate constants for the interconversion of the native and denatured proteins. Reprinted in part with permission from J. K. Myers and T. G. Oas, *Biochemistry* **38**, 6761 (1999). ©1999 by American Chemical Society.

tion of the protein present in the native and denatured states can be derived from the line shape analysis as it is directly proportional to the strength of the resonance for each state. Thus, the individual rate constants can be directly determined. The rate constants vary linearly with the urea concentration but are in the range of

$10^2 - 10^3\,s^{-1}$. When extrapolated to zero urea concentration, the rate constants for folding and unfolding are $3600\,s^{-1}$ and $27\,s^{-1}$ at 37°C, respectively, for the particular variant of the λ repressor studied (9).

Why does this protein fold so fast? It is relatively small, only 79 amino acid residues, but some proteins of this size fold much slower. A unique feature of this protein is that the only significant secondary structural elements are α-helices, which are known to wind and unwind in microseconds or less. In contrast, β-sheet structures form and break down considerably slower. The detailed folding mechanism has been explored further by site-specific mutagenesis of amino acid residues. The results obtained suggest that minor modifications, such as changing two glycines to alanines or disruption of a single hydrogen bond, can have significant effects on the folding mechanism. The results further suggest that formation of one of the five helical stretches may be a crucial slow step in the folding mechanism.

Knowledge of the kinetics of fast folding proteins provides unique information about the folding mechanism, and NMR provides one of the few methods available for studying these very fast reactions.

12.5 RNA FOLDING

The folding of RNA into functional structures is of obvious importance to biology. The mechanism of folding appears to be more complex than that for proteins (cf. Ref. 12). A number of structures exist that have similar free energies so that molecules can become trapped in nonfunctional structures. Furthermore, because RNA molecules are highly charged, the structures formed are very dependent on the ionic environment, particularly on the concentration of Mg^{2+}. A frequently invoked mechanism is that two major structural changes occur in the folding reaction. First, stable secondary structures form on the microsecond time scale. Secondary structure is defined as local structural elements such as helices, etc. due to base pairing. The second stage is formation of tertiary folding that brings the secondary structural elements together. This is similar to the mechanism of protein folding discussed above. The study described below suggests that this picture is too simple in many cases.

We have previously discussed ribozymes (Chapters 5 and 7). The folding of a selfsplicing RNA from *Tetrahymena* has served as a prototypical example of RNA folding (13,14). This group I intron consists of two large domains, labeled P4–P6 and P1–P2/P3–P9. The stable P4–P6 domain folds independently, and its crystal structure has been determined by X-ray crystallography (15,16). Within this structure are three helices, labeled P5abc, and the structure of this RNA has been studied with NMR (17). The P5abc RNA can fold independently, and its structure is amenable to determination by NMR.

To arrive at a structure, the one-dimensional proton NMR spectrum was obtained for the imino protons, protons that are involved in the hydrogen bonding of base pairs. An A–U base pair has one imino proton, and a G–C base pair has two imino protons. A two-dimensional NOESY study was then carried out to determine which imino protons were close to each other. Interpretation of these spectra required assignment of each of the resonance peaks in the one-dimensional spectrum to specific bases in the RNA. The imino protons of guanine and uracil give rise to very sharp NMR resonances in the 10–15 ppm range. This suggests that a single conformation is present and no aggregation is occurring. In the NOESY spectrum, the diagonal peaks corresponded to the one-dimensional spectrum, and each cross-peak connected two diagonal peaks, which implies that the two protons are within 5 Å of each other. Neighboring base pairs give rise to NOEs and helped establish the assignments of specific resonance peaks and the secondary structure of the RNA.

The results are summarized in Figure 12-5, where the secondary structure of the 56-nucleotide RNA is shown. The arrows indicate the connectivity established by the NOESY experiments, often called a "NOE walk." Included in the structure are disks that indicate the locations of the imino protons, all of which are involved in base pairs.

The NMR results give a quite clear picture of the secondary structure. Surprisingly, when this structure was compared with that determined with X-ray crystallography, the two structures were found to be different. In the crystals, the hydrogen bonding of six base pairs is broken (two G–C, one A–U, and three G–U) and four new G–C base pairs are formed, as well as two nonstandard A–U pairs. In addition, a tetraloop is disrupted, and other changes occur. The calculated free energy of the secondary structure of the RNA crystal structure is actually higher than that of the NMR secondary structure.

However, this difference in free energy is more than compensated for by interactions within the tertiary structure. These structural differences can be reconciled if Mg^{2+} is added to the solution. The addition of this metal results in a conversion to a new structure that is identical to that found in the crystal structure. In fact, two sets of resonances are seen, indicating that the two structures are slowly interconverting. In this case, slow means with a rate constant less than 30 s^{-1}. (This is determined by the differences in chemical shifts between the two structures.) In terms of the structure, the interactions of RNA with Mg^{2+} alter both the secondary structure and the tertiary structure. This change in structure is shown schematically in Figure 12-6. In the fully folded RNA, a cluster of five Mg^{2+} ions are coordinated to phosphate oxygens, indicating that the tertiary folding occurs on a Mg^{2+} core.

The significance of these results in understanding RNA folding is that the preformed secondary structure cannot be assumed to be the direct precursor of the final tertiary structure (17,18). In other cases, this has been found to be a valid assumption. It also indicates that many stable conformations are available for RNA within a range of salt and Mg^{2+} concentrations. This makes the delineation of the physiological folding mechanism for RNA more difficult than that for proteins where a more restricted number of stable conformations probably exist.

In a follow-up study, site-specific mutations of the P5abc RNA fragment were made to explore the balance between tertiary and secondary structures (19). In particular, point mutations designed to disrupt the secondary structure greatly affected

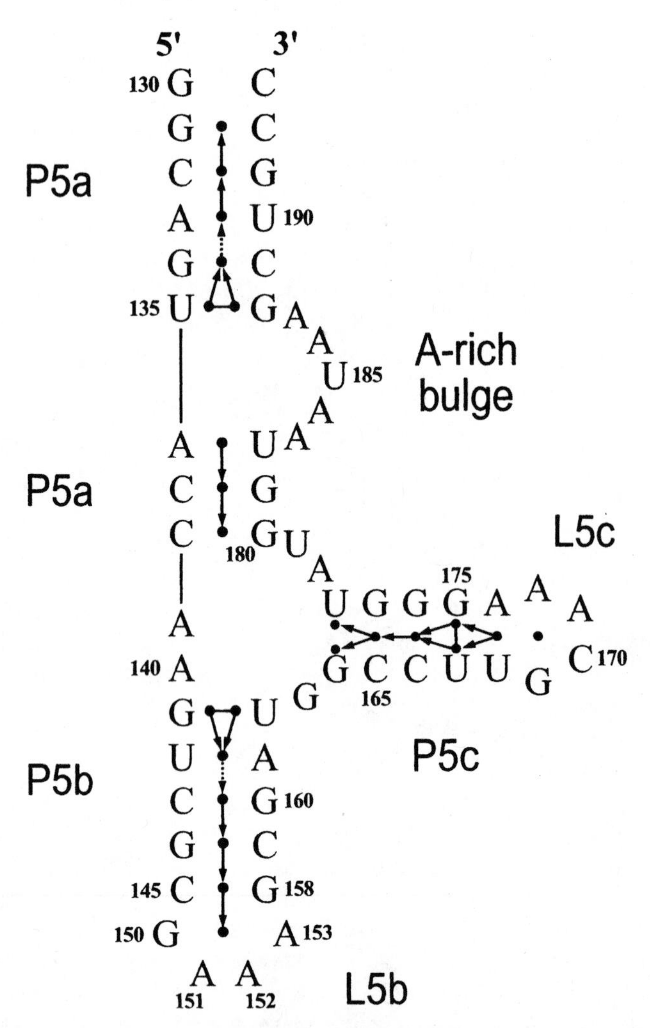

FIGURE 12-5. Secondary structure of the 56-nucleotide RNA ribozyme fragment. The disk between each base pair represents the imino protons that are observed by NMR. The arrows represent the connectivity established by the NOESY experiments. The dotted arrow NOE was not observed because the resonances could not be resolved. Reproduced with permission from M. Wu and I. Tinoco, Jr., *Proc. Natl. Acad. Sci. USA* **95**, 11555 (1998). © 1998 National Academy of Sciences, USA.

the Mg^{2+}-dependent folding into the new structure. Moreover, two single-point mutations are sufficient to prevent the rearrangement of the secondary structure observed in the crystal structure. However, if the same experiments are done with the P4–P6 domain, formation of the tertiary structure is sufficient to alter the secondary structure of P5abc, as observed with the native P5abc RNA, even with the point mutations. The message of this work is that the balance between ter-

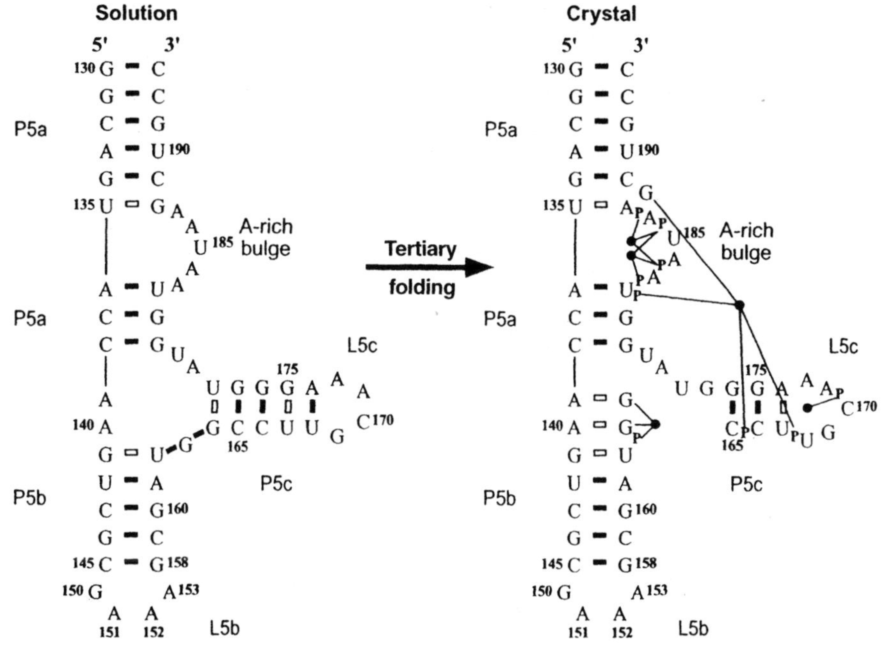

FIGURE 12-6. Comparison of the structures of the RNA ribozyme fragment determined by NMR and crystallography. The tertiary folding is caused by the five Mg^{2+} ions shown in the crystal structure as filled circles. The solid bars between bases represent Watson–Crick pairing, and the open bars represent non-Watson–Crick base pairing. Note the changes in base pairing caused by the presence of Mg^{2+}. Reproduced with permission from M. Wu and I. Tinoco, Jr., *Proc. Natl. Acad. Sci. USA* **95**, 11555 (1998). © 1998 National Academy of Sciences, USA.

tiary and secondary structure is very delicate in RNA. It depends on both the ionic conditions and the context of the RNA domains. This is likely to be a general feature of RNA structure/folding.

12.6 LACTOSE PERMEASE

Transport proteins are integral membrane proteins that are responsible for the flow of many metabolites across the cell membrane. Two broad classes of transport mechanisms exist, *facilitated diffusion* and *active transport*. Facilitated diffusion depends on a concentration gradient, and the molecules being transported flow from a higher to a lower concentration. Selectivity is due to the size of the pore and/or gating of the channel by a stimulus. For example, Gram-negative bacteria contain several porins, 34–38 kDa proteins, in their outer membranes. The porins permit molecules and ions with a molecular mass of about 600 to enter. The molecules that are transported include maltodextrins, sugar phosphates, and chelated iron. Glucose and HCO_3^- are also generally transported across membranes by facilitated diffusion. In active transport, material is carried across the membrane against a concentration gradient, from low to high concentration. Active transport must be coupled to an electrochemical gradient created by a free energy favorable process such as the hydrolysis of ATP.

Lactose permease (LacY) is responsible for all of the translocation reactions carried out by the galactoside transport system in *E. coli.* It couples the free energy associated with the collapse of a proton gradient, that is, transport of protons, with the energetically uphill translocation of galactosides against a concentration gradient. The lactose permease from *E. coli* has been extensively studied (cf. Ref. 20), and ESR has been one of the tools used to elucidate its structure in the membrane. The protein has been purified, reconstituted into proteoliposomes, and shown to be solely responsible for the galactoside transport.

The crystal structure of lactose permease is now known (21), but prior to the structure determination, the primary features of the molecule had been determined by less direct methods such as mutagenesis, chemical cross-linking, and ESR. Lactose permease has a molecular weight of about 45,500, and the structure contains 12 α-helices passing through the membrane, linked by loops on each side of the membrane. The structure is shown schematically in Figure 12-7. A hydrophilic cavity is formed between the helices that alternately faces the inside and outside of the cell as the sugar is transported. We will not be concerned about the detailed mechanism of transport here, but rather with the structural information that was obtained from a series of ESR studies. In simplistic terms, the mechanism consists of a series of sugar and proton bindings that result in the conformation of the protein switching between conformations that expose the hydrophilic cavity to the appropriate side of

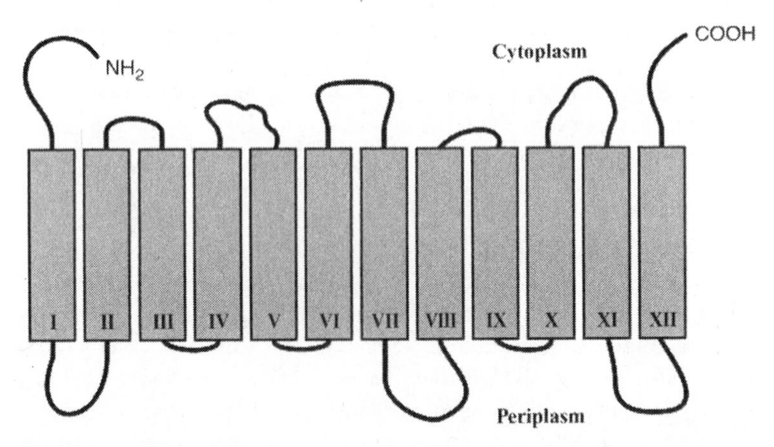

FIGURE 12-7. Secondary structure model of lac permcase. The permease has a hydrophilic N-terminus, followed by 12 α-helical hydrophobic domains that are connected by hydrophilic loops, and a hydrophilic C-terminus. Adapted with permission from J. Wu, J. Voss, W. L. Hubbell, and H. R. Kaback, *Proc. Natl. Acad. Sci. USA* **93,** 10123 (1996). © 1996 National Academy of Sciences, USA.

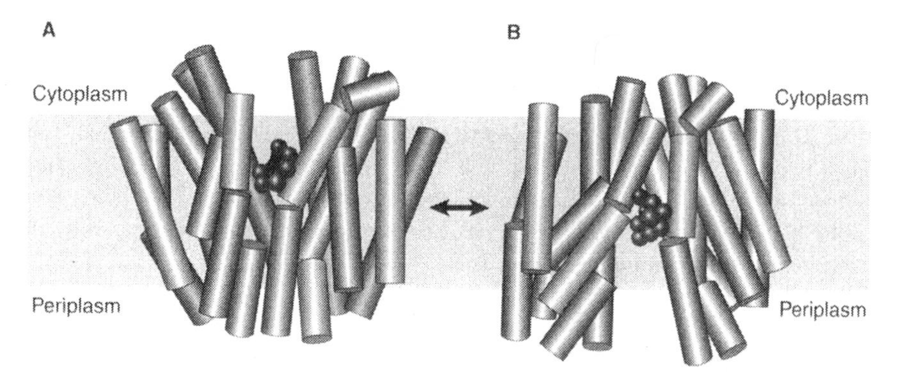

FIGURE 12-8. Schematic representation of the inward and outward-facing conformations of lac permease. The mechanism of lactose transport involves alternating these conformations. Lactose is shown as a space-filling molecular structure. Adapted with permission from J. Abramson, I. Smirnova, V. Kasho, G. Verner, H. R. Kaback, and S. Iwata, *Science* **301**, 610 (2003). © 2003 AAAS.

the membrane. The structural models for the inward and outward facing conformations are shown schematically in Figure 12-8: note the complex arrangement and distortions of the helical rods that make up the structure.

Site-directed spin labeling of lactose permease can be carried out by introduction of a single cysteine residue into a protein with no cysteines, and then labeling the thiol with a nitroxide spin label. The protein with no cysteines (prepared by site-specific mutagenesis) retains its transport activity. In the initial set of experiments (22), three derivatives were prepared with cysteines at amino acid positions 148 (helix V) and 228 (helix VII), 148 and 226 (helix VII), or 148 and 275 (helix VIII). A nitroxide spin label was then covalently linked to the cysteines. The ESR spectra showed relatively broad lines for all three pairs, suggesting that the nitroxides were relatively immobile due to strong interactions with their environment, nearby protein and membrane. Additional broadening could be detected at low temperatures for the 148/228 and 148/275 pairs due to close proximity of the nitroxides (spin–spin interactions). Analysis of this additional broadening indicates that the spin labels are within about 15 Å of each other. These results suggest that position 148 in helix V is in close proximity to position 228 in helix VII and position 275 in helix VIII. Chemical cross-linking experiment indicates that position 148 is closer to helix VII than to helix VIII. On the other hand, the lack of spin–spin interactions between positions 148 and 226 suggests that cysteine 226 is on the opposite face of helix VII from cysteine 228.

In a second study (23), a high-affinity Cu^{2+} binding site was created on lactose permease by replacing Arg 301 (helix IX) and Glu 325 (helix X) with His residues. In addition, a series of proteins with single cysteines at various positions in helices II, V, or VII were prepared, and nitroxide spin labels were attached to the cysteines. The ESR spectra of the nitroxides varied, depending on their locations. Those interacting within the protein structure were broad, indicating restricted rotation, whereas those directed away from the interior of the protein were much sharper, indicating freer rotation. When Cu^{2+} was added, some of the lines broadened due to interaction of the unpaired electron of the metal ion with the spin label. This permitted a mapping of the cysteines with respect to the metal ion. When these results were combined with those from chemical cross-linking, fluorescence, and other techniques, a model for the helix packing could be developed. Additional experiments of this type utilized the interaction of Gd(III) with spin labels to provide confirmation that helix V lies close to both helices VII and VIII (24).

Single cysteines also were introduced at positions 126 – 156, and the cysteines were derivatized with nitroxide spin labels (25). The dynamics of these labels varied from highly mobile to highly immobilized, as judged by the broadness of the spectra. These results can be interpreted in terms of structure in that enhanced mobility can be interpreted as increased accessibility of the side chain. In addition, spectral broadening due to addition of paramagnetic species (potassium chromium oxalate and oxygen) provided further information about the accessibility of the spin labels to the solvent. These experiments provide information about both the structure and the dynamics of the protein side chains.

These extensive ESR studies provide detailed structural information about a membrane-bound protein. Such information is difficult to obtain because membrane-bound proteins are not readily accessible to the large cadre of techniques that can be applied to soluble proteins. In this case, a crystal structure has been obtained that is in reasonable accord with the more indirect methods used. The studies of lactose permease serve as a useful paradigm for the investigation of the structure of membrane-bound proteins.

12.7 CONCLUSION

These selected examples of the application of magnetic resonance methods to biology are a small fraction of the many interesting investigations that have been carried out. Note that in each case, important questions were being asked about structure and function. In point of fact, the biological problem is the most significant aspect of these studies, and magnetic resonance is one of many spectroscopic/ biochemical tools that can be used for the elucidation of these problems.

REFERENCES

1. A. Giordano and M. L. Avantaggiati, *J. Cell Phys.* **181**, 218 (1999).
2. R. H. Goodman and S. M. Smolik, *Genes Dev.* **14**, 1553 (2000).
3. R. N. De Guzman, H. Y. Liu, M. Martinez-Yamout, H. J. Dyson, and P. E. Wright, *J. Mol. Biol.* **303**, 243 (2000).
4. R. Duncan, L. Bazar, G. Michelotti, T. Tomonaga, H. Krutzsch, M. Avigan, and D. Levens, *Genes Dev.* **8**, 465 (1994).

5. C. G. Burd and G. Dreyfuss, *Science* **265**, 615 (1994).
6. D. T. Braddock, J. M. Louis, J. L. Baber, D. Levens, and G. M. Clore, *Nature* **415**, 1051 (2002).
7. J. K. Myers and T. G. Oas, *Annu. Rev. Biochem.* **71**, 783 (2002).
8. R. T. Sauer, S. R. Jordan, and C. O. Pabo, *J. Mol. Biol.* **227**, 177 (1990).
9. G. S. Huang and T. G. Oas, *Proc. Natl. Acad. Sci. USA* **92**, 6878 (1995).
10. R. E. Burton, G. S. Huang, M. A. Daugherty, P. W. Fullbright, and T. G. Oas, *J. Mol. Biol.* **263**, 311 (1996).
11. J. K. Myers and T. G. Oas, *Biochemistry* **38**, 6761 (1999).
12. P. Brion and E. Westhof, *Annu. Rev. Biophys. Biomol. Struct.* **26**, 113 (1997).
13. P. P. Zarrinker and J. R. Williamson, *Nat. Struct. Biol.* **3**, 432 (1996).
14. B. Scalvi, M. Sullivan, M. R. Chance, M. Brenowitz, and S. A. Woodson, *Science* **279**, 1940 (1998).
15. J. H. Cate, A. R. Gooding, E. Podell, K. Zhou, B. L. Golden, C. E. Kundrot, T. R. Cech, and J. A. Doudna, *Science* **273**, 1678 (1996).
16. J. H. Cate, A. R. Gooding, E. Podell, K. Zhou, B. L. Golden, A. A. Szewczak, T. R. Cech, and J. A. Doudna, *Science* **273**, 1696 (1996).
17. M. Wu and I. Tinoco, Jr., *Proc. Natl. Acad. Sci. USA* **95**, 11555 (1998).
18. D. Thirumalai, *Proc. Natl. Acad. Sci. USA* **95**, 11506 (1998).
19. S. K. Silverman, M. Zheng, M. Wu, I. Tinoco, Jr., and T. R. Cech, *RNA* **5**, 1665 (1999).
20. J. Abramson, I. Smirnova, V. Kasho, G. Verner, S. Iwata, and H. R. Kaback, *FEBS Lett.* **555**, 96 (2003).
21. J. Abramson, I. Smirnova, V. Kasho, G. Verner, H. R. Kaback, and S. Iwata, *Science* **301**, 610 (2003).
22. J. Wu, J. Voss, W. L. Hubbell, and H. R. Kaback, *Proc. Natl. Acad. Sci. USA* **93**, 10123 (1996).
23. J. Voss, W. L. Hubbell, and H. R. Kaback, *Biochemistry* **37**, 211 (1998).
24. J. Voss, J. Wu, W. L. Hubbell, V. Jacques, C. F. Meares, and H. R. Kaback, *Biochemistry* **40**, 3184 (2001).
25. M. Zhao, K. C. Zen, J. Hernandez-Borrell, C. Altenbach, W. L. Hubbell, and H. R. Kaback, *Biochemistry* **38**, 15970 (1999).

SPECIAL TOPICS

CHAPTER 13

Ligand Binding to Macromolecules

13.1 INTRODUCTION

The binding of ligands to macromolecules is a key element in virtually all biological processes. *Ligands* can be small molecules, such as metabolites, or large molecules, such as proteins and nucleic acids. Ligands bind to a variety of *receptors*, such as enzymes, antibodies, DNA, and membrane-bound proteins. For example, the binding of substrates to enzymes initiates the catalytic reaction. The binding of hormones, such as insulin, to receptors regulates metabolic events, and the binding of repressors and activators to DNA regulates gene transcription. The uptake and release of oxygen by hemoglobin is essential for life. Indeed, compilation of a comprehensive list of biological processes in which ligand binding plays a key role would be a formidable task.

In this chapter, we shall discuss how to analyze ligand binding to macromolecules quantitatively for both simple and complex systems. We will also consider experimental methods that are used to study ligand binding. The application and importance of this analysis for biology will be illustrated through specific examples. The treatment presented will be adequate for most situations: more complete (and more complex) discussions of this topic are available (1–3).

13.2 BINDING OF SMALL MOLECULES TO MULTIPLE IDENTICAL BINDING SITES

The binding of a small molecule to identical sites on a macromolecule is a common occurrence. For example, enzymes frequently have several binding sites for substrates on a single molecule. Proteins have multiple binding sites for protons that are often essentially identical, for example, carboxyl or amino groups. Let us assume the simplest case, namely, a single ligand, L, binding to a single site on a protein, P:

$$\mathrm{L} + \mathrm{P} \rightleftharpoons \mathrm{PL} \tag{13-1}$$

Physical Chemistry for the Biological Sciences by Gordon G. Hammes

This equilibrium can be characterized by the equilibrium constant, K:

$$K = [\mathrm{PL}]/([\mathrm{P}][\mathrm{L}]) \tag{13-2}$$

Binding equilibria involving macromolecules are conveniently characterized by a binding isotherm, the moles of ligand bound per mole of protein, r. For the above case,

$$r = \frac{[\mathrm{PL}]}{[\mathrm{P}] + [\mathrm{PL}]} = \frac{K[\mathrm{L}]}{1 + K[\mathrm{L}]} \tag{13-3}$$

A plot of r versus [L] is shown in Figure 13-1a: It is a hyperbolic curve with a limiting value of 1 at high ligand concentrations, and when $r = 0.5$, $[\mathrm{L}] = 1/K$.

If there are n identical binding sites on the protein, the binding isotherm for the macromolecule is simply the sum of those for each of the sites:

$$r = \frac{nK[\mathrm{L}]}{1 + K[\mathrm{L}]} \tag{13-4}$$

A plot of r versus [L] has the same shape as before, but now the limiting value of r at high ligand concentrations is n, the number of identical binding sites on the macromolecule, and when $r = n/2$, $[\mathrm{L}] = 1/K$. Thus, a study of ligand binding to a protein containing n identical binding sites permits determination of the number of binding sites and the equilibrium binding constant. In practice, alternative plots of the data are frequently used that yield a straight line plot and therefore can be analyzed more easily. With the availability of nonlinear least squares programs on desktop computers, this is not really necessary. However, it should be kept in mind that statistical analyses always fit the data—this does not necessarily mean that the fit is a good one. The quality of the fit must carefully be examined to be sure that the equations used and the fitting procedures are appropriate. In addition, it is extremely important that a very wide range of ligand concentration is used.

The most obvious recasting of Eq. 13-4 is to take its reciprocal:

$$1/r = 1/n + 1/(nK[\mathrm{L}]) \tag{13-5}$$

A plot of $1/r$ versus 1/[L] is a straight line with an intercept on the [L] axis of $1/n$ and a slope of $1/(nK)$. Another possibility is to multiply Eq. 13-5 by [L]:

$$[\mathrm{L}]/r = [\mathrm{L}]/n + 1/(nK) \tag{13-6}$$

If $[\mathrm{L}]/r$ is plotted versus [L], a straight line is obtained with a slope of $1/n$ and an intercept of $1/(nK)$. A common alternative plot is a Scatchard plot (4), named after George Scatchard, a pioneer in the study of small molecule binding to proteins. Rearrangement of Eq. 13-4 gives

$$r/[\mathrm{L}] = nK - rK \tag{13-7}$$

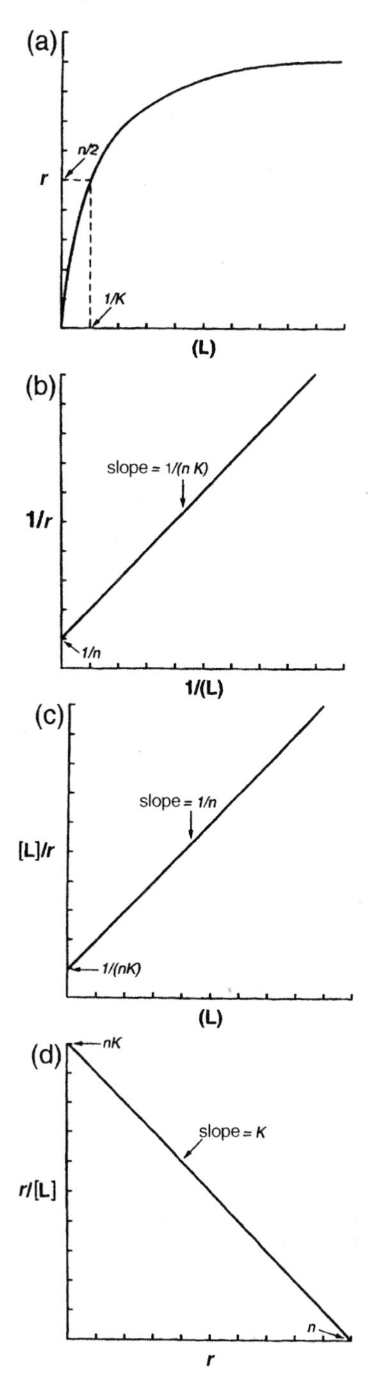

FIGURE 13-1. Plots of binding isotherms for n identical sites on a macromolecule according to (a) Eq. 13-4, (b) Eq. 13-5, (c) Eq. 13-6, and (d) Eq. 13-7. When $n = 1$, the curve in (a) obeys Eq. 13-3.

A plot of $r/[L]$ versus r is a straight line, with the intercept on the x axis ($r/[L] = 0$) being equal to the number of binding sites, n, and the intercept on the y axis being equal to nK. Examples of these straight line plots are included in Figure 13-1.

13.3 MACROSCOPIC AND MICROSCOPIC EQUILIBRIUM CONSTANTS

Before we proceed further with the analysis of ligand binding to macromolecules, it is important to understand the distinction between macroscopic and microscopic equilibrium constants. Consider, for example, a dibasic acid such as the amino acid glycine. Four possible protonation states are possible:

$$GH_2^+ = {}^+H_3NCH_2COOH$$
$$GH = H_2NCH_2COOH$$
$$GH' = {}^+H_3NCH_2COO^-$$
$$GH^- = H_2NCH_2COO^-$$

If a pH titration is carried out, the states GH and GH′ cannot be distinguished as they contain the same number of protons. The pH titration can be used to determine the *macroscopic* ionization constants:

$$K_1 = ([GH] + [GH'])[H^+]/[GH_2{}^+] \tag{13-8}$$
$$K_2 = [G^-][H^+]/([GH] + [GH']) \tag{13-9}$$

The two pK values determined by pH titration are p$K_1 = 2.35$ and p$K_2 = 9.78$ (25°C and zero salt concentration).

If we consider the *microscopic* states of glycine, four *microscopic ionization constants, k_i, are needed to characterize the system:*

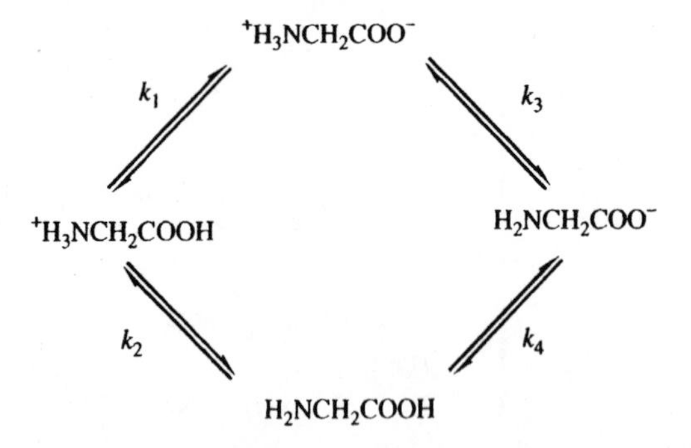

Note that these four microscopic ionization constants are not independent because this is a closed cycle. If the four microscopic ionization constants are written in terms of the concentrations, it can be seen that

$$k_1k_3 = k_2k_4 \tag{13-10}$$

This is an example of the principle of detailed balance. Whenever a closed cycle of reactions occurs, a relationship between the individual equilibrium constants such as Eq. 13-10 exists. The relationship between the macroscopic and microscopic ionization constants can easily be seen by reference to Eqs. 13-8 and 13-9, namely,

$$K_1 = k_1 + k_2 \tag{13-11}$$

and

$$K_2 = k_3 k_4 / (k_3 + k_4) \tag{13-12}$$

Since the two macroscopic ionization constants and Eq. 13-10 provide only three relationships between the microscopic ionization constants, it is apparent that the microscopic ionization constants cannot be determined from a pH titration. In order to calculate the microscopic ionization constants, additional information is needed. In this case, it might be assumed that k_2 is equal to the ionization constant determined by pH titration of the methyl ester of glycine, namely, $pk_2 = 7.70$. This assumes that the ionization constant for protonation of the amino group is the same when either a proton or a methyl group is bound to the carboxyl, which is not unreasonable. With this assumption, the other three microscopic pK values can be calculated: $pk_1 = 2.35$, $pk_3 = 9.78$, and $pk_4 = 4.43$. These results indicate that the bottom state in the ionization scheme above is present at very low concentrations throughout a pH titration.

This is a simple illustration of the distinction between macroscopic and microscopic equilibrium constants. Experimental measurements usually only give information about macroscopic binding constants although exceptions exist. For example, the state of protonation of the nitrogen can be monitored by nuclear magnetic resonance (NMR) so that, in principle, a titration curve can be obtained for the amino group independently of that for the dibasic acid.

13.4 STATISTICAL EFFECTS IN LIGAND BINDING TO MACROMOLECULES

Let us return again to the binding of a ligand to multiple identical binding sites on a macromolecule and consider the matter of macroscopic and microscopic equilibrium constants. As a simple example, consider a macromolecule, P, with two identical binding sites; symbolically, one of the binding sites will be on the left of P and the other on the right of P. The macroscopic equilibrium constants are

$$K_1 = ([PL] + [LP])/[L][P] \tag{13-13}$$

and

$$K_2 = [PL_2]/\{[L]([PL] + [LP])\} \tag{13-14}$$

In relating this discussion to the previous consideration of glycine, it should be noted that the glycine protolytic equilibria are characterized by equilibrium *dissociation* constants whereas equilibrium *association* constants are used here.

The equilibrium binding isotherm can be written as

$$r = \frac{[PL] + [LP] + 2[PL_2]}{[P] + [PL] + [LP] + [PL_2]} \tag{13-15}$$

If both the numerator and the denominator are divided by [P], it can easily be seen that

$$r = \frac{K_1[L] + 2K_1K_2[L]^2}{1 + K_1[L] + K_1K_2[L]^2} \tag{13-16}$$

This does not look the same as Eq. 13-4. This is because the equilibrium constant in Eq.13-4 is the microscopic equilibrium constant. In this case, it is easy to see that $K_1 = 2K$ and $K_2 = K/2$. If these relationships are put into Eq. 13-16,

$$r = \frac{2K[L](1 + K[L])}{(1 + K[L])^2}$$

$$r = \frac{2K[L]}{1 + K[L]}$$

Thus, in the case of multiple identical binding sites, the binding isotherm is the same whether macroscopic or microscopic equilibrium constants are considered. The relationship between microscopic and macroscopic constants for this particular case is particularly simple. When both sites are empty, there are two possible sites available for the ligand so that the macroscopic equilibrium constant is twice as large as the microscopic constant. When both sites are occupied by ligand, there are two sites from which the ligand can dissociate; therefore, the macroscopic equilibrium constant is one-half of the microscopic constant.

If more than two identical binding sites are present, the relationship between the macroscopic and microscopic equilibrium constants can be determined in a similar manner. If a macromolecule has n identical binding sites, the relationship between the macroscopic equilibrium constant for binding to the ith site and the microscopic equilibrium constant is

$$K_i = \frac{\text{number of free sites on P before binding}}{\text{number of occupied sites on P after binding}} K$$
$$K_i = [(n - i + 1)/i]K, \quad i \geq 1 \tag{13-17}$$

We now return to consideration of a macromolecule with n identical binding sites.

The macroscopic equilibria are

$$\begin{aligned} L + P &\rightleftharpoons PL \\ L + PL &\rightleftharpoons PL_2 \\ \vdots \quad & \quad \vdots \\ L + PL_{n-1} &\rightleftharpoons PL_n \end{aligned} \tag{13-18}$$

The corresponding macroscopic equilibrium constants are

$$\begin{aligned} K_1 &= [PL]/([P][L]) \\ \vdots \quad & \quad \vdots \\ K_n &= [PL_n]/([L][PL_{n-1}]) \end{aligned} \tag{13-19}$$

The binding isotherm can be written as

$$r = \frac{[PL] + 2[PL_2] + \cdots + n[PL_n]}{[P] + [PL] + [PL_2] + \cdots + [PL_n]} \tag{13-20}$$

If we divide the numerator and denominator by [P] and use the definitions of the macroscopic association constants for the ratios $[PL_i]/[P]$, Eq. 13-20 becomes

$$r = \frac{K_1[L] + 2K_1K_2[L]^2 + \cdots + nK_1K_2\cdots K_n[L]^n}{1 + K_1[L] + K_1K_2[L]^2 + \cdots + K_1K_2\cdots K_n[L]^n} \tag{13-21}$$

This equation describes the binding of a ligand to n *different* binding sites as no relationship has been assumed between the binding constants. If the sites are identical, the macroscopic constants are related to the microscopic or intrinsic binding constant by Eq. 13-17. If this relationship is inserted into Eq.13-21, it becomes

$$r = \frac{nK[L] + \dfrac{2n(n-1)}{2!}K^2[L]^2 + \cdots + nK_n[L]^n}{1 + nK[L] + \dfrac{n(n-1)}{2!}K^2[L]^2 + \cdots + K_n[L]^n}$$

$$r = \frac{nK[L](1 + K[L])^{n-1}}{(1 + K[L])^n} = \frac{nK[L]}{1 + K[L]} \tag{13-22}$$

The binomial theorem has been used to obtain the final result; that is, the series of terms in the denominator that contain successive powers of [L] is recognized as the expansion of $(1 + K[L])^n$. Similarly, the expansion of $(1 + K[L])^{n-1}$ can be factored out of the numerator. This result is exactly the same as Eq.13-4, which was obtained in a more intuitive manner.

To conclude this section, we consider the situation where a macromolecule has more than one set of independent identical binding sites for a ligand. In this case, the equilibrium binding isotherm is simply a sum of the terms given in Eq. 13-4, with the number of terms in sum being equal to the number of sets of binding sites:

$$r = \sum \frac{n_i K_i [\mathrm{L}]}{1 + K_i [\mathrm{L}]} \tag{13-23}$$

Here n_i and K_i are the number of sites in each set and the intrinsic equilibrium constant for each set, respectively. Unless independent information is available about the structure of the macromolecule, it is usually preferable to use Eq. 13-21 to fit the data as no assumptions are made about the nature or number of the binding sites in this case.

Virtually any ligand binding isotherm can be fit to Eq. 13-21, but this is not a meaningful exercise unless the result can be related to the structure and/or the function of the macromolecule. For example, the binding isotherm for the binding of laurate ion to human serum albumin is shown in Figure 13-2 (5,6). The concentration range covered is so large that the concentration axis is logarithmic. The corresponding Scatchard plot is also shown in Figure 13-2. It is clear that saturation of the binding sites on serum albumin is never reached. The data in the figures, nevertheless, can be fit well with the assumption that $n = 10$. Clearly, this does not establish that 10 binding sites exist, nor regrettably does it provide information about the structure of

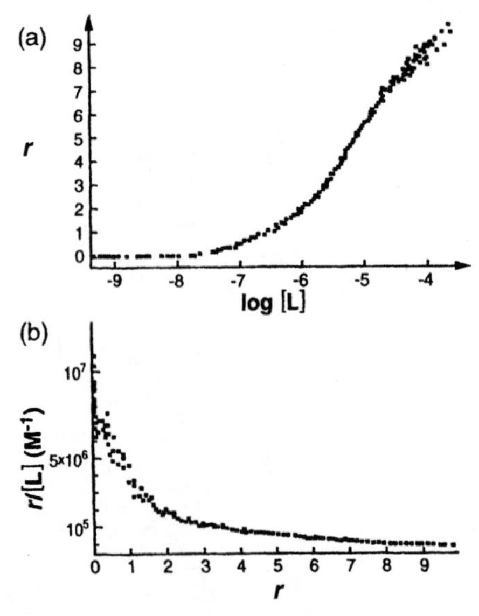

FIGURE 13-2. (a) Plot of r versus log [L] for the binding of laurate ion by human serum albumin. (b) The same data plotted as r/[L] versus r (Scatchard plot). Adapted from I.M. Klotz, *Ligand-Receptor Complexes*, John Wiley & Sons, Inc. New York, 1997. Copyright © 1997 John Wiley & Sons, Inc. Reprinted by permission of John Wiley & Sons, Inc.

the macromolecule. We will not dwell on this matter here. We will, however, return to the topic of multiple binding sites when discussing the concept of cooperativity.

13.5 EXPERIMENTAL DETERMINATION OF LIGAND BINDING ISOTHERMS

We will now make a digression to discuss briefly some of the experimental aspects of ligand binding studies. Many different experimental methods exist for determining the binding of ligands to macromolecules. The experimental methods used fall into two classes: (1) direct determination of the unbound ligand concentration; and (2) measurement of a change in physical or biological property of the ligand or macromolecule when the ligand binds. Generally, stoichiometry and binding constants can be determined reliably only if concentrations of the unbound ligand, the bound ligand, and unbound macromolecule can be determined.

The most straightforward method is equilibrium dialysis. This method is pictured schematically in Figure 13-3. With this method, a solution is separated into two parts by a semipermeable membrane, which will not permit the macromolecule to cross but will permit the ligand to pass freely. The macromolecule is put on one side of the membrane and the ligand on both sides. The system is then permitted to come to equilibrium. At equilibrium, the concentration of the unbound ligand is the same on both sides of the membrane. (Strictly speaking, their thermodynamic activities are equal, but we will not worry about the difference between activity and concentration here.) If the total amount of ligand is known, and the unbound amount of ligand is known, the amount of bound ligand can be determined from mass balance. The concentration of the macromolecule can be determined independently so that the binding isotherm can be calculated directly.

If equilibrium dialysis is to provide reliable data for analysis of the binding isotherm, it is usually necessary for the concentrations of bound and unbound ligand and macromolecule to be comparable. If essentially all of the ligand is

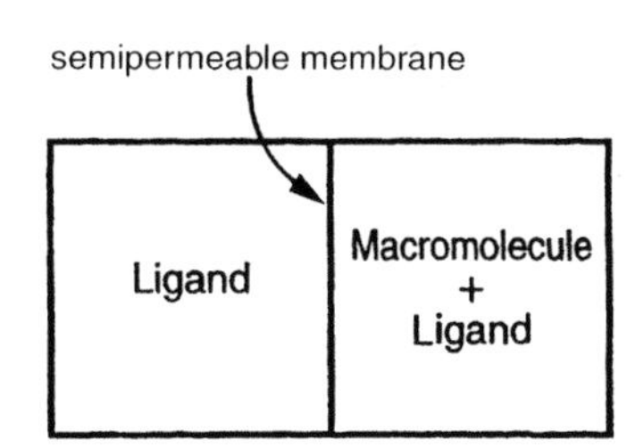

FIGURE 13-3. Schematic representation of an equilibrium dialysis experiment. The macromolecule is on one side of a semipermeable membrane and cannot pass through the membrane. The ligand can pass through the membrane and therefore is on both sides of the membrane. The concentration of the ligand on the side of the membrane that does not contain the macromolecule is equal to the concentration of the unbound ligand.

bound, the amounts of unbound ligand and/or protein become very small and difficult to determine. On the other hand, if essentially no ligand is bound, calculation of the bound ligand becomes problematical. In the former case, the stoichiometry of binding can often be determined, but not the binding constant(s). In the latter case, it is virtually impossible to determine either the stoichiometry or binding constant(s). Another way of stating this is that the concentrations of all the species should be of the same order of magnitude as the reciprocal of the binding constant so that experimental points can be determined over a wide range of r. This limitation becomes restrictive for very weak binding, primarily in terms of the large amount of material required. In the case of very tight binding, it may be difficult to work with the very dilute solutions required.

Equilibrium dialysis is an example of the more general method of determining the distribution between phases. This could, for example, be the equilibration between two nonmiscible solvents that partition the reactants, although this is rarely used. Various gel exclusion media such as Agarose are, however, often used. Such gels will exclude macromolecules from their beads, with the size of the macromolecule excluded depending on the property of the specific gel. The gel will include the ligand so that the gel bead serves as a semipermeable membrane, with equilibration occurring between the inside and the outside of the bead. If a macromolecule is passed through a column equilibrated with a given ligand concentration, the unbound ligand concentration is equal to this concentration. The amount of ligand bound and the macromolecule concentration can be determined by analysis of the column effluent. Many variations on this type of experimental protocol exist.

In some cases, the concentration of the unbound ligand is determined directly in the absence of a semipermeable membrane, for example, by a biological assay. In this case, the implicit assumption is that the rate of equilibration of the ligand and macromolecule occurs more slowly than the time for analysis. This is not necessarily the case, so such methods must be used with extreme caution.

Perturbations in the physical properties of the ligand are often used to determine binding isotherms. However, some care and caution must be taken in doing so. The most frequent situation is when the optical absorption properties of the ligand are different for the bound and unbound ligand, or for the macromolecule with and without ligand. The technique of difference spectroscopy is often used (cf. Chapter 8). With this method, the difference in the optical absorption of a solution containing the ligand and a solution containing the same concentration of ligand plus the macromolecule is determined. If the concentration of the total ligand in the first solutions is $[\mathrm{L_T}]$ and the concentrations of the free (unbound) ligand and the bound ligand are $[\mathrm{L_f}]$ and $[\mathrm{L_b}]$, respectively, this difference, ΔA, can be written as

$$\begin{aligned} \Delta A &= \varepsilon_\mathrm{f}[\mathrm{L_T}] - \varepsilon_\mathrm{f}[\mathrm{L_f}] - \varepsilon_\mathrm{b}[\mathrm{L_b}] \\ \Delta A &= (\varepsilon_\mathrm{f} - \varepsilon_\mathrm{b})[\mathrm{L_b}] \end{aligned} \tag{13-24}$$

where ε_i are extinction coefficients and the relationship $[\mathrm{L_T}] = [\mathrm{L_f}] + [\mathrm{L_b}]$ has been used. This derivation assumes that the extinction coefficient of bound ligand is the same at all binding sites, which may or may not be valid. Consequently, this method is not as direct as the partitioning methods previously described. The difference extinction coefficient, $\varepsilon_f - \varepsilon_b$, can be determined by extrapolation to high ligand concentrations where the macromolecule is saturated with the ligand. The difference absorbance measurements will then permit the determination of the bound ligand concentration, and the concentrations of the other species can be determined by mass balance. This method also works if the spectral change occurs in the macromolecule rather than in the ligand. The only difference is that the comparison solution contains macromolecule rather than protein, and the concentrations and extinction coefficients of the protein appear in Eq. 13-24. Perhaps not so obvious is the fact that this method also works if spectral changes occur in both the ligand and the macromolecule when the ligand binds. The derivation of ΔA for this case is given in Chapter 8.

As a specific example of a difference spectrum titration, the difference absorbance is shown as a function of the ligand concentration for the binding of 2′-cytidine monophosphate to ribonuclease A in Figure 13-4 (7). Other optical properties such as fluorescence and circular dichroism can be used in a similar manner. Finally, it is some times possible to monitor a specific site on the macromolecule with physical methods such as NMR or other types of spectroscopy. If changes occur when the ligand binds, the concentrations of empty and occupied sites can be determined and translated into binding isotherms.

The determination of high-quality binding isotherms is not trivial, and this discussion does not do full justice to the topic. One of the difficulties not discussed that should be mentioned is the occurrence of nonspecific binding. It is commonplace

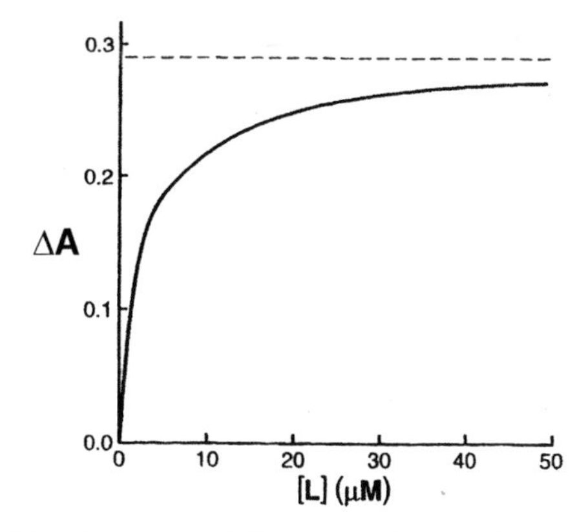

FIGURE 13-4. Plot of the absorbance difference at 288 nm, ΔA, accompanying the binding of 2′-CMP to ribonuclease A versus the concentration of unbound 2′-CMP at pH 5.5, 25°C. The total enzyme concentration is 10^{-4} M. The difference extinction coefficient is 2.88×10^3 cm^{-1} M^{-1} and the binding constant is 2.96×10^5 M^{-1}. The curve was calculated from the data in D.G. Anderson, G.G. Hammes, and Frederick G. Walz, Jr., *Biochemistry* **7**, 1637 (1968). The dashed line is the difference absorbance when all of the enzyme has been converted to enzyme-2′-CMP.

for nonspecific binding to occur along with binding of the ligand to the site(s) of interest. Macromolecules have many different side chains that can attract ligands, for example, by charge–charge interactions. This nonspecific binding is characterized by binding isotherms that do not level off at a specific value of n–due to a large number of such sites characterized by very weak (small) binding constants. More complete treatises concerned with the experimental aspect of the binding of ligands to macromolecules are available (cf. Ref. 3).

13.6 BINDING OF CRO REPRESSOR PROTEIN TO DNA

As an example of a ligand binding study, we will consider some of the results obtained in the studies of the binding of the Cro repressor protein to DNA. DNA transcription is regulated by the binding of proteins to DNA. The Cro repressor is a protein that plays a regulatory role in λ phage. The binding of this protein to DNA has been examined by many different methods (8–10), and the three-dimensional structure of a Cro protein–DNA complex has been determined (11). The Cro protein binds as a dimer, and amino acids form DNA sequence specific hydrogen bonds with exposed parts of DNA bases. Both the thermodynamics and the kinetics of Cro protein binding to DNA have been studied extensively.

The binding of Cro protein to DNA is difficult to study because the binding is very tight. In order to study the binding, a very sensitive filter binding assay was used. Under certain conditions, linear duplex DNA passes through nitrocellulose filters, while a protein or protein–DNA complex is retained. Radioactive (^{32}P) DNA was prepared and mixed with Cro protein at very low concentrations (as low as <1 ng/ml). The reaction mixture was filtered within a few seconds, and from the radioactivity bound to the filter, the concentration of Cro protein–DNA could be calculated. Since the total concentrations of Cro protein and DNA are known, the equilibrium concentrations of all species can be calculated. The binding of Cro protein to various operator regions of λ phage DNA was studied. The operator is 17 base pairs long, but the DNAs investigated contained an extra 2 base pairs at each end. In addition, a consensus operator was constructed that had the consensus sequence of the six λ operator regions examined. The consensus sequence is identical to the OR3 operator, except for a few base pairs on the interior of the sequence.

The equilibrium association constants obtained are given in Table 13-1. The stoichiometry of the reaction was determined to be one to one. The association constants are very large, with the largest being for the consensus and OR3 operators. The OR3 operator was previously determined to be the preferred binding site for Cro protein. The binding of Cro operator to nonspecific DNA was also determined and found to be a linear function of the length of the DNA with a binding constant of $6.8 \times 10^5\ \mathrm{M}^{-1}$ per base pair, considerably weaker than the specific binding, as expected. The rate constants characterizing the reaction between the DNA operator sequences and the Cro the protein were also determined. The dissociation rate constants are included in Table 13-1. The association rate constants that

TABLE 13-1. Binding Constants and Dissociation Rate Constants at 273 K for Cro Protein–λDNA

Operator	K (M^{-1})	k_d (s^{-1})
Consensus	8.3×10^{11}	7.7×10^{-5}
OR3	5.0×10^{11}	1.7×10^{-4}
OR2	8.3×10^{9}	1.2×10^{-2}
OR1	1.2×10^{11}	1.7×10^{-3}
OL1	6.7×10^{10}	2.2×10^{-3}
OL2	3.7×10^{10}	2.6×10^{-3}
OL3	1.9×10^{10}	9.2×10^{-3}

Source: J. G. Kim, Y. Takeda, B. W. Matthews, and W. F. Anderson, *J. Mol. Biol.* **196**, 149 (1987).

could be measured were all approximately $3 \times 10^8\ M^{-1}\ s^{-1}$, the value expected for a diffusion controlled reaction. Every time a Cro protein and the operator DNA collide, a complex is formed.

The binding of Cro protein to long DNAs containing the operator region was also studied. The length of DNA varied from 73 to 2410 base pairs. For all except the shortest DNA, the equilibrium constant was about the same, $6.7 \times 10^{10}\ M^{-1}$. However, the association and dissociation rate constants increase as the length of DNA increased, leveling off at values of about $4.5 \times 10^9\ M^{-1}\ s^{-1}$ and $1.7 \times 10^{-2}\ s^{-1}$, respectively. This very high second-order rate constant suggests that every time Cro protein collides with DNA, it binds very tightly. This result is puzzling as it would be expected that Cro protein would have to sample various parts of the DNA until it found the operator and bound tightly. An explanation for these results is that Cro protein binds to the DNA on every collision and then diffuses rapidly along the DNA chain until it encounters the operator region. The rapid sliding of DNA binding protein along the DNA chain until it finds the correct place for a specific interaction appears to occur in other systems, but other mechanisms may also be operative.

Calorimetry was also carried out on the binding reaction (10). The results obtained indicate that the association of Cro protein with nonspecific DNA at 15°C is characterized by $\Delta H^\circ = 4.4$ kcal/mol, $\Delta S^\circ = 49$ cal/(mol K), $\Delta G^\circ = -9.7$ kcal/mol, and $\Delta C_P \approx 0$. The parameters obtained with the OR3 DNA are quite different, $\Delta H^\circ = 0.8$ kcal/mol, $\Delta S^\circ = 59$ cal/(mol K), $\Delta G^\circ = -16.1$ kcal/mol, and $\Delta C_P = -360$ cal/(mol K). In both cases, the favorable free energy change is entropy driven.

The specific molecular interactions between the DNA and the Cro protein have been probed by studying the binding of Cro protein to OR1 DNA, 21 base pairs as above, with systematic base substitutions along the entire DNA chain (9). Experiments were also done with point mutations in the protein of amino acids thought to be involved in the binding interaction. The changes in free energy were then interpreted as due to specific interactions between the protein and the DNA, and a structural map of the interactions was proposed. A high-resolution crystal structure of a Cro protein–DNA complex determined some years later provided a definitive description of the molecular interactions (11). Many features proposed on the basis of the binding/mutation studies proved to be correct although some were not. As often stated in this text, molecular interpretations of thermodynamic studies must be viewed with caution. However, in retrospect, all of the mutation studies can be understood in terms of the crystal structure. A view of the overall complex is presented in Figure 13-5 (see color plates). Each Cro protein monomer consists of three α-helices and three β-strands. Only the helix–turn-helix portion of the protein makes direct contact with the DNA bases: the α_3-helix is inserted into the major groove of DNA, and its interactions with DNA bases account for the tight operator binding. A schematic map of the DNA base–Cro protein interactions is shown in Figure 13-6. Multiple hydrogen bonds are formed, although other types of interactions are also important. Both the protein and DNA conformations are altered when the complex is formed. Many other interesting details of the molecular interactions can be inferred from the crystal structure but will not be considered here.

This study is an elegant demonstration of the type of information that can be obtained from thermodynamic, kinetic, and structural studies of ligand binding.

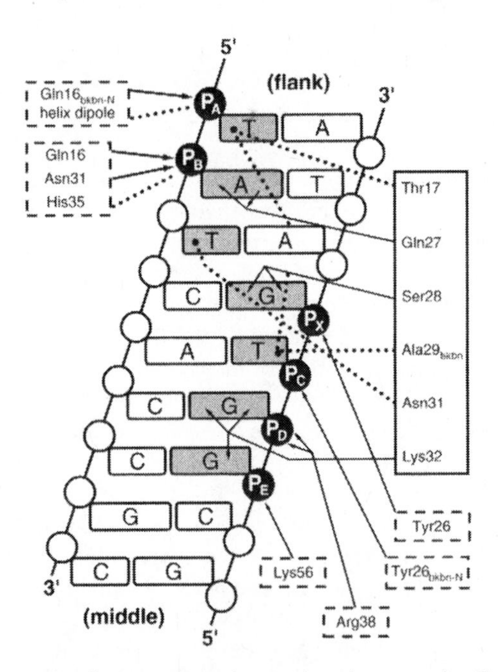

FIGURE 13-6. Schematic diagram of the interactions between the Cro protein and the λ operator DNA. The interactions for one of the polypeptide chains are shown as the interactions with the other polypeptide chain of the dimeric Cro protein are symmetric. Hydrogen bonds are shown as continuous lines with arrows pointing from the donor to the acceptor. Broken lines are van der Waals contacts, bkbn indicates a contact with the protein backbone, and the dotted lines are presumed electrostatic interactions. Reproduced from R. A. Albright and B. W. Matthews, crystal structure of lambda-Cro bound to a consesus operator, *J. Mol. Biol.* **280**, 137 (1998). Copyright 1998, with permission from Elsevier.

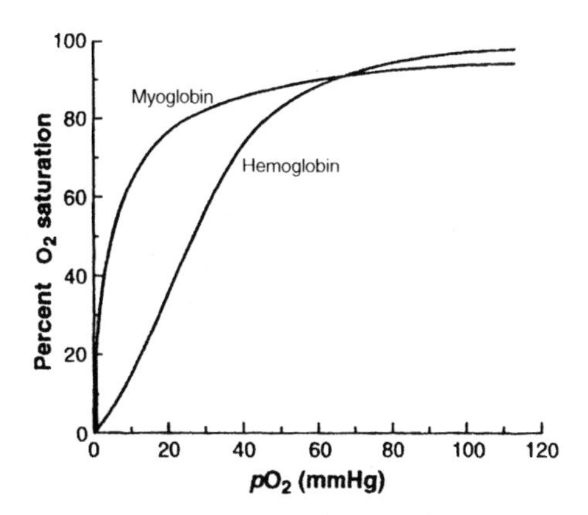

FIGURE 13-7. Binding of oxygen by myoglobin and hemoglobin at pH 7 and 38°C. The pressure of O_2 is a measure of the concentration of unbound O_2. Adapted from F. Daniels and R. A. Alberty, *Physical Chemistry*, 4th edition, John Wiley & Sons, Inc., New York, 1975.

13.7 COOPERATIVITY IN LIGAND BINDING

We now return to the subject of macromolecules with multiple ligand binding sites. One of the most famous and best studied examples is hemoglobin. In Figure 13-7, the binding isotherm is shown for the binding of oxygen to myoglobin and hemoglobin. Hemoglobin contains four identical binding sites for oxygen. Myoglobin contains a single binding site for oxygen. The concentration of unbound oxygen is expressed in pressure units of mmHg and rather than r; the percent saturation (r/n) is shown so that the binding isotherms can be compared directly even though $n = 1$ for myoglobin and 4 for hemoglobin. The binding isotherm for myoglobin can be fit to Eq. 13-4 with $K = 0.23\,(\text{mmHg})^{-1}\{K = [\text{MyO}_2]/([\text{My}][\text{O}_2])$, where My is myoglobin$\}$. The data for hemoglobin clearly cannot be fit so simply, as the binding isotherm is not hyperbolic. Equation 13-21 can be used, however. Before doing so, it is useful to take into account the statistical effects associated with identical binding sites, that is, express Eq. 13-21 in terms of intrinsic binding constants rather than macroscopic binding constants by use of Eq. 13-17. With this substitution, Eq. 13-21 becomes

$$r = \frac{4K_1[\text{O}_2] + 12K_1K_2[\text{O}_2]^2 + 12K_1K_2K_3[\text{O}_2]^3 + 4K_1K_2K_3K_4[\text{O}_2]^4}{1 + 4K_1[\text{O}_2] + 6K_1K_2[\text{O}_2]^2 + 4K_1K_2K_3[\text{O}_2]^3 + K_1K_2K_3K_4[\text{O}_2]^4} \qquad (13\text{-}25)$$

If this equation is used to fit the data obtained at pH 7.4, 25°C, 0.1M NaCl, the binding constants obtained are 0.024, 0.077, 0.083, and 7.1 $(\text{mmHg})^{-1}$ (12). Note that the intrinsic binding constant becomes larger as each oxygen is bound. This is an example of cooperative binding. When an oxygen binds to hemoglobin, it increases the affinity of hemoglobin for the next oxygen. The result is a sigmoidal binding isotherm rather than a hyperbolic isotherm. This has important physiological consequences as it permits oxygen to be picked up and released over a very narrow range of oxygen pressure. Cooperative binding such as this is found frequently because it permits the biological activity to be regulated over a very narrow range of concentration.

When the binding of oxygen by hemoglobin was first studied, the binding isotherm was fit by the empirical equation

$$r/n = \frac{[\mathrm{L}]^{\alpha}/K^{\alpha}}{1 + [\mathrm{L}]^{\alpha}/K^{\alpha}} \tag{13-26}$$

where the equilibrium constant is expressed as a dissociation constant and α is an empirical parameter called the Hill coefficient obtained from experiment by plotting $\ln[(r/n)/(1-r/n)]$ versus $\ln[\mathrm{L}]$ (13). {Note that $(r/n)/(1-r/n) = [\mathrm{L}]^{\alpha}/K^{\alpha}$.} As might be expected, such plots are linear over a limited range of ligand concentration. In the case of hemoglobin, α is about 2.5 and depends on the specific experimental conditions. The use of this equation does not have a physical meaning. It assumes that the binding equilibrium is

$$\text{Hemoglobin} + \alpha\mathrm{O}_2 \rightleftharpoons \text{Hemoglobin}(\mathrm{O}_2)_{\alpha}$$

Obviously, the number of binding sites on hemoglobin must be an integer. Nevertheless, cooperative binding is frequently characterized by a Hill coefficient. The closer the Hill coefficient is to the actual number of binding sites, the more cooperative the binding.

It is also possible to have binding isotherms in which the binding constants decrease as successive ligands bind. This is usually termed *negative cooperativity* or *anticooperativity*, both oxymorons of a sort. It should be remembered that if macroscopic equilibrium binding constants are used, the binding constants decrease as each ligand is added to the macromolecule even if the sites are equivalent. This is the statistical effect embodied in Eq. 13-17. Anti- or negative cooperativity therefore means that the binding constants decrease more than the statistical effect expected for equivalent sites. An example of negative cooperativity is shown in Figure 13-8 for the binding of cytidine 3′-triphosphate (CTP) to the enzyme aspartate transcarbamoylase (14). This enzyme catalyzes the carbamoylation of aspartic acid at a branch point in metabolism that eventually leads to the synthesis of pyrimidines. When the concentration of CTP becomes high, it inhibits aspartate transcarbamoylase and shuts down the metabolic pathway for its synthesis. As can be seen from the Scatchard plot presented, the data clearly do not conform to the straight line expected for independent equivalent binding sites. The total number of binding sites appears to be six, although extrapolation to this number is not precise. In this case, the data were fit to a model of two sets of three independent binding sites. The intrinsic binding constants obtained were $7.1 \times 10^5\,\mathrm{M}^{-1}$ and $4.4 \times 10^3\,\mathrm{M}^{-1}$. The data could be fit equally well to Eq. 13-21, but the data were not sufficient to define all six constants well. Aspartate transcarbamoylase is known to contain six identical binding sites for CTP and has three dimers of identical polypeptide chains. When CTP binds to one of the sites on a dimer, it weakens

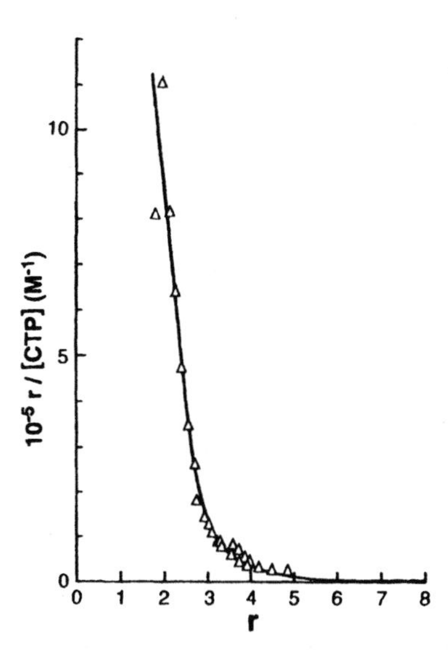

FIGURE 13-8. Scatchard plot for the binding of CTP to aspartate transcarbamoylase at 4°C and pH 7.3 in the presence of 2 mM carbamoyl phosphate and 10 mM succinate. Adapted from S. Matsumoto and G. G. Hammes, *Biochemistry* **12**, 1388 (1973). Copyright © 1973 American Chemical Society.

the binding for the second CTP. Interestingly, the binding of aspartate to this enzyme exhibits positive cooperativity, and the extent of the cooperativity is modulated by the binding of CTP, as will be discussed later.

The presence of positive or negative cooperativity can readily be diagnosed from the binding isotherm. The shapes of the various plots commonly used are shown in Figure 13-9 for no cooperativity, positive cooperativity, and negative cooperativity. The type of cooperativity occurring can readily be diagnosed from these plots. It is particularly obvious in Scatchard plots as no cooperativity gives a straight line, a maximum in the plot is observed for positive cooperativity, and a concave curve is found for negative cooperativity. For some studies, log[L] is used in order that a wide range of concentrations can be represented in a single plot (Fig.13-9b).

When positive and negative cooperativities are observed, one must determine if this is due to a preexisting difference between the binding sites, or if the binding of one ligand alters the binding affinity of the remaining sites for the ligand. Usually, but not always, the latter is true, as for the cases cited above. The possibility also exists that both positive and negative cooperativity could occur in a single binding isotherm.

13.8 MODELS FOR COOPERATIVITY

Thus far, we have been concerned with fitting data to binding isotherms and developing criteria with regard to whether the binding is cooperative. However,

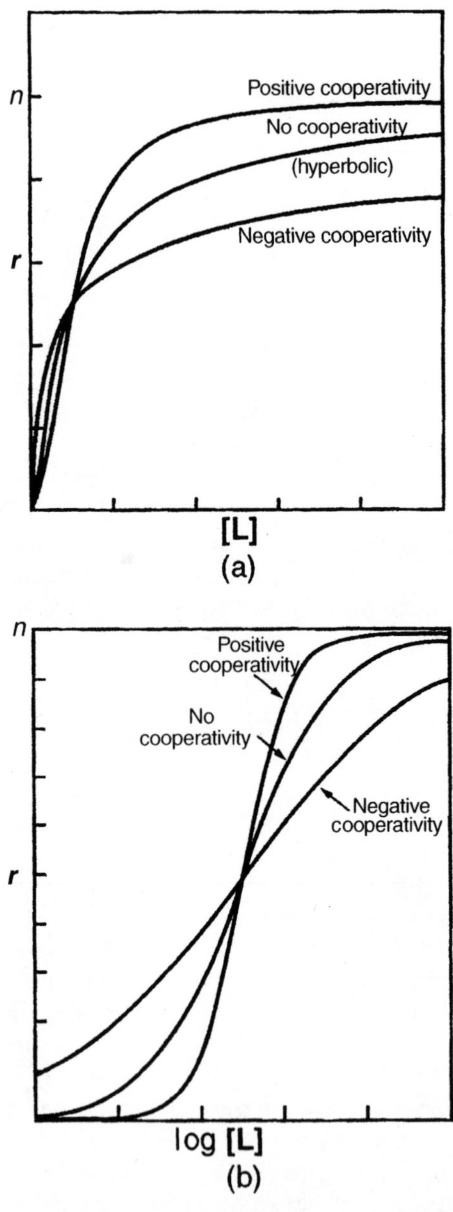

FIGURE 13-9. Schematic representations of equilibrium binding data demonstrating no cooperativity, positive cooperativity, and negative cooperativity. In these figures, r is the moles of ligand bound per mole of protein and [L] is the concentration of unbound ligand.

(Continued)

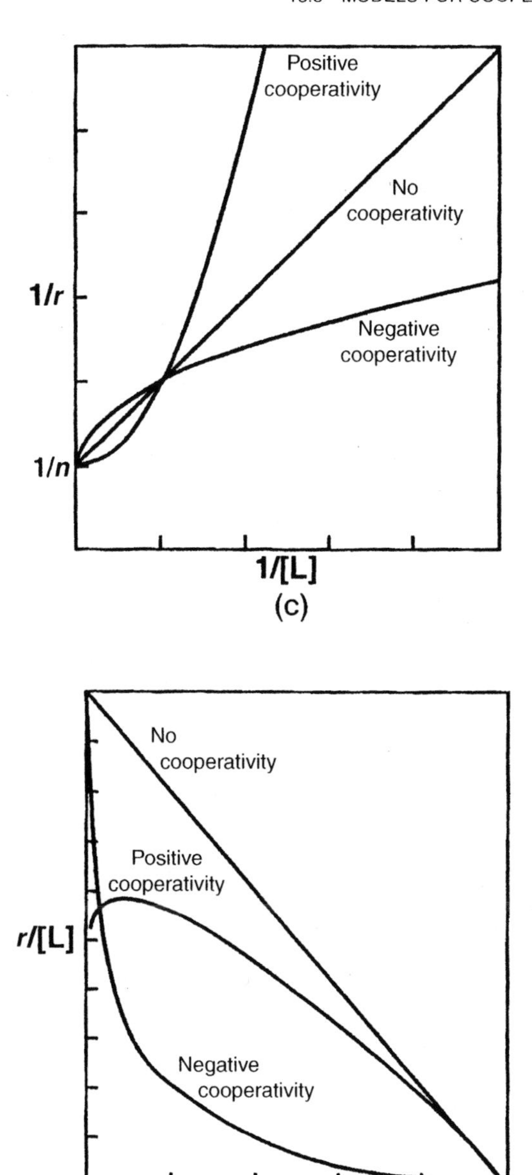

FIGURE 13-9. *(Continued)*

the ultimate goal is to relate the experimental data to molecular structure. This requires the development of theoretical models that relate structure to ligand binding isotherms. Two limiting models have been developed to explain cooperative ligand binding to proteins. Both models are based on the general hypothesis that

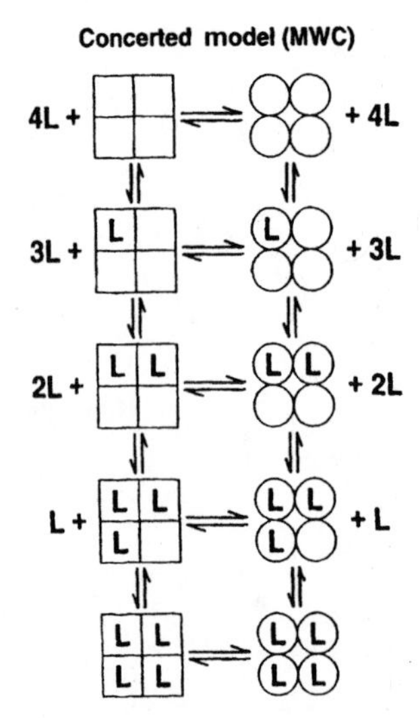

FIGURE 13-10. Schematic representation of the Monod–Wyman–Changeux model for a four-subunit protein. The squares and circles designate different subunit conformations of the protein, and L is the ligand.

cooperativity is the result of alterations in the interactions between polypeptide chains through conformational changes in the macromolecule. One of the models assumes a concerted or global conformational change, whereas the other assumes a sequential change in the conformation of each polypeptide chain or chains that contain a ligand binding site.

We first consider the concerted model that has been developed by Monod, Wyman, and Changeux (15). This model (MWC) is based on three assumptions: (1) the protein consists of two or more identical subunits, each containing a site for the ligand; (2) at least two conformational states, usually designated as R and T states, are in equilibrium and differ in their affinities for the ligand; and (3) the conformational changes of all subunits occur in a concerted manner (conservation of structural symmetry). A schematic illustration of the MWC model for a four-subunit protein is shown in Figure 13-10. A sigmoidal binding isotherm can be generated from this model in the following way. In the absence of ligand, the protein exists largely in the T state (the square conformation), but the substrate binds preferentially to the R state (the circular conformation). When the ligand binds, it shifts the protein from the T to the R state. Thus, at low ligand concentrations, the protein is primarily in the T state, whereas at high ligand concentration, the protein is largely in the R state. This shift in equilibria can give rise to a sigmoidal binding isotherm, or positive cooperativity. This model cannot, however, explain anti(negative) cooperativity.

The quantitative development of this model is complex, although not difficult, and will only be given in outline form here. The scheme for ligand binding can be represented as

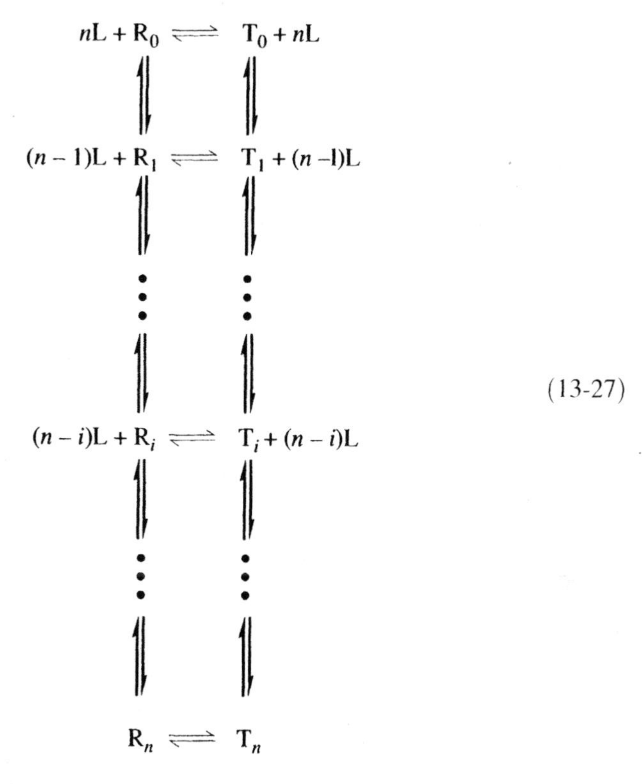

(13-27)

where R_0 and T_0 are the two different conformational states in the absence of ligand, and R_i and T_i designate their complexes with i molecules of L. Three constants are needed to specify the equilibrium binding isotherms: the intrinsic dissociation constants for ligand binding to the R and T states, and the equilibrium constant for the ratio of the R_0 to T_0 states. These constants can be written as

$$\begin{aligned} L_0 &= [T_0]/[R_0] \\ K_R &= [(n-i+1)/i][R_{i-1}][L]/[R_i] \\ K_T &= [(n-i+1)/i][T_{i-1}][L]/[T_i] \end{aligned} \tag{13-28}$$

The fraction of sites occupied by the ligand, Y, can be expressed as

$$\begin{aligned} Y &= \frac{r}{n} = \frac{([R_1] + 2[R_2] + \cdots + n[R_n]) + ([T_1] + 2[T_2] + \cdots + n[T_n])}{n\{([R_0] + [R_1] + \cdots + [R_n]) + ([T_0] + [T_1] + \cdots + [T_n])\}} \\ Y &= \frac{L_0 c\alpha(1 + c\alpha)^{n-1} + \alpha(1+\alpha)^{n-1}}{L_0(1 + c\alpha)^n + (1+\alpha)^n} \end{aligned} \tag{13-29}$$

where $\alpha = [L]/K_R$ and $c = K_R/K_T$. The transformation of the first part of Eq.13-29 to the second part requires the use of the binomial theorem and will not be detailed here. The nature of the binding isotherm depends on the values of L_0 and c. A

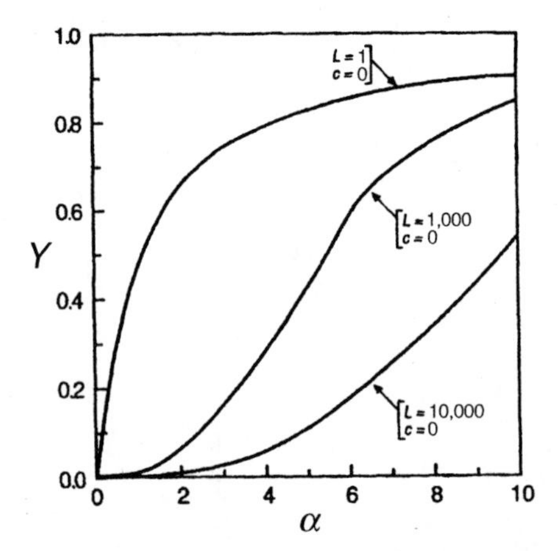

FIGURE 13-11. Plot of the fraction of sites bound by ligand, Y, versus $\alpha(= [\mathrm{L}]/K_{\mathrm{R}})$ for various values of c and L according to Eq. 13-29. A sigmoidal binding isotherm is generated when L is large and c is small.

hyperbolic isotherm is obtained when the ligand binds equally well to both conformations, $c = 1$, and when both L_0 and c are either very large or very small. However, when L_0 is large and c is small, sigmoidal binding isotherms occur as illustrated in Figure 13-11. When c is very small, Eq. 13-29 becomes

$$Y = \frac{\alpha(1+\alpha)^{n-1}}{L_0 + (1+\alpha)^n} \qquad (13\text{-}30)$$

With this limiting case, it can easily be seen that a hyperbolic isotherm is found when L_0 is small, whereas a sigmoidal isotherm is predicted when L_0 is large.

The basic assumptions of an alternative model developed by Koshland, Nemethy, and Filmer (KNF) are the following (16): (1) two conformational states, A and B, are available to each subunit; (2) only the subunit to which the ligand is bound changes its conformation; and (3) the ligand-induced conformational change in one subunit alters its interactions with the neighboring subunits. The strength of the subunit interactions may be increased, decreased, or stay the same. The result of this change in subunit interactions is that the binding of the next ligand can be weaker, stronger, or the same as the binding of the previous ligand. Clearly, this model can produce either positive or negative cooperativity—or a hyperbolic binding isotherm. This sequential model is shown schematically in Figure 13-12 for a four-subunit protein in a square configuration.

Calculation of the binding isotherm for the KNF model is complex and will not be presented here. The basic parameters are the intrinsic binding constant, an equilibrium constant characterizing the conformational change that occurs, and constants characterizing the subunit interactions, AB and BB. (The AA interaction is taken as the reference state, so it does not appear in the calculation.) The binding isotherm that results is identical in form to Eq. 13-21, except that the appropriate

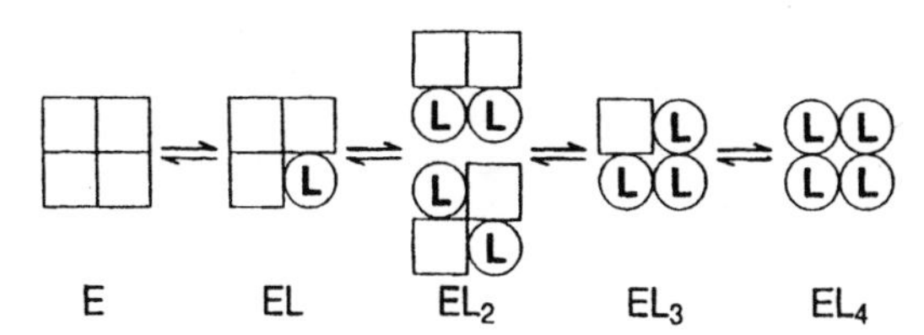

FIGURE 13-12. Schematic representation of the Koshland–Nemethy–Filmer model for a four subunit (square) protein. The squares and circles designate different subunit conformations, and L is the ligand. Note that two structures are shown for the intermediate with two ligands bound as the subunit interactions are different for these two species.

statistical corrections are included so that the intrinsic binding constant appears. The effective binding constant multiplying each successive power of L may increase, decrease, or stay the same, depending on the nature of the subunit interactions, so that positive, negative, or mixed cooperativity is possible.

The MWC and KNF models are limiting cases of a more general scheme shown in Figure 13-13. This figure illustrates a general mechanism involving a tetrameric protein and only two conformational states for each subunit. The real situation is somewhat more complex, as the permutations of the ligand among the subunits for a given conformational state are not shown. The extreme right- and left-hand columns enclosed by dashed lines represent the MWC mechanism, whereas the diagonal, enclosed by dotted lines, represents the KNF model. Thus, these two models are limiting cases of an even more complex scheme. As might be suspected, distinguishing among these models when positive cooperativity occurs is not a simple task.

13.9 KINETIC STUDIES OF COOPERATIVE BINDING

Thus far, we have confined the discussion to ligand binding at equilibrium. The binding process and the cooperativity that may occur play an important role in many biological processes, as evidenced by hemoglobin, membrane receptor binding, and operator binding to DNA. The role of cooperativity in regulating such reactions has been well established. However, many enzyme reactions are also regulated through cooperative binding processes. In such cases, the rate of the enzymatic reaction is usually measured, often the initial velocity, and it is observed that a plot of the initial velocity versus the substrate concentration exhibits cooperativity. An example already discussed is aspartate transcarbamoylase: A plot of the initial velocity versus the aspartate concentration is sigmoidal so that small changes in the concentration of aspartate can cause significant changes in the rate of the enzymatic reaction.

For many enzymes, plots of the initial velocity versus the substrate concentration are essentially identical in form with the binding isotherm. This suggests that the binding steps prior to the rate-determining step are rapid and reversible and that the

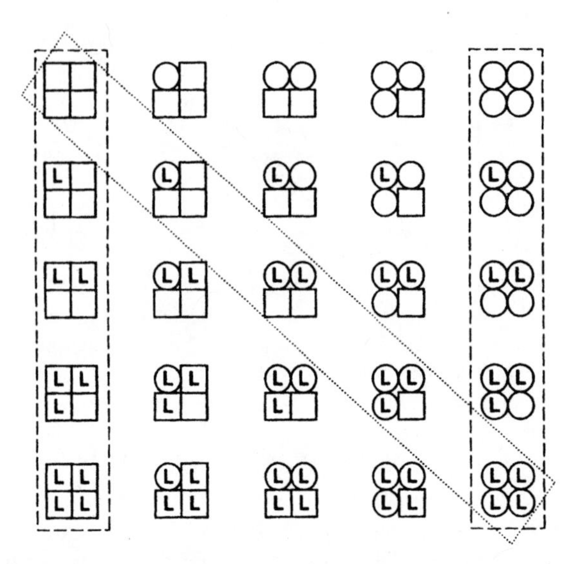

FIGURE 13-13. Schematic illustration of a general allosteric model for the binding of a ligand to a four-subunit protein. The squares and circles designate different subunit conformations. The portion enclosed by dashed lines is the Monod–Wyman–Changeux model, whereas that enclosed by dotted lines is the Koshland–Nemethy–Filmer model. For the sake of simplicity, the permutations of the ligand among the subunits and the free ligand are omitted.

turnover numbers (catalytic efficiency) for all of the binding sites are identical. If this is the situation, the initial velocity, v, can readily be related to the binding isotherm:

$$v = V_m r/n \tag{13-31}$$

where V_m is the maximum velocity of the enzyme reaction expressed as the product of the molar concentration of enzyme and the turnover number. This simple analysis seems to be sufficient for many cases, and if it is, the treatment of the rate data parallels what was done for equilibrium binding, including the methods of plotting and fitting the data. As might be expected, both positive and negative cooperativity are observed.

The interpretation of kinetic data can, however, be more complex. The turnover numbers for different sites could be different. For example, in terms of the MWC model, the R and T forms might have different catalytic activities although usually the assumption is made that only the R form is enzymatically active. For the KNF model, each of the ligand binding sites might have a different catalytic activity. In fact, these complications are rarely included, or justified, in the data analysis. Finally, it should be noted that apparent cooperativity in the rate of an enzymatic reaction can arise purely from special kinetic situations and may have nothing to do with cooperative binding of substrate. A few such situations have been well documented, but we will not consider such complications further. This is just a reminder that kinetic measurements are not a substitute for equilibrium binding studies. They may provide similar information in some cases, but they are inherently more difficult to interpret. Of course, conversely, kinetic studies can provide dynamic information that cannot be obtained from equilibrium measurements.

13.10 ALLOSTERISM

Thus far, we have considered only cases where a single ligand binds to a macromolecule, and important biological control can occur through cooperative interactions for a single ligand. These are *homotropic* interactions. However, in many instances regulation occurs through the binding of a second ligand that influences the binding of the first ligand. These are called *heterotropic* interactions. Regulatory control by reaction of a second ligand is termed *allosterism*, and the second binding site is called an *allosteric* site. Two specific examples will be discussed to illustrate the principles involved: hemoglobin and aspartate transcarbamoylase.

The association of oxygen with hemoglobin is strongly pH dependent as shown in Figure 13-14. Note that the oxygenation isotherm becomes more sigmoidal as the pH is lowered. Thus, the binding of protons to hemoglobin clearly affects oxygen binding. In fact, the addition of protons decreases the amount of oxygen bound, and vice versa. This reciprocal relationship is called the Bohr effect. The effect of proton binding is typical allosteric regulation, in this case of oxygen binding. How is this accommodated in the models we have discussed previously? For the MWC model, allosteric effectors are assumed to bind selectively to the R or T conformation. An inhibitor, I, such as the proton in the case under consideration,

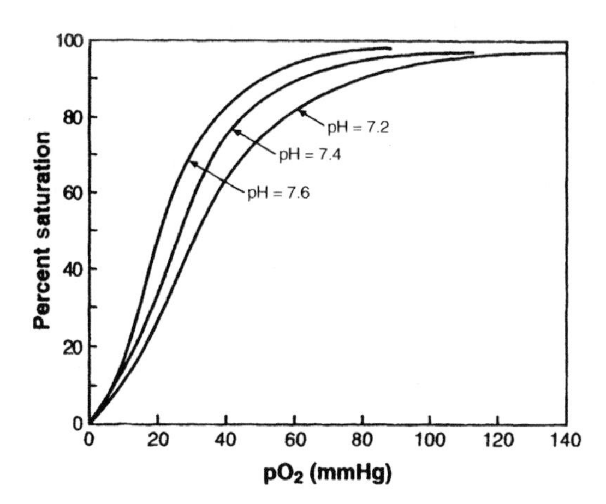

FIGURE 13-14. The effect of pH on the oxygenation of hemoglobin (the Bohr effect). The percent of saturation by oxygen is plotted versus the pressure of O_2. Adapted from R. E. Benesch and R. Benesch, The mechanism of interaction of red cell organic phosphates with hemoglobin, *Adv. Protein Chem.* **28**, 211 (1974). Copyright 1974, with permission from Elsevier.

would bind selectively to the T conformation. This shifts the equilibrium from the R to the T state, which effectively changes the equilibrium constant L_0 to

$$L_0' = L_0(1 + [\mathrm{I}]/K_\mathrm{I})^n \tag{13-32}$$

where K_I is the intrinsic dissociation constant for the binding of inhibitor, and it has been assumed that a single proton binds to each of the four subunits. Insertion of this relationship into Eq. 13-29 can quantitatively account for the Bohr effect. The KNF model can also explain the observations by assuming that binding of the inhibitor alters the subunit interactions and conformational changes, effectively altering the binding constants for oxygen as successive ligands are bound. Organic phosphates are also strong effectors of oxygen binding and bind preferentially to deoxyhemoglobin.

An activator, A, can enhance ligand binding by binding selectively to the R conformation. This effectively changes the equilibrium constant L_0 to

$$L_0' = \frac{L_0}{(1 + [\mathrm{A}]/K_\mathrm{A})^n} \tag{13-33}$$

where K_A is the intrinsic dissociation constant for the binding of activator.

Which of the models for allosterism best accommodates the known data for hemoglobin? The nature of the conformational change occurring when oxygen binds is known from structural studies of deoxy–and oxyhemoglobin. The hemoglobin studied contains four polypeptide chains, two α chains and two β chains. The α and β chains are similar but not identical. When oxygen binds, salt bridges are broken between the polypeptide chains, and all four chains rotate slightly to accommodate the movement of iron into the plane of the heme. The iron moves only a few tenths of an angstrom, and two of the polypeptides move about 7 Å closer (17). These structural changes and the binding data fit quite well to the MWC model. However, there is evidence that in addition to this major conformational change, sequential conformational changes occur in the subunits. Therefore, it is likely the case that both models are needed to accommodate the data, and multiple conformational changes occur. The structure of hemoglobin and the proposed major concerted conformational change are shown schematically in Figure 7-9 (see color plates).

As previously indicated, aspartate transcarbamoylase is a key enzyme in the pathway for pyrimidine biosynthesis. It is subject to inhibition by CTP and to activation by ATP. The allosteric binding of CTP makes the dependence of the rate on aspartate concentration more sigmoidal, whereas binding of the activator, ATP, makes the curve less sigmoidal (18). This is a prototype for feedback inhibition in metabolism and is shown schematically in Figure 13-15. An additional wrinkle in the regulatory process is that the binding of both CTP and ATP to the enzyme displays negative cooperativity, and these two ligands compete for the same binding site. The effect of ATP and CTP on the binding of aspartate can readily be accommodated by the MWC model with the assumption that CTP binds

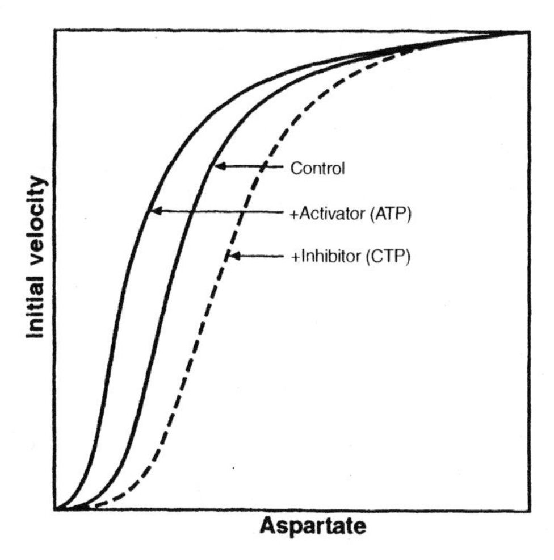

FIGURE 13-15. Schematic representation of the dependence of the initial velocity of the reaction catalyzed by aspartate transcarbamoylase on the concentration of aspartate, a substrate. The effects of an allosteric inhibitor, CTP, and an allosteric activator, ATP, are shown.

selectively to the T conformation and ATP to the R conformation. This assumes that aspartate binds selectively to the R conformation. This model, however, cannot accommodate the negative cooperativity observed in the binding of CTP and ATP. The KNF model can also accommodate these results through alterations in the interactions between subunits.

Again, the structure of aspartate transcarbamoylase is known. It consists of two trimers of catalytic sites, with three dimers of regulatory sites at the interface of the trimers (19). A threefold rotational axis is present, and it has been shown that conversion of the putative R to T form involves rotation around this axis and alteration of the interactions between the regulatory and catalytic subunits. The MWC model accommodates much of the data, but the data also require sequential conformational changes in the subunits. Thus, the conclusion is similar to that for hemoglobin. A major conformational change consistent with R and T conformations appears to occur, but additional conformational changes more localized and sequential in nature also occur. The regulatory process, therefore, seems to require multiple conformations and the interplay between global and local conformational changes. The use of multiple conformational changes enhances the versatility and sensitivity of the regulatory process. The structure of aspartate transcarbamoylase and the nature of the concerted conformational change are shown in Figure 13-16 (see color plates). The top of the figure is the T state with the catalytic trimers in blue and white at the top and bottom of the structure. Two of the regulatory dimers are at the sides of the structure in yellow. The third dimer is at the back of the structure. The bottom structure is the R state. The movement of the subunits with respect to each other can be clearly seen.

This concludes our discussion of ligand binding in biology. We have developed the theoretical and experimental frameworks and have provided several examples of applications to biological systems. This, and further reading of the literature, should permit the interested reader to develop specific applications as needed.

REFERENCES

1. G. G. Hammes, *Enzyme Catalysis and Regulation*, Academic Press, New York, 1982.
2. C. R. Cantor and P. R. Schimmel, *Biophysical Chemistry*, Parts I, II, and III, Freeman, San Francisco, CA, 1980.
3. I. M. Klotz, *Ligand–Receptor Complexes*, John Wiley & Sons, New York, 1997.
4. G. Scatchard, *Ann. N. Y. Acad. Sci.* **51**, 660 (1949).
5. A. O. Pedersen, B. Hust, S. Andersen, F. Nielsen, and R. Brodersen, *Eur. J. Biochem.* **154**, 545 (1986).
6. R. Brodersen, B. Honoré, A. O. Pedersen, and I. M. Klotz, *Trends Pharm. Sci.* **9**, 252 (1988).
7. D. G. Anderson, G. G. Hammes, and Frederick G. Walz, Jr., *Biochemistry* **7**, 1637 (1968).
8. J. G. Kim, Y. Takeda, B. W. Matthews, and W. F. Anderson, *J. Mol. Biol.* **196**, 149 (1987).
9. Y. Takeda, A. Sarai, and V. M. Rivera, *Proc. Natl. Acad. Sci. USA* **86**, 439 (1989).
10. Y. Takeda, P. D. Ross, and C. P. Mudd, *Proc. Natl. Acad. Sci. USA* **89**, 8180 (1992).
11. R. A. Albright and B. W. Matthews, *J. Mol. Biol.* **280**, 137 (1998).
12. I. Tyuma, K. Imai, and K. Shimizu, *Biochemistry* **12**, 1491 (1973).
13. A. V. Hill, *J. Physiol.* **40**, iv (1910).
14. S. Matsumoto and G. G. Hammes, *Biochemistry* **12**, 1388 (1973).
15. J. Monod, J. Wyman, and J.-P. Changeux, *J. Mol. Biol.* **12**, 88 (1965).
16. D. E. Koshland, Jr., G. Nemethy, and G. Filmer, *Biochemistry* **5**, 365 (1966).
17. J. M. Friedman, *Science* **228**, 1273 (1985).
18. J. C. Gerhart and A. B. Pardee, *J. Biol. Chem.* **237**, 819 (1962).
19. W. N. Lipscomb, *Adv. Enzymol.* **68**, 67 (1994).

PROBLEMS

13.1. The following data were obtained for the binding of ADP to an ATPase.

How many binding sites are present per mole of enzyme and what is the intrinsic binding constant? The concentration in the table is unbound ADP. (Nonspecific binding occurs, which is not uncommon. You will have to decide how to handle this problem.)

Total ligand concentration (μM)	
Side without macromolecule	Side with macromolecule
0.158	0.436
0.395	0.960
0.890	1.83
2.37	3.78
4.00	5.60
6.18	7.91
8.12	9.90

13-2. Some typical equilibrium dialysis data for the binding of a ligand to a macromolecule are given below. The total concentration of the macromolecule is 0.500 μM.

Determine the number of binding sites on the macromolecule and the intrinsic binding constant.

13-3. The following initial velocities, v, were measured for the carbamoylation of aspartic acid by carbamoyl phosphate as catalyzed by the enzyme aspartate transcarbamoylase. The concentration of carbamoyl phosphate was 1 mM, and the concentration of aspartate was varied.

v (arbitrary units)	Aspartate (mM)
0.90	2.0
1.60	3.0
2.25	4.0
3.20	5.0
3.65	6.0
4.70	8.0
5.05	10.0
5.25	12.0
5.80	15.0
6.00	17.0
6.05	20.0

a. Assume that the initial velocity is related to r by Eq. 13-31 and construct a Hill plot of the data. The slope provides a lower bound to the number of aspartate binding sites.

b. In fact, six aspartate binding sites are present per mole of protein. Using this information and the data, make a table of r and the corresponding aspartate concentration. Make a plot of r/[aspartate] versus r.

c. What type of cooperativity is occurring? Which of the two limiting models discussed (MWC and KNF) is consistent with the data?

13-4. Consider a macromolecule that has two different conformations, M and M′. The two conformations bind a single ligand, L, per molecule, but the binding equilibrium is different for each conformation. This can be represented as

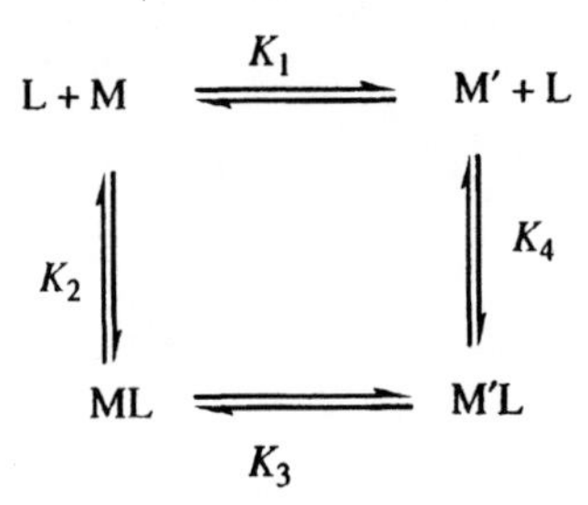

r	[ADP] (μM)
0.500	0.719
0.694	1.23
1.06	2.36
1.22	3.30
1.32	4.55
1.63	8.81
1.69	16.32
1.91	27.15
2.21	40.62
2.15	79.75
2.40	115.5

where K_i are the equilibrium constants for the individual reactions. Calculate the binding isotherm, r, for this macromolecule in terms of the equilibrium constants and the concentration of the unbound ligand. What type of cooperativity, if any, is displayed by this system? How many of the individual equilibrium constants can be determined from the binding isotherm? What relationship, if any, exists between the four constants?

CHAPTER 14

Hydrodynamics of Macromolecules

14.1 INTRODUCTION

Hydrodynamics is the science of the movement and interactions of particles in fluids, both in the absence and in the presence of external forces such as centrifugal fields (centrifugation) and electrical fields (electrophoresis). Hydrodynamic measurements are used frequently to characterize the behavior of macromolecules in solutions and can provide important molecular information. In this chapter, we consider some of the fundamental aspects of hydrodynamics in relatively simple terms and analyze some of the methods used to obtain molecular information about macromolecules. The mathematical details are not presented: more detailed discussions are available (cf. Refs. 1–4).

14.2 FRICTIONAL COEFFICIENT

If a molecule moves through a solution, it is subjected to a frictional drag because of interactions with the solvent. If the movement is slow enough so as not to cause turbulence, the frictional drag is directly proportional to the velocity of the molecule, and the proportionality constant is called the *frictional coefficient*, f. The actual calculation of this frictional coefficient is quite complex, and we present only the results. If the molecule is assumed to be a sphere of radius R_S, the frictional coefficient is

$$f = 6\pi\eta R_S \tag{14-1}$$

where η is the viscosity of the liquid. This relationship is for the translational frictional coefficient. The frictional coefficient for rotational motion can also be calculated:

$$f_{rot} = 6\eta V \tag{14-2}$$

Physical Chemistry for the Biological Sciences by Gordon G. Hammes

where V is the volume of the sphere, $(4/3)\pi R_S{}^3$. Two immediate questions arise with regard to these definitions, namely, what is the "volume," and what if the macromolecule cannot be approximated as a sphere.

Defining the volume turns out to be more difficult than might be imagined. What can be measured precisely is the partial specific volume of the solute, $\bar{V}_2$. In thermodynamic terms, the partial specific volume of a solute is the change in solution volume when a small increment of solute is added to the solution, extrapolated to infinite dilution. In practice, the volume of the solution, V_{total}, is determined as a function of the weight concentration of the solute, g. At a given constant weight, temperature, and pressure, $\bar{V}_2 = \mathrm{d}V_{\text{total}}/\mathrm{d}g$, and the observed value of $\bar{V}_2$ at various values of g is extrapolated to zero concentration to eliminate the effect of interactions between macromolecules. In practice, $\bar{V}_2$ is rarely measured directly because of the large amount of solute required. Instead, it is estimated from the amino acid composition of the protein. The partial specific volumes of the amino acids have been carefully measured so that the partial specific volume of the protein can be calculated by summing the partial specific volumes of the amino acids on a weight basis. For proteins, $\bar{V}_2$ ranges from 0.69 to 0.76 cm^3/g, and a value of 0.73 cm^3/g is used in the absence of amino acid data. Nucleic acids are more dense, and the value of $\bar{V}_2$ is more sensitive to temperature and solvent, particularly the specific counter ion, because of the strong electrostatic interactions between the polyphosphate chain and metals alter the hydration of the macromolecule. A value of about 0.5 cm^3/g is obtained if the counter ion is Na^+.

Since $\bar{V}_2$ is calculated on a weight basis, the observed volume of molecule is $(M/N_0)\bar{V}_2$, where M is the molecular weight and N_0 is Avogadro's number. This determination, however, does not take into account the hydration sphere associated with the molecule. The hydration of the macromolecule can be included by adding the volume of hydration $\delta\bar{V}_1$, where δ can be estimated by experimental measurements not considered here and $\bar{V}_1$ is the specific volume of the solvent. Thus, the hydrated volume of the molecule, V_h, is

$$V_h = (M/N_0)(\bar{V}_2 + \delta\bar{V}_1) \tag{14-3}$$

This should be regarded as an empirical estimate of the hydrated volume. The first term is often considered the anhydrous volume of the macromolecule and the second term the water of hydration. However, this is an oversimplification: $\bar{V}_2$ is also dependent on the solvent. As further validation of this approximate model, the thickness of the water layer around the macromolecule can be calculated to be about one layer. Because of the uncertainty in the extent of hydration of a macromolecule and its dependence on the details of the molecular surfaces, the hydrodynamic radius is sometimes calculated from $\bar{V}_2$, that is not explicitly including hydration. This radius is often called the hydrodynamic or Stokes' radius. George Stokes (1819–1903) was one of the pioneers in the field of hydrodynamics, and Eq. 14-1 is often called Stokes' law.

Most macromolecules are not spheres: this is obvious for nucleic acids and even in the case of proteins a better representation would be some type of irregularly

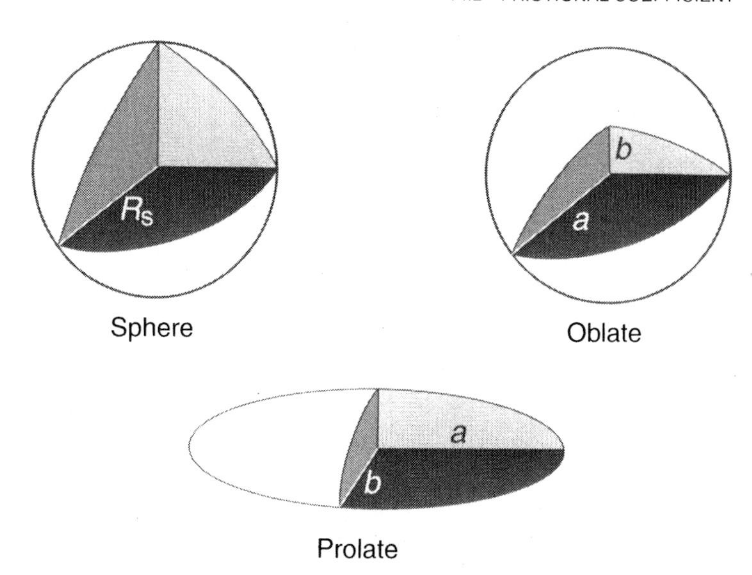

FIGURE 14-1. The most common hydrodynamic shapes used as models for molecules: the sphere and oblate and prolate ellipsoids. The cross sections are shown to illustrate the shapes. The a and b axes are defined as the length and width of the entire ellipsoid. Adapted from Ref. 1.

shaped rigid body. Nevertheless, a sphere is the most common shape analyzed because of the relative simplicity of the mathematics. In fact, even when molecules are known to be nonspherical, an "apparent" Stokes' radius is often calculated with Eq. 14-1. Two other shapes have been analyzed in detail, prolate and oblate ellipsoids of revolution. These shapes are shown in Figure 14-1, along with a sphere. A prolate ellipsoid is cigar shaped and is generated by rotation about its long axis. The two parameters characterizing it are the long axis, a, and the short axis, b. Obviously for a sphere, $a = b$. An oblate ellipsoid is generated by rotation around its short axis, b, and resembles a disk in shape. As might be guessed, these ellipsoids of revolution have larger translational frictional coefficients than spheres.

Tables of the ratio of the frictional coefficient of the ellipse to that of a sphere are available as a function of the axial ratio, a/b (prolate ellipsoids) or b/a (oblate ellipsoids). An abbreviated compilation is given in Table 14-1. This frictional coefficient ratio, f/f_{sphere}, varies from 1 to 6 for a prolate ellipsoid and from 1 to 4 for an oblate ellipsoid when the axial ratio goes from 1 to 200. As will be discussed later, hydrodynamic measurements can be used to infer information about the shape of macromolecules.

Rotational frictional coefficients are more complex since rotation can occur around either the long or short axis of an ellipsoid, so that two frictional coefficients are needed for each ellipsoid. This case will not be dealt with further except to note the following. For oblate ellipsoids, the rotational friction coefficient is similar about both axes and larger than that of the sphere. For prolate ellipsoids, the two friction coefficients are quite different, with rotation around the long axis having a

TABLE 14-1. Frictional Coefficient Ratios and Intrinsic Viscosity Shape Factors for Ellipsoids of Revolution

	Prolate		Oblate	
Axial ratio	ν	f/f_{sphere}	ν	f/f_{sphere}
1	2.500	1.000	2.500	1.000
5	5.806	1.250	4.708	1.224
10	13.634	1.543	8.043	1.458
20	38.53	1.996	14.80	1.782
50	176.81	2.946	35.16	2.375
100	593.7	4.067	69.10	2.974
200	2052.9	5.708	137.01	3.735

Data from Ref. 1 and H. A. Scheraga, *Protein Structure*, Academic Press, New York, 1961.

smaller friction coefficient than the equivalent sphere and rotation around the short axis having a larger friction coefficient. This is qualitatively what might be expected.

Frictional coefficients for more complex shapes have been calculated but are not considered here.

14.3 DIFFUSION

The motion of molecules in solution in the absence of external forces is due to diffusion. In a solution at equilibrium, the motion of molecules is random and caused by thermal energy: as the temperature goes up, the speed of the molecules increases. Diffusion cannot be easily observed in solutions at equilibrium: sophisticated light scattering techniques are required. The diffusion of macromolecules can be studied more directly by creating a boundary between a buffer and a solution of the macromolecule. As time passes the boundary spreads, due to diffusion. The movement of molecules with time is governed by Fick's laws. The first law states that when a concentration gradient exists, $\mathrm{d}c/\mathrm{d}x$, the number of molecules, n, passing through an area, A, per unit time, $\mathrm{d}n/\mathrm{d}t$, is proportional to the concentration gradient, with the proportionality constant defining the *diffusion constant*, D. This can be written in one dimension, x, as

$$\mathrm{d}n/\mathrm{d}t = -DA(\mathrm{d}c/\mathrm{d}x) \tag{14-4}$$

Strictly speaking, these should be partial derivatives, that is, the derivative with respect to time assumes that x is constant and the derivative with respect to x assumes that the time is constant. The minus sign arises because the flow of molecules is opposite in sign to the concentration gradient, and D is defined as a positive number. From dimensional analysis, D can be seen to have the dimensions of cm^2/s.

Fick's first law is easy to understand on an intuitive basis. It states that the net rate of flow of molecules is proportional to the concentration gradient, dc/dx: the larger the concentration gradient, the faster the flow. If there is no concentration gradient ($dc/dx = 0$), no net flow occurs. Fick's second law is not as apparent. It is based on the first law and the conservation of mass and relates the time dependence of the concentration to the concentration gradient. Often it is more convenient to measure the concentration rather than the flux (dn/dt). Fick's second law in one dimension can be written as

$$dc/dt = D\,d^2c/dx^2 \tag{14-5}$$

Here again, the derivative on the left is carried out with x held constant and that on the right with t held constant.

For the boundary spreading experiments in which diffusion is studied, the concentration of the macromolecule as it spreads can be detected by various optical techniques. A schematic depiction of the boundary spreading is shown in Figure 14-2. An analysis of the boundary spreading as a function of time according to Fick's laws permits the determination of the diffusion constant.

On the surface, this appears to be an easy experimental measurement, but in actuality it is quite difficult. Ideally, an infinitely sharp boundary is set up at time zero, and boundary spreading is measured as a function of time in the absence of forces other than the concentration gradient. Diffusion is slow for macromolecules so that the experiments take a long time, typically days. During the experiment, the equipment must be mechanically stable to prevent distortion of the boundary, and the temperature must be very carefully controlled to prevent thermal gradients. In addition, the net charge of the macromolecule must be kept as close to neutrality as

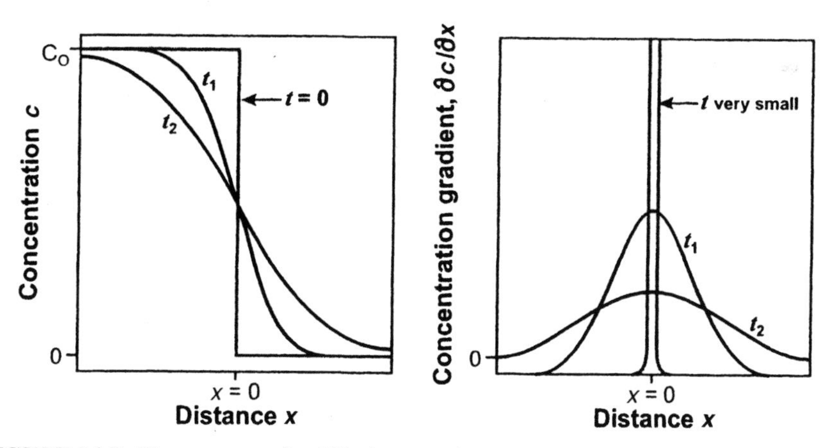

FIGURE 14-2. The progress of a diffusion experiment that begins with a sharp boundary. The concentration is shown at three different times on the left and the concentration gradient on the right. Reprinted with permission from C. Tanford, *Physical Chemistry of Macromolecules*, Wiley, New York, 1961, p. 354.

TABLE 14-2. Selected Diffusion Constants in Water at 293 K

Sample	Molecular weight	$D \times 10^7$ (cm^2/s)	$\bar{V}_2$ (cm^3/g)	f/f_{sphere}
Sucrose	342	45.9	–	–
Ribonuclease	13,683	11.9	0.728	1.14
Lysozyme	14,100	10.4	0.688	1.32
Bovine serum albumin	66,500	6.1	0.734	1.31
Hemoglobin	68,000	6.9	0.749	1.14
Tropomyosin	93,000	2.2	0.71	3.22
Myosin	493,000	1.10	0.728	3.65
Collagen	345,000	0.69	0.695	6.8
DNA	6,000,000	0.13	0.53	15
Tobacco mosaic virus	40×10^6	0.44	0.73	2.19

Data from Refs. 1 and 2.

possible to minimize electrostatic gradients. In practice, most modern day studies of diffusion are done in an analytical centrifuge with a synthetic boundary cell in which the boundary is created by layering buffer on the top of the solution. At low centrifuge speeds, little movement of the molecules occurs due to centrifugal force. Some typical diffusion constants for macromolecules are given in Table 14-2.

Complex hydrodynamic calculations show that the diffusion constant, D, can be written as

$$D = (k_B T)/f \tag{14-6}$$

where k_B is Boltzman's constant, T is the temperature, and f is the translational frictional coefficient. Thus, the frictional coefficient can be directly calculated from the diffusion constant. If the molecule is assumed to be a sphere, f is given by Eq. 14-1, and Eq. 14-6 can be written as

$$D = (k_B T)/(6\pi\eta R_S) \tag{14-7}$$

The radius, R_S, can be related to the partial specific volume, $\bar{V}_2$, by the equation

$$R_S = [(3M\bar{V}_2)/(4\pi N_0)]^{1/3} \tag{14-8}$$

and

$$D = \left(\frac{k_B T}{6\pi\eta}\right)\left(\frac{4\pi N_0}{3M\bar{V}_2}\right)^{1/3} \tag{14-9}$$

The calculated diffusion constant for a given molecular weight will differ from the experimentally determined diffusion constant if the macromolecule is not spherical in shape. This is usually summarized by the ratio f/f_{sphere}, where f is the experimentally determined frictional coefficient and f_{sphere} is the calculated frictional coefficient for a sphere with a given molecular weight and partial specific volume (Eq. 14-8). This ratio is included in Table 14-1. This ratio reflects both the shape and the hydration of the molecule. Typically, the effect of hydration is much less than the effect of shape. As expected, the ratio is near 1 for globular proteins (the first four entries for proteins), but deviates considerably for collagen, tropomyosin, and myosin (molecules known to be elongated) and DNA. These molecules are more accurately described as prolate ellipsoids rather than as spheres. An "apparent" radius can be calculated directly from Eqs. 14-1 and 14-7. If the molecule is not spherical, the apparent radius will be greater than the radius calculated from Eq. 14-9. Thus, the measurement of diffusion constants can give direct information about the hydrodynamic shape of a macromolecule.

14.4 CENTRIFUGATION

The most common laboratory technique that directly measures the hydrodynamic properties of molecules is centrifugation. Centrifugation involves generating a centrifugal field by rapidly spinning a sample in a rotor. The separation of a solid precipitate from a solution requires relatively slow spinning of the rotor. In order to cause macromolecules to move in a centrifical field, very rapid spinning is required in a special instrument called an ultracentrifuge. The centrifugal force on a particle of mass m is $m\omega^2 r$, where ω is the angular velocity of the rotor in radians/s and r is the distance of the particle from the center of the rotor. The outward motion of the particle in the field will be partially opposed by the buoyancy of the molecule, i.e., the mass of the solvent displaced by the macromolecule as it moves through the solution. The buoyant force is $\omega^2 r$ times the mass of the displaced solution which is $m\bar{V}_2\rho$, where ρ is the density of the solvent. The net force on the particle, therefore, is the difference between these two opposing factors, namely $\omega^2 rm(1 - \bar{V}_2\rho)$. The frictional force of the particle moving past each other with a relative velocity v is fv. When a particle moves in a centrifical field in a closed cell, the net force is zero so that the centrifugal force on the particle is balanced by the frictional force and buoyancy:

$$f\mathrm{v} = \omega^2 rm(1 - \bar{V}_2\rho) \tag{14-10}$$

and the velocity of the particle in the field is

$$\mathrm{v} = \omega^2 rm(1 - \bar{V}_2\rho)/f \tag{14-11}$$

As expected, the more massive the particle, the more rapidly it sediments. Similarly, the larger the frictional coefficient, the slower the particle sediments. Also, the less dense the molecule and the more dense the solvent, the slower the particle sediments. In order to obtain a parameter independent of the speed of the sedimentation, it is customary to determine the *sedimentation coefficient*, s, the velocity per unit field:

$$s = \mathrm{v}/(\omega^2 r) = m(1 - \bar{V}_2\rho)/f \tag{14-12}$$

The sedimentation coefficient has the dimensions of seconds and for typical macromolecules is in the range of 10^{-11}–10^{-13} s. In order to eliminate the necessity of writing these exponentials, the sedimentation coefficient is usually reported in terms of the svedberg, S, with $S = 10^{-13}$ s. Theodor Svedberg (1884–1971) was the inventor of the ultracentrifuge and was awarded the Nobel Prize for this accomplishment.

14.5 VELOCITY SEDIMENTATION

Before proceeding further with discussion of the sedimentation coefficient, the instrumentation used for analytical ultracentrifugation is worth discussing. The rotor spins at speeds greater than 50,000 revolutions per minute (rpm) [$\omega = 2\pi$ rpm/60 radians/s], with forces greater than 600,000 times the force of gravity being produced. The rotor, therefore, must be mechanically stable and very carefully balanced. The cell holding the solution must be designed so that the optical properties of the solution can be observed during the centrifugation, typically through ultraviolet absorption. The design of the cell is shown in Figure 14-3. It

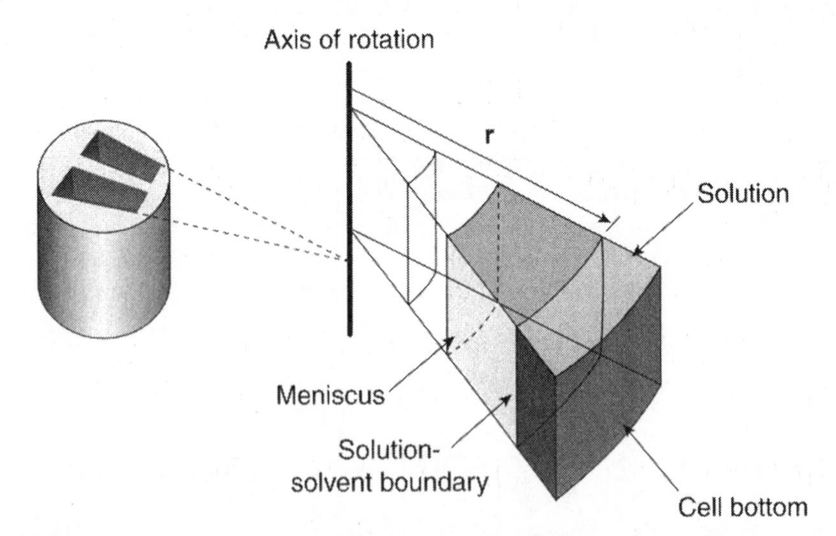

FIGURE 14-3. Schematic representation of the ultracentrifuge sample holder. The drawing on the left is a two-sector holder. One sector is used for the sample and the other for the buffer as an optical blank. Sample holders with more than two sectors are also used. The drawing on the left is a blow-up of the sample sector only. The axis of rotation is shown to indicate that the sectors are designed as pie slices, with the axis of rotation at the center of the pie.

consists of a centerpiece sandwiched by two quartz or sapphire windows. The centerpiece has two wedge-shaped sectors that are extensions of pie-shaped lines with the apex at the center of the rotor. With the windows in place, two cells are defined: one is filled with the solution of interest and the other with the solvent. The ultracentrifuge is equipped with an optical system that measures the absorbance of light at a range of distances, r, from the center of rotation. (Refractive index measurements also are possible.)

When the cell is spun, a boundary is formed between the fluid in each cell and an air bubble, creating a meniscus in each cell. Initially, the concentration of the solute is uniform throughout the fluid volume. As the centrifugation proceeds, a progressive movement of the macromolecule away from the meniscus occurs, and a moving boundary is formed between the solute and the trailing solvent. The net result is that the macromolecule moves to the rear of the cell. A typical plot of the concentration of the macromolecule versus the distance from the center of the rotor is shown in Figure 14-4. Note that the sample becomes diluted as the centrifugation proceeds because the cell is wider at the bottom (radial dilution). The boundary

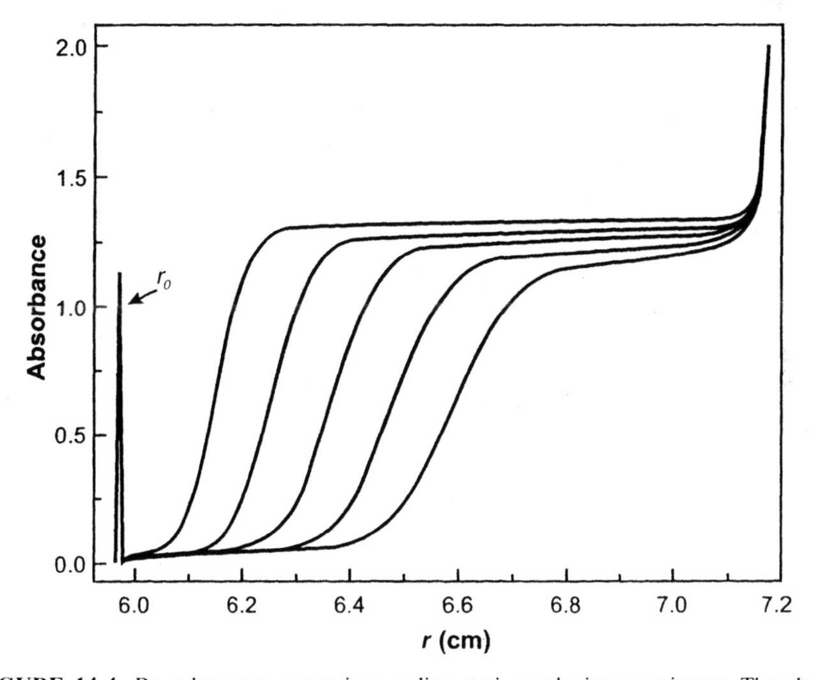

FIGURE 14-4. Boundary movement in a sedimentation velocity experiment. The absorbance is plotted versus r, the distance from the axis of rotation. The boundary moves to the bottom of the centrifuge cell as the centrifugation proceeds with a loss of sharpness due to diffusion. The meniscus at the top of the cell is at the position r_0. The concentration decreases because the sector volume increases from the top of the cell to the bottom. The MoaC protein ($s = 5.21$ S) was centrifuged at 50,000 rpm and 20°C and the absorbance was scanned at 20-min intervals at 230 nm. Figure courtesy of Dr. Harvey Sage.

stays sharp as long as significant diffusion does not occur during the duration of the experiment, typically a few hours. If significant broadening of the boundary occurs during the experiment, the diffusion constant can be calculated by analyzing the shape of the boundary as a function of time. Because diffusion is relatively slow for large macromolecules, diffusion constants are usually measured independently over a longer time period as previously discussed.

The sedimentation coefficient can be calculated from a plot of the position of the center of the boundary versus time. In practice, sedimentation coefficients are normally tabulated at a standard temperature of 20°C in water. If the sedimentation is measured at a different temperature, T, and solvent, the assumption is made that the only correction needed is to the frictional coefficient and specific volume of the macromolecule (Eq. 14-11). If the frictional coefficient is assumed to depend only on the viscosity (Eq. 14-1):

$$s_{20,\mathrm{w}} = s_{\mathrm{T}} \left(\frac{\eta_{\mathrm{T}}}{\eta_{20,\mathrm{w}}} \right) \frac{(1 - \bar{V}_2 \rho)_{20,\mathrm{w}}}{(1 - \bar{V}_2 \rho)_{\mathrm{T}}} \tag{14-13}$$

To make this conversion, the density and viscosity of the solvent used for the sedimentation experiment must be known, and the implicit assumption is that the shape and solvation of the macromolecule are the same in water and the solvent. In addition, the sedimentation coefficient is sometimes concentration dependent because of interactions between the macromolecules. In this case, the experimental sedimentation constant is extrapolated to zero concentration to obtain the value of $s_{20,\mathrm{w}}$. From Eq. 14-12, it is apparent that if the molecular weight of the species is known, the frictional coefficient can be calculated. As with diffusion, this provides information about the shape of the macromolecule. Sedimentation coefficients for some typical proteins are presented in Table 14-3, along with the ratio of the frictional coefficients, f/f_{sphere}, calculated from the sedimentation coefficients and diffusion constant.

The molecular weight of the macromolecule can be calculated from sedimentation velocity experiments if both the diffusion constant and the sedimentation coefficient

TABLE 14-3. Selected Protein Sedimentation Coefficients in Water at 293 K

Sample	Molecular weight	s (S)	$\bar{V}_2$ ($\mathrm{cm^3/g}$)	f/f_{sphere}
Ribonuclease A	12,400	1.85	0.728	1.29
Lysozyme	14,100	1.91	0.688	1.22
Bovine serum albumin	66,500	4.31	0.734	1.33
Hemoglobin	68,000	4.31	0.749	1.28
Tropomyosin	93,000	2.6	0.71	2.65
Myosin	570,000	6.43	0.728	3.63
Tobacco mosaic virus	40×10^6	192	0.73	2.65

Data from Ref. 1.

have been measured. Equations 14-6 and 14-12 can be combined to eliminate the frictional coefficient. The result is

$$M = \frac{RTs}{D(1 - \bar{V}_2\rho)} \tag{14-14}$$

(Note that $m = M/N_0$ and $R = N_0 k_B$.)

14.6 EQUILIBRIUM CENTRIFUGATION

The best method for determining molecular weights with ultracentrifugation is by equilibrium centrifugation. These measurements provide direct determination of the absolute molecular weight. As the macromolecule sediments in the centrifuge, a concentration gradient is created in the solution as the macromolecule accumulates at the back of the cell. This will cause the macromolecule to diffuse back in the direction where the concentration is depleted. Thus, diffusion causes the molecule to move to the top of the cell, and centrifugation causes it to move to the bottom of the cell. Eventually, these two opposing factors will balance, and a stable concentration distribution will be established within the cell. Typically, this takes many hours, even days, to occur. The final equilibrium distribution, however, has nothing to do with the shape of the macromolecule. Since the system is at equilibrium, all that matters is the energy of the macromolecule in the centrifugal field. From our previous considerations, it is apparent that the force acting on the macromolecule is $(M/N_0)(1 - \bar{V}_2\rho)\omega^2 r$, where r is the distance from the center of rotation. The energy is the negative of the integral of the force times dr. Thus, the energy per mole, E, at a given position in the centrifical field is

$$E = -M(1 - \bar{V}_2\rho)\omega^2 r^2/2 \tag{14-15}$$

The concentration distribution can be calculated from the Boltzmann equation that prescribes the probability distribution of the concentration as a function of energy, namely, for two concentrations c_1 and c_2 with energies E_1 and E_2, respectively,

$$c_2/c_1 = \exp[(E_1 - E_2)/RT] \tag{14-16}$$

Substituting Eq. 14-16 into Eq. 14-15, we obtain

$$\ln(c_2/c_1) = \frac{M(1 - \bar{V}_2\rho)\omega^2(r_2^2 - r_1^2)}{2RT} \tag{14-17}$$

This result predicts that a plot of ln c versus r^2 is a straight line and that the slope of this line permits the absolute molecular weight of the macromolecule to be determined. From Eq. 14-17, the molecular weight is

$$M = \frac{2RT}{(1 - \bar{V}_2\rho)\omega^2}(\text{slope of } \ln c \text{ versus } r^2) \tag{14-18}$$

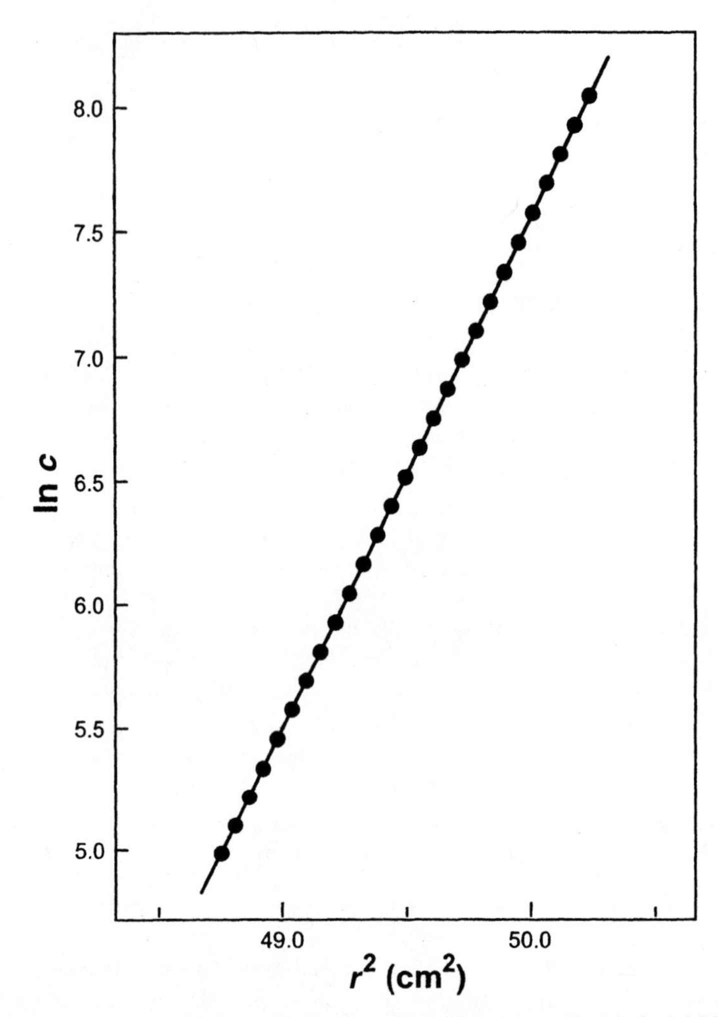

FIGURE 14-5. Sedimentation equilibrium data for bovine serum albumin in 0.1 M NaCl at 27,960 rpm, 20°C. The initial protein concentration was 650 μg/ml. The natural logarithm of the concentration is plotted versus the square of the distance from the axis of rotation (Eq. 14-17). Reprinted from D. Freifelder, *Physical Biochemistry: Applications to Biochemistry and Molecular Biology*, 2nd edition, W.H Freeman, New York, 1982, p. 421. © 1976, 1982 by Freeman. Used with permission.

A typical plot of ln c vs r^2 is shown in Figure 14-5 for bovine serum albumin.

Analytical centrifugation is one of the primary tools for determining the molecular weight of a macromolecule. It is particularly important if the protein contains multiple subunits or if a complex between nucleic acids and proteins occurs. Ultracentrifugation provides a means of determining the molecular weight of the complex formed. In addition, if multiple species are in equilibrium, for example, with concentrations of both complexed and uncomplexed species present, information can be obtained about the equilibria involved.

14.7 PREPARATIVE CENTRIFUGATION

Thus far, we have discussed ultracentrifugation as an analytical tool. However, it is also useful as a preparative method. This use is based on the obvious fact that macromolecules of different molecular weights centrifuge at different rates, as in velocity sedimentation. In principle, one might envisage simply centrifuging a mixture of macromolecules in a centrifuge tube and then punching a hole at the bottom of the centrifuge tube to elute the solution from the tube, with different molecular weight macromolecules eluting at different places in the elution pattern. This is the general idea of preparative or zonal centrifugation, but effective separation is not usually possible in a homogeneous solution because of gravitational instability. If a mixture of macromolecules is layered onto the top of a centrifuge tube, the density of the sample is higher in the layered band than in the solution below it, and the band would collapse. In zonal centrifugation, this problem is circumvented by creating a density gradient in the centrifuge tube before layering the sample onto the solution. This can be done with materials such as salt, sucrose, and glycerol. The density is highest at the bottom of the centrifuge tube. Particles will sediment through the gradient with a gradually decreasing velocity, each moving with an overall speed proportional to its s value. Thus, if the centrifugation is stopped before they reach the bottom of the tube, a separation of the macromolecules will occur, and the separate molecules can be isolated by collecting the contents of the centrifuge tube in fractions. If internal standards with known s values are included in the mixture of macromolecules, the values of the sedimentation coefficients for the individual macromolecules can be estimated.

The time course of zonal centrifugation for a mixture of three macromolecules is shown schematically in Figure 14-6. The separation of components depends on the molecular weights, the partial specific volumes of the macromolecules, and their shape.

14.8 DENSITY CENTRIFUGATION

Density centrifugation is related to zonal centrifugation and is also an effective preparative method. The best known example of this is the separation of nucleic acids by CsCl gradient centrifugation. Rather than creating a preformed gradient in a centrifuge tube, a CsCl solution is centrifuged until a CsCl gradient is formed. This is because small molecules will tend to distribute in the tube just as large molecules, and this creates an effective gradient if the solution is reasonably dense (a saturated solution of CsCl has a density of 1.9 g/cm^3). CsCl is commonly used because the solution density increases considerably as the concentration of CsCl is increased. However, other materials can work equally well, as long as a broad range of densities

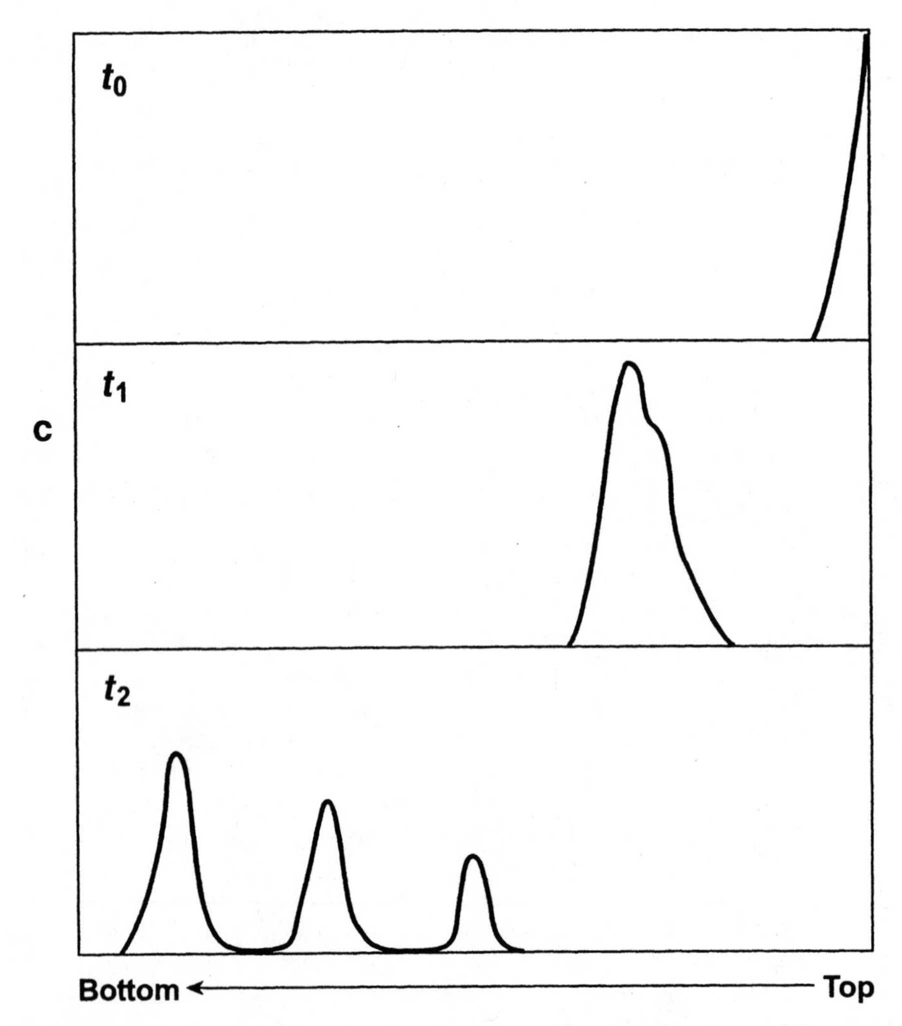

FIGURE 14-6. Schematic representation of zonal centrifugation of a mixture of three proteins. The proteins separate according to their *s* values as the centrifugation proceeds. In a real experiment, the area under the curves would remain constant.

can be achieved. If a macromolecule is included in the solution, it will tend to form a tight band at a solvent density the same is its own, thus providing an effective method of separation because of the difference in buoyant density of the macromolecules. This has proved especially effective for the separation of different DNA molecules.

The most famous experiment done with density centrifugation was the original demonstration that DNA replicates by making a new complementary strand for each original strand in the parent DNA. This was done by growing bacteria with DNA having ^{14}N at all positions in media containing ^{15}N. In the original double-stranded DNA, each strand contained ^{14}N. In the first generation, the double-stranded DNA had one strand with ^{14}N and one strand with ^{15}N. This could be easily seen with density gradient centrifugation as the ^{14}N:^{15}N DNA moved further from the center of rotation than the ^{14}N:^{14}N DNA because it is heavier. In the second generation, some ^{15}N:^{15}N DNA forms which moves even further down the centrifuge tube. This result, of course, is one of the basic tenets of modern molecular biology.

14.9 VISCOSITY

Viscosity plays a central role in hydrodynamics, as seen, for example, in the definition of the frictional coefficient. However, it is rarely used to determine the molecular properties of macromolecules because it provides limited information, and better tools are available. The measurement of viscosity is relatively easy. With a simple viscometer (e.g., Ostwald viscometer), the time, t, required for a defined volume of liquid to flow through a capillary is measured, and the flow time for an unknown liquid is compared with that of liquid whose viscosity is known. The relative viscosities are related through the equation

$$\eta_u/\eta_k = (\rho_u t_k)/(\rho_k t_u) \tag{14-19}$$

where ρ is the density and the subscripts k and u are the known and unknown liquids, respectively. The most common unit used for the viscosity coefficient is the poise [g/(cm s)] or centipoise (10^{-2} poise). The SI units are kg/(m s) = P s, where P is the pascal.

The viscosity of a solution of macromolecules is strongly dependent on the concentration of the macromolecule. For a given concentration, the specific viscosity, η_{sp}, is defined as

$$\eta_{sp} = (\eta' - \eta)/\eta \tag{14-20}$$

where η' is the viscosity of the solution and η is the viscosity of the solvent. Finally, to obtain a viscosity that is intrinsic to the macromolecule and independent of the concentration of the macromolecule, c, the *intrinsic viscosity*, $[\eta]$, is defined as the value of η_{sp}/c extrapolated to zero concentration. Note that the intrinsic viscosity has the units of reciprocal concentration as the specific viscosity is dimensionless. This is usually taken as g/cm^3. The intrinsic viscosities of a few selected proteins are given in Table 14-4, along with the frictional coefficient ratio f/f_{sphere} determined by diffusion constant measurements.

The intrinsic viscosity can be related to the shape of the macromolecule through the relationship

$$[\eta] = \nu(\bar{V}_2 + \delta\bar{V}_1) \tag{14-21}$$

In this equation, ν is a dimensionless factor that is dependent only on the shape of the macromolecule. For a sphere, ν is 2.5, and it is substantially larger for nonspherical

TABLE 14-4. Intrinsic Viscosities of Proteins in Aqueous Salt (20–25°C)

Sample	Molecular weight	$[\eta]$ (cm^3/g)	f/f^a_{sphere}
Ribonuclease	13,683	3.30	1.14
Bovine serum albumin	65,000	3.7	1.31
Hemoglobin	68,000	3.6	1.14
Tropomysoin	93,000	52	3.22
Myosin	493,000	217	3.65
Collagen	345,000	1150	6.8

Data from Ref. 2.
[a]From diffusion constant measurements (Table 14-2).

shapes. Values of ν are included in Table 14-1. From Eq. 14-21, it can be seen that the intrinsic viscosity is a measure of the shape and/or volume of a macromolecule. Thus, measurements of the intrinsic viscosity can be used to monitor processes such as the assembly of multimolecular species or large shape changes such as protein denaturation.

Viscosity measurements, for example, can be used to show that the activity of DNA polymerase increases the length of DNA. When the first polymerases were isolated, the activity of the enzyme was demonstrated by the incorporation of radioactive nucleotides into DNA. However, this could be due to synthesis of new DNA or due to exchange reactions with existing DNA. The viscosity of DNA as it grows longer increases significantly so that viscosity measurements demonstrated that new DNA was being made.

14.10 ELECTROPHORESIS

Biological macromolecules are usually charged so that if an electric field is applied across a solution, the molecules will move in the field. Positively charged molecules will move to the cathode (negative electrode) and negatively charged molecules to the anode (positive electrode). The fundamental relationship for the movement of a molecule in a nonconducting solvent is

$$u = Ze\mathbf{E}/f \tag{14-22}$$

where u is the velocity of the molecule, Z is the number of charges on the molecule, e is the charge on an electron, $\mathbf{E}$ is the electric field, and f is the frictional coefficient. (If mks units are used, u is in m/s, Z is dimensionless, e is in coulombs, $\mathbf{E}$ is in volts/m, and f is in kg/s.) In principle, this relationship can be used to determine the charge of a molecule, as was done for determining the charge on an electron. however, biological macromolecules are in water solutions with other ions that influence the ionic atmosphere around the macromolecule. This makes it virtually impossible to obtain quantitative information about macromolecules with electrophoretic methods. However, electrophoresis is a valuable analytical tool, and some of its application will be briefly described.

The first major application of electrophoresis was the separation of the protein components in blood through electrophoresis in aqueous solutions. This ground-breaking experiment demonstrated that blood contained many different proteins that could be separated and purified. Most current applications, however, utilize electrophoresis in gels. The gel is a polymer dispersed in aqueous solution that forms a three-dimensional network. Typical examples are agarose gels, a polysaccharide from agar in water, and polyacrylamide gels. Most of the gel is water, typically 90%, but the three-dimensional network of the polymer constricts the flow of molecules significantly. Furthermore, in some cases the macromolecules may be too large to enter the network. The extent to which flow is inhibited can be controlled by the degree of crosslinking of the gel. If a macromolecule is put into a gel and an electric field applied, the macromolecule will migrate in the gel. Exactly how far the macromolecule migrates is determined by the hydrodynamics of the macromolecule (size and shape) and its charge (Eq. 14-22).

DNA fingerprinting is an example of the analytical power of gel electrophoresis. All humans have similar DNA, but the exact base sequences are very specific for individuals. Much of human DNA does not code for proteins (>90%) and consists of repeating sequences of bases. The number and nature of these sequences differ for each individual, and such a region is used for DNA fingerprinting. These sequences can be isolated, and restriction enzymes exist that cut double-stranded DNA at specific places determined by the nucleotides in the sequence. A large number of restriction enzymes are readily available, and they can be used to cleave the fingerprinting region of the DNA to obtain different fragmentation patterns. The fragments are then separated by electrophoresis on agarose gels. The DNA separates on the gel according to its size because every residue has the same charge due to the phosphate portion of the DNA. The pattern of bands obtained with only a few restriction enzymes is unique for each person (except identical twins) and therefore can be used to distinguish between people.

The actual detection of the gel band is accomplished by denaturing the DNA on the gel, either with heat or chemically, and making radioactive DNA oligonucleotides complementary to the individual gel bands. The gel bands are transferred ("blotted") to a nitrocellulose membrane, usually by simple capillary action. The membrane is then soaked with the radioactive nucleotides. After washing, the radioactivity of the blotted bands is measured. This is termed a Southern blot, after the person who invented this technique, Edward M. Southern. The use of this analysis in criminal cases is quite extensive. Not only can this method identify individuals with an extremely high probability, but it can also identify related people. The probability of the analysis being correct depends on how many restriction maps are used—the odds become overwhelmingly large with a relatively few restriction enzyme maps. Two other gel electrophoresis blotting methods are commonly used, one with radioactive RNA oligonucleotides and the other with antibodies. These are called Northern and Western blots, respectively—a scientific joke to those who know the origin of these terms.

Nucleic acids have one charge per nucleotide so that separation of oligonucleotides according to molecular weight is relatively easy with electrophoresis. The situation is quite different for proteins as the number of charges varies between individual proteins and with pH, salt concentration, etc. Furthermore, the shapes of proteins also vary, so they may have quite different frictional coefficients. Nevertheless, gel electrophoresis of native proteins may provide a convenient separation method, even though the migration of the protein in the gel cannot be quantitatively interpreted. Sodium dodecyl sulfate–polyacrylamide gel electrophoresis (SDS-PAGE) gets around this problem by the addition of SDS to the protein. Proteins bind about the same amount of SDS per monomer. As SDS is highly charged, the charge per unit mass of protein is about the same. Furthermore, if disulfide bonds are broken by reduction with a reagent such as 2-mercaptoethanol, the protein is denatured by SDS into a similar shape for all proteins. If SDS-PAGE is carried out for reduced proteins reacted with SDS (typically about 10^{-3} M), the extent of movement of the proteins is determined entirely by their molecular weight. The protein is usually detected by staining the gel with a dye, e.g., Coomassie blue, after electrophoresis. Under controlled conditions, the intensity of the stain can be related to the concentration in the gel.

The logarithm of the molecular weight is plotted versus the mobility of the protein for a variety of proteins and gels in Figure 14-7. The straight line relationship is a convenient method for determining the approximate molecular weight of a denatured protein. This straight line can be represented by the equation

$$\log M = A - Bx \tag{14-23}$$

where M is the molecular weight, x is the distance the protein has migrated in the gel, and A and B are constants determined empirically. In practice, a set of standard proteins is included in the gel along with the protein of unknown molecular weight in order to determine A and B. However, this is an empirical method and is not a substitute for determining the exact molecular weight by a method such as equilibrium centrifugation. If the protein itself is highly charged, has an unusual shape when denatured with SDS, or does not bind the "typical" amount of SDS, SDS-PAGE can give misleading results.

These two examples illustrate the importance of gel electrophoresis as an analytical technique in modern biology. These methods are dependent on the hydrodynamic properties of the macromolecules, even though these properties are used in an entirely empirical way. More extensive discussions of electrophoresis are available elsewhere (1,3).

14.11 PEPTIDE-INDUCED CONFORMATIONAL CHANGE OF A MAJOR HISTOCOMPATIBILITY COMPLEX PROTEIN

Major histocompatibility complex proteins (MHC) are cell surface proteins that are essential for the cell-mediated immune response. Class II MHC proteins bind

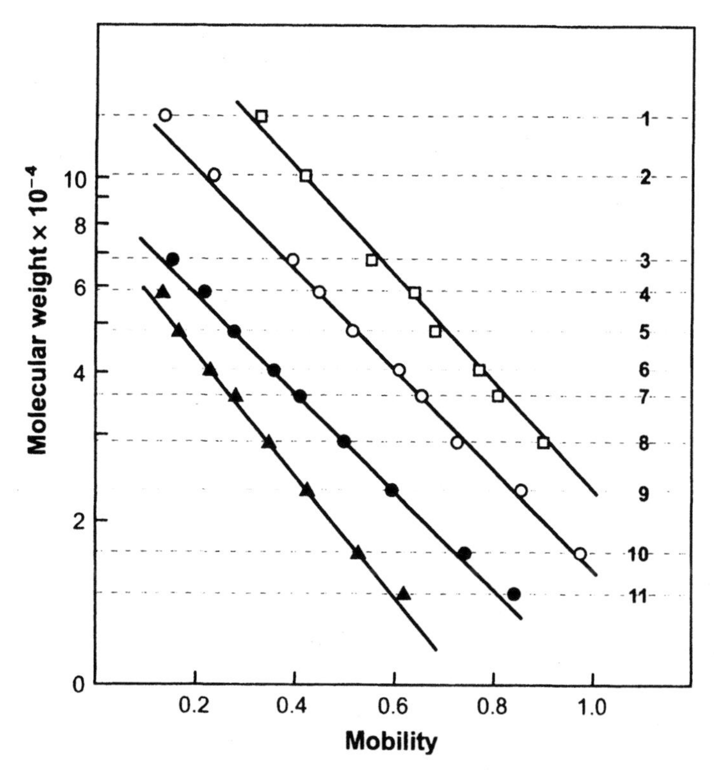

FIGURE 14-7. Electrophoretic mobilities of standard proteins in an SDS-PAGE experiment. The numbers refer to the standard proteins used, and the lines are different acrylamide concentrations. The acrylamide concentrations are 15% (▲), 10% (•), 7.5% (○), and 5% (□). As expected, the higher the acrylamide concentration, the slower the migration. The standard proteins are β-galactosidase, phosphorylase A, serum albumin, catalase, fumarase, aldolase, glyceraldehyde phosphate dehydrogenase, carbonic anhydrase, trypsin, myoglobin, and lysozyme. Reprinted from K. Weber, J. R. Pringle, and M. Osborn, Measurement of molecular weights by electrophoresis on SDS–acrylamide gel, *Methods Enzymol.* **26**, 3 (1972). Copyright 1972, with permission from Elsevier.

peptides that are produced by proteolysis and present them to the surface of the T cell, thereby initiating an immune response (5). Complexes of these proteins with peptides have been extensively studied by a variety of techniques, including X-ray crystallography, thermodynamics, and kinetics. Here we briefly describe a hydrodynamic investigation that demonstrated a major conformational change of an isolated MHC protein induced by the binding of a peptide (6). In this work, a soluble portion of a specific MHC protein, DR1, was cloned and expressed in *E. coli*. DR1 consists of two different polypeptide chains that are tightly associated. The hydrodynamic properties of the protein and the protein–peptide complex were then measured by a variety of techniques: gel filtration, light scattering, and ultracentrifugation. The results obtained are summarized in Table 14-5 for DR1 and a DR1–peptide complex.

TABLE 14-5. Hydrodynamic Data for DR1 and a DR1–Peptide Complex

	DR1	DR1–peptide
Gel filtration molecular weight	48,000	40,000
s (S)	2.67	3.23
$D\,(\times 10^{-7}\ \mathrm{cm}^2/\mathrm{s})$, light scattering	6.17	7.40
$D\,(\times 10^{-7}\ \mathrm{cm}^2/\mathrm{s})$, sedimentation velocity	5.73	6.74
$D\,(\times 10^{-7}\ \mathrm{cm}^2/\mathrm{s})$, calculated		7.68
f/f_{sphere}, light scattering	1.3	1.1
R_S (Å), light scattering	35	29
R_S (Å), sedimentation velocity	37.4	31.8
R_S (Å), calculated		27.9

D is the translational diffusion constant, s is the sedimentation constant, $R_S$is Stokes' radius, and f/f_{sphere} is the ratio of frictional coefficients. The calculated values are based on the crystal structure of the DR1–peptide complex and the estimated hydrated volume of the DR1–peptide. Data from Ref. 8.

Gel filtration is an empirical method for determining molecular weight. Molecular sieve gels have pores in the individual beads that allow proteins to enter if they are small enough. The proteins partition between the inside and the outside of the beads. In general, the larger proteins partition more favorably on the outside of the beads, and the smaller proteins partition more favorably on the inside of the beads. A variety of gel materials are available with different pore sizes so that graded partitioning between the inside and the outside is available for a wide range of molecular weights. If proteins are passed through a gel filtration column in which all of the proteins can be partitioned inside the gel beads to some extent, they will elute from the column in the order of their molecular weight, with the highest molecular weight appearing first. If the column is calibrated with proteins of known molecular weights, the molecular weight of an unknown protein can be determined. Gel filtration, however, must be used with caution for determining molecular weights, however, as the shape of the macromolecule and its hydration are also important factors in determining the elution volume of a given protein. In fact, gel filtration separates proteins on the basis of their effective hydrodynamic radius rather than simply by molecular weight.

The diffusion constant obtained from the sedimentation coefficient was calculated by the use of Eq. 14-14, Stokes' radius with Eq. 14-7, and the frictional coefficient ratio as in Table 14-2. Table 14-5 includes a diffusion constant and a Stokes' radius calculated from the crystal structure of the DR1–peptide. This was done by calculating a radius based on the specific volume of the protein and the estimated water of hydration. The radius obtained from this procedure was multiplied by the frictional coefficient ratio for an oblate ellipsoid estimated from the crystal structure. In other words, $R_S = \left(\frac{3V_h}{4\pi}\right)^{1/3}(f/f_{\mathrm{sphere}})$, where V_h is the hydrated volume.

The results obtained from the three methods are similar: binding of the two peptides causes an apparent decrease in the molecular weight and the Stokes' radius, and an increase in the diffusion constant. This cannot be due to an actual change in molecular weight as the peptide does not significantly change the molecular weight of DR1. This was confirmed by sedimentation equilibrium experiments. The molecular masses were found to be 45,000 and 47,000 for DR1 and DR1–peptide, respectively. The most obvious way to explain these results is that the DR1–peptide complex has a more compact shape than the DR1 protein alone. Circular dichroism measurements suggested that structural changes occur when DR1 binds peptide. Thus, the evidence is quite convincing that a significant conformational change occurs when the peptide binds.

What is the biological significance of this conformational change? This is a matter of speculation, but the argument is made that this conformational change is important for presentation of the peptide to the T cell and perhaps for regulation of the immune response.

14.12 ULTRACENTRIFUGE ANALYSIS OF PROTEIN–DNA INTERACTIONS

Translin is a nucleic acid binding protein that appears to recognize sequences found at chromosomal break points. Chromosomal breakage and rejoining is associated with the development of malignancies (7). Consequently, the interaction of translin with DNA is of considerable biological significance. Translin is a protein of 228 amino acid residues and a molecular weight of 26,180 (calculated from the DNA sequence of its gene).

However, it exists and functions as a multisubunit protein. Unltracentrifugation analysis was carried out to determine its molecular weight and shape. In addition, the binding of oligonucleotides to translin was also investigated (8).

Velocity sedimentation analysis reveled a major species with an s value of 8.5 S. A broad distribution of a small amount of heavier material was present that is attributed to larger aggregates. The major species was attributed to an octamer with a frictional coefficient ratio of 1.35. In order to confirm the presence of an octamer and to examine more carefully the distribution of molecular species, equilibrium centrifugation experiments were also carried out. The results confirmed the presence of an octamer, along with a small amount of high molecular weight material. No evidence of lower molecular weight species was found, suggesting that the octamer is a very stable species. The shape of the octamer was derived from the frictional coefficient and electron microscopy. The octamer can be approximated by an oblate ellipsoid with an axial ratio of 7.5:1.

The interaction of translin with a 24-mer oligodeoxynucleotide was also studied. The oligonucleotide was labeled with the fluorescent molecule fluorescein. With this label, it was possible to observe both the free and unbound nucleotides by measuring the absorbance at 490 nm. Data were also collected at 280 and 260 nm: as discussed in Chapter 8, proteins have a maximum absorbance at 280 nm and nucleic acids have a maximum absorbance at 260 nm. However, at these two wavelengths both the protein and the nucleic acid contribute to the absorbance. Equilibrium sedimentation experiments were carried out for several concentrations of DNA and translin, and these data were simultaneously fit to a model in which the nucleic acid was assumed to form a one-to-one complex with the octameric translin. The equilibrium dissociation constant obtained was 84 nM. This is very tight binding, attesting to the specificity of this interaction.

The overall model that emerges from this work (and electron microscopy) is that translin forms an annular oblate ellipsoid structure of eight subunits that binds tightly to DNA at chromosomal break points.

As is apparent from the above discussion, hydrodynamic considerations form the basis of several important experimental methods used in modern biological research. Also, determination of the hydrodynamic properties of macromolecules provides important information about their structure and shape which in turn provides insight into their function.

REFERENCES

1. C. R. Cantor and P. R. Schimmel, *Biophysical Chemistry*, W. H. Freeman, San Francisco, CA, 1980.
2. C. Tanford, *Physical Chemistry of Macromolecules*, Wiley, New York, 1961.
3. D. Freifelder, *Physical Biochemistry: Applications to Biochemistry and Molecular Biology, 2nd edition, Freeman, New York, 1982.*
4. I. Tinoco, Jr., K. Sauer, J. C. Wang, and J. D. Puglisi, *Physical Chemistry: Principles and Applications in Biological Sciences*, 4th edition, Prentice-Hall, Englewood Cliffs, NJ, 2002.
5. C. Watts, *Annu. Rev. Immunol.* **15**, 821 (1997).
6. J. A. Zarutskie, A. K. Sato, M. M. Rushe, I. C. Chan, A. Lomakin, G. B. Benedek, and L. J. Stern, *Biochemistry* **38**, 5878 (1999).
7. T. H. Rabbits, *Nature* **372**, 143 (1994).
8. S. P. Lee, E. Fuior, M. S. Lewis, and M. K. Han, *Biochemistry* **40**, 14081 (2001).

PROBLEMS

14-1. The value of s can usually be calculated by measuring the rate of change of the midpoint of the boundary, $x_{1/2}$, with respect to time (Fig 14-4). From Eq. 14-12, $s = \mathrm{v}/(\omega^2 x) = (\mathrm{d}x/\mathrm{d}t)/(\omega^2 x) = (\mathrm{d}\ln x/\mathrm{d}t)/\omega^2$. Thus, the slope of a plot of ln x versus time and the value of ω allow s to be determined. The following data were obtained for a protein of molecular weight 74,000 with a specific volume of 0.737 cm^3/g at 20°C. The density of the buffer is 1.010 g/cm^3, the viscosity is 1.002 centipoise $[10^{-2}\ \mathrm{g/(cm\ s)}]$, and the centrifugation was carried out at 52,000 rpm.

Time (min)	$x_{1/2}$ (cm)
0	5.8591
30	5.9541
60	6.0763
90	6.2010
120	6.3282
150	6.4581
180	6.5905

a. Calculate the sedimentation coefficient, s.

b. Calculate the frictional coefficient.

c. Calculate the diffusion constant.

d. Calculate the frictional coefficient ratio, f/f_{sphere}.

14-2. High molecular weight DNA can be modeled as a prolate ellipsoid of revolution. The frictional coefficient can be approximated as $f = 6\pi\eta a/[\ln(2a/b)]$, where η is the viscosity of the solvent, a is the long axis of the ellipsoid, and b is the short axis. DNA with a molecular weight of 1 million is about 5200 Å long and 22 Å in diameter.

a. Calculate the diffusion constant and sedimentation coefficient in 0.1 M NaCl for this DNA at 20°C. The viscosity of the solvent at this temperature is 1.016 centipoise, its density is 1.0025 g/cm^3, and the specific volume of DNA is 0.556 cm^3/g.

b. The calculated values are only in fair agreement with experimental values. Provide a possible explanation for this discrepancy.

14-3. Equilibrium centrifugation is carried out on a protein at pH 7.0 and pH 10.5 at 20°C. The partial specific volume of the protein is 0.749 cm^3/g at this temperature and the density of water is 0.9982 g/cm^3. The following results were obtained.

r^2 (cm^2) 40,000 rpm	c (μM) pH 7.0, 30,000 rpm	c (μM) pH 10.5,
49.00	0.431	0.333
49.10	0.611	0.388
49.20	0.865	0.453
49.30	1.22	0.529
49.40	1.72	0.616
49.50	2.44	0.720
49.60	3.46	0.840
49.70	4.89	0.980
49.80	6.91	1.14
49.90	9.78	1.33
50.00	14.2	1.55

a. Calculate the molecular weight for the protein at the two pH values.

b. Provide an explanation for the different molecular weights obtained at the two different pH values.

14-4. a. Ribonuclease is a small molecular weight protein consisting of a single polypeptide chain. When it is thermally denatured, the relative viscosity of the solution increases. What is a likely explanation for this observation in terms of the molecular structures of the native and denatured states? What would you predict for the changes in the diffusion constant and sedimentation coefficients accompanying denaturation?

b. In contrast, when double-stranded DNA is thermally denatured into single-stranded DNA, the viscosity of the solution decreases. Explain this observation in terms of the molecular structures of single- and double-stranded DNA. If the melting temperature of the DNA is 30°C, sketch a plot of the intrinsic viscosity versus temperature. Again, what would you predict for the changes in diffusion and sedimentation coefficients for the DNA. (Assume that the degree of polymerization of the DNA is unchanged. DNA melting and the melting temperature are discussed in Chapters 3 and 5.)

14-5. The ribosome from *E. coli* can be isolated as two particles, the 30S and 50S subunits. (The nomenclature for ribosomes uses the sedimentation coefficients determined in early experiments.) Both particles contain proteins and RNA. The 30S subunit contains 16S RNA and proteins. A question that was asked was whether the 16S RNA has the same configuration when it is isolated as when it is in the 30S subunit. One approach to this question is to determine the effective hydrodynamic radii of the isolated 16S RNA and the 30S subunit. The following data were obtained at 20°C [M. F. Tam, J. A. Dodd, and W. E. Hill, *J. Biol. Chem.* **256**, 6430 (1981)].

	$D\,(\times 10^7\ \text{cm}^2/\text{s})$	s (S)	$\bar{V}_2\ (\text{cm}^3/\text{g})$
16S RNA	1.721	21.0	0.541
30S subunit	1.97	31.8	0.590

a. Calculate the molecular weights of 16S RNA and the 30S subunit. The density of water under these conditions is 0.9982 g/cm^3.

b. Calculate the effective Stokes' radii of the 16S RNA and the 30S subunit from the diffusion constant and sedimentation coefficient. The viscosity of water at this temperature is 1.005 centipoise. What do you conclude about the conformation of the isolated 16S RNA versus its conformation in the 30S subunit?

c. Calculate the ratio f/f_{sphere} for 16S RNA and 30S subunit from the diffusion constant. What can you conclude about the shape of 16S RNA and 30S subunit.

(The calculations in this problem are somewhat simplified relative to the original publication but the conclusions are the same.)

14-6. Aspartate transcarbamoylase is an enzyme that is important for the regulation of pyrimidines (Chapter 13). It has an unusual subunit structure that could have been determined by a combination of gel chromatography and SDS-PAGE. The hypothetical results of such experiments are summarized below.

A gel filtration column was calibrated with the following proteins. The elution volumes can be assumed to be a linear function of the logarithm of the molecular weight.

Protein	Molecular weight	Elution volume (ml)
Cytochrome *c*	12,400	208
Myoglobin	17,800	194
Chymotrypsinogen	25,000	185
Bovine serum albumin	66,200	152
Lactate dehydrogenase	140,000	127
α-conarachin	295,000	102
Apoferritin	475,000	90

a. When aspartate transcarbamoylase was put through this column, it eluted with a volume of 101 ml. What is the molecular weight of aspartate transcarbamoylase?

b. The enzyme was reacted with a reagent that modifies sulfhydral groups and again passed through the gel filtration column. This time two peaks emerged with elution volumes of 138 and 174 ml. What are the molecular weights of these components?

c. Finally, aspartate transcarbamoylase was subjected to SDS-PAGE. The following proteins were used to calibrate the gel.

Protein (mm)	Molecular weight	Gel migration
Lysozyme	14,400	88.0
β-Lactoglobulin	18,400	78.2
Carbonic anhydrase	29,000	70.0
Lactate dehydrogenase	35,000	55.2
Ovalbumin	45,000	46.0
Bovine serum albumin	66,200	30.2
β-galactosidase	116,000	10.0

Two bands were observed for aspartate transcarbamoylase with gel migrations of 58.0 and 82.0 mm. What are the molecular weights of these components?

d. On the basis of the above results, what can you say about the subunit structure of aspartate transcarbamoylase?

CHAPTER 15

Mass Spectrometry

15.1 INTRODUCTION

The use of mass spectrometry in biology has become sufficiently prevalent that the basic concepts and their applications merit consideration in this text. The basic variable parameter in spectroscopy is wavelength, whereas in mass spectrometry it is m/z, the ratio of the mass of a particle to its charge. In very simple terms, a mass spectrometry experiment can be divided into three main steps: ionization, mass analysis, and ion detection. Although modern mass spectrometry experiments are more complicated than suggested by this simple analysis, these three steps are always involved. Importantly, the only environment in which ions are stable for a sufficient time to analyze readily is a vacuum, and a good vacuum is an essential part of a mass spectrometer. This chapter constitutes an introduction to the field. More comprehensive treatments can be found in Refs. 1–3.

15.2 MASS ANALYSIS

The basic principles underlying mass spectrometry can be understood by considering the motion of a charged particle in electric and magnetic fields.

The kinetic energy of a charged particle in an electric field is given by the relationship

$$\frac{1}{2}m\mathrm{v}^2 = zV \tag{15-1}$$

where m is the mass, v is the velocity, z is the charge, and V is the applied voltage. The trajectory of an ion in a magnetic field, H, is an arc of radius r, as shown in Figure 15-1. The conservation of angular momentum requires that

$$m\mathrm{v}^2/r = z\mathrm{v}H \tag{15-2}$$

Physical Chemistry for the Biological Sciences by Gordon G. Hammes

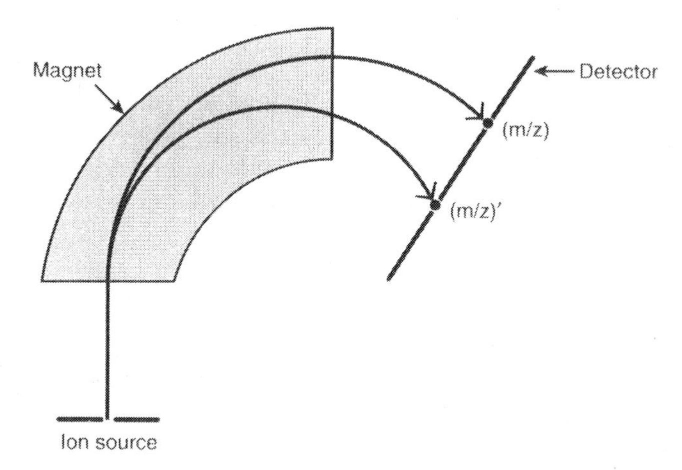

FIGURE 15-1. Schematic representation of the trajectory of ions with different m/z values in a magnetic field. In this illustration, $m/z > (m/z)'$.

Eliminating the velocity between these two equations gives

$$z/m = H^2r^2/(2V) \tag{15-3}$$

This relationship shows that in constant magnetic and electric fields, the trajectory will be the same for all ions with the same ratio m/z. Thus, if the magnetic and electric fields are kept constant, the mass spectrum can be collected by determining the positions of the ions after they have passed through the magnetic field. Alternatively, the magnetic field can be scanned to determine the mass spectrum.

The ultimate goal of the mass analysis is to determine m/z with the smallest possible error and to analyze a broad range of m/z. Unfortunately, these goals are often in conflict so that several different methods of mass analysis are used. The resolution of a mass spectrometer is defined as $m/\Delta m$. For example, a mass resolution of 1000 can distinguish between ions with m/z of 1000 and 1001.

The simplest type of mass spectrometer utilizes only a magnetic analyzer. As is evident from Eq. 15-2, the path of an ion in a magnetic field is different for every value of m/z. If Eq. 15-2 is solved for r, we find that

$$r = (1/H)(2mV/z)^{1/2} \tag{15-4}$$

The path for two ions with different m/z passing through a magnetic field is included in Figure 15-1. The resolution of the separation can be determined by taking the natural logarithm of Eq. 15-4 and differentiating:

$$\Delta r/r = (1/2)(\Delta m/m) + (1/2)(\Delta V/V) \tag{15-5}$$

Thus, in order to sharply focus the ions, thereby decreasing the error in m, it is important for all ions to be homogeneous in energy, that is, the last term must be small. This can be partially accomplished by using a high voltage (V), typically 8000 V, but usually the magnetic focusing is coupled with an electrostatic analyzer in order to produce ions that are homogeneous in energy (small ΔV). This can be done by subjecting the ions to a constant voltage either prior or subsequent to the magnetic field. Instruments of this type, double-focusing magnetic sector mass spectrometers, have high resolution and a mass range up to about 15,000 Da. They are also very expensive.

The quadrupole mass spectrometer is the most widely used, primarily because of its relatively low cost. The separation of ions is accomplished by utilizing electric fields only. The mass analyzer consists of four metal rods, as shown in Figure 15-2. The trajectory of the ions is between the four rods. The rods are electrically connected in pairs, A to A and B to B in the figure. A constant voltage of opposite sign is applied to the A and B rods. An oscillating voltage is superimposed on the constant voltage, with the phase differing by 180° between A and B. The quadrupole serves as a mass filter: only ions with a specified value of m/z can get through the filter. Other ions collide with the rods and do not reach the detector. To obtain a mass spectrum, the applied electric fields are varied, thus allowing ions with different values of m/z to be detected. The resolution of quadrupole instruments is not as good as double-focusing magnetic sector instruments, 10,000 versus 100,000, but it is significantly less expensive and can tolerate relatively high pressures. The latter feature is important if the ions are generated by electrospray, a frequently used technique that will be discussed later.

The quadrupole mass spectrometer has been coupled with an "ion trap." With these instruments, the ions are held within the quadrupole and manipulated before

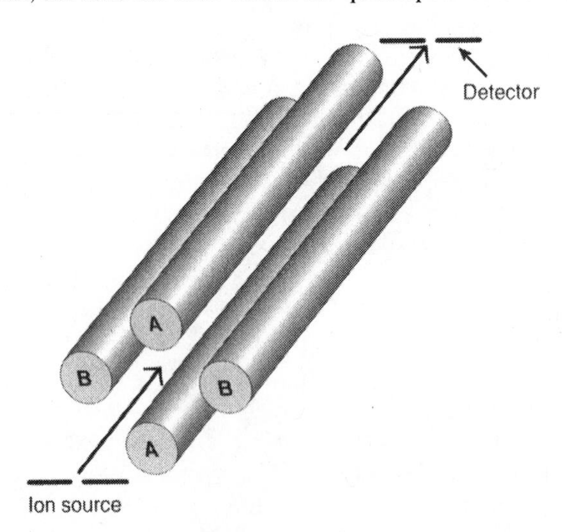

FIGURE 15-2. Schematic representation of a quadropole mass analyzer. The ions pass through the middle of four parallel rods. The A rods are connected and have the same DC and superimposed radiofrequency voltages. The B rods are also connected and have a DC voltage opposite in sign to the A rods and the radiofrequency voltage phase shifted 180° relative to the A rods.

proceeding to the detector. By holding the ions in the trap, both time and space dimensions are available. This improves the resolution and sensitivity. In simple terms, the ions are physically trapped between electrodes and subjected to both constant and oscillating electric fields such that ions of specific m/z precess within the trap. The magnitudes of the fields are increased, thereby causing ions of specific m/z to be ejected from the trap.

The time-of-flight mass spectrometer selects ions by measuring the time of arrival of the ion at the detector. Equation 15-1 can be rearranged to give

$$\mathrm{v} = (2Vz/m)^{1/2} \tag{15-6}$$

This equation predicts that the lighter the ion the faster it will arrive at the detector. In order for this method to work, the ions must enter the flight tube at the same time. This is accomplished by generating ions in short bursts. The difference in the time of arrival of ions is not great, typically in the microsecond range. Thus, a complete spectrum can be determined in very short times. In some cases, reflectors are used to improve the sensitivity and resolution. This is done by slowing the ions with a series of electric field "lenses" until they essentially stop and then accelerating them in the opposite direction. The reflection increases the path length traveled, thus providing better separation of the time-of-flight, and the lenses focus the ions with a specific m/z by reducing the spread in kinetic energies for a given ion. The time-of-flight instrument has the advantage of essentially unlimited mass range and high sensitivity.

Other methods exist for mass analysis but are not discussed in detail here. Most notably, Fourier transform methods using ion cyclotron resonance have been developed that permit very high precision mass determinations. The ions are inserted into a small volume in the cyclotron, and a large magnetic field is applied so that the ions precess in circular orbits according to Eq.15-2. The ions are constrained to the cell by an electric field applied to front and rear plates of the sample cell. If a pulsed electric field is applied at a frequency matching the precession frequency of the ions, energy is absorbed, analogous to a magnetic resonance experiment. The ions then transmit a radio frequency current at the detector plates that contains the frequency components of each of the ions. This is converted to a free ion decay signal (analogous to the free induction decay in NMR) that can be Fourier transformed to the mass spectrum of all of the ions, thus permitting detection over a wide mass range.

A summary of the various mass analyzers, along with their approximate range of m/z and resolution, is given in Table 15-1.

15.3 TANDEM MASS SPECTROMETRY (MS/MS)

Tandem mass spectrometry couples two (or more) mass analyzers to obtain additional information about the sample in question. Three steps are typically involved in MS/MS analysis: mass selection, fragmentation, and mass analysis. The first step

TABLE 15-1. Characteristics of Mass Analyzers

Method	Mass range (Da)	Resolution
Magnetic sector	15,000	200,000
Quadrupole	4000	4000
Quadrupole ion trap	100,000	30,000
Time-of-flight	Unlimited	15,000
Fourier transform ion cyclotron resonance	$>10^6$	$>10^6$

Adapted from reference 1.

is the selection of a specific ion for further study. In the second step, the selected ion then undergoes fragmentation, usually through collisions with neutral gas atoms. The ion fragments are then analyzed by a second mass analyzer. Because modern ionization methods produce very little fragmentation, the ion selected is often similar to the ion of the parent compound. Selection of a specific ion and determination of its fragmentation pattern may be important for its molecular identification.

A variety of different MS/MS instruments are available. For example, a triple quadrupole instrument uses the first quadrupole for ion selection, the second for collision-induced dissociation, and the third for analysis of the fragments produced. Ion traps can also be incorporated at a relatively low cost. Similarly, multiple sector instruments are available that use a series of magnetic and electric fields for analysis. As might be expected, multiple sector instruments are quite expensive. Time-of-flight and Fourier transform cyclotron resonance instruments are also available. The use of MS/MS is particularly useful for biological systems to identify and characterize substances uniquely.

15.4 ION DETECTORS

The two most common methods of detecting ions after they have been sorted by the mass spectrometer are electron multipliers and photomultipliers. With electron multipliers, the ion strikes a dynode that emits secondary electrons. Typical dynode surfaces are BeO, GaP, and CsSb. These secondary electrons are accelerated by a voltage and attracted to a second dynode that emits more electrons. This process continues through a series of dynodes, resulting in a cascade of electrons. The resulting current can be read with standard technology. Electron multipliers are very sensitive: a signal amplification factor of 10^6 can be readily obtained.

Photomultiplier detectors operate in a similar fashion except that the ion first strikes a phosphorous screen. The phosphorous screen releases photons, which are detected by a photomultiplier. The photomultiplier also has a series of dynodes and causes a cascade of electrons when the light strikes it. This type of detection is commonly called scintillation counting and is often used to measure radioactivity quantitatively. The amplification factor is similar to electron multipliers. A major advantage of photomultipliers is that they have significantly longer lifetimes than electron multipliers. The lifetime of electron multipliers is limited by contamination/damage of the surface that the ions strike. Nonetheless, electron multipliers are currently the most common devices used.

Detection of the ion signal is often done by a point detector so that only a single type of ion is detected, that is, a single m/z. However, array detectors are also available. Array detectors consist of a linear arrangement of detectors so that multiple ions can be detected simultaneously.

15.5 IONIZATION OF THE SAMPLE

The first step in the analysis is to ionize the sample into the vacuum of the mass spectrometer. A variety of methods are used to ionize the sample, and we will deal with only a few of them. Currently, the two most widely used methods for macromolecules are MALDI (Matrix Assisted Laser Desorption/Ionization) and ESI (ElectroSpray Ionization), but a few other methods that are commonly used for relatively small molecules will also be discussed.

Electron ionization vaporizes the sample into the mass spectrometer and then bombards the vapor with a beam of high-energy electrons, 50–100 eV. This is typically accomplished by thermal evaporation from a probe containing the sample. The high-energy electrons are produced from a filament and acceleration of the electrons through a large electric field. In order to produce ions, the electrons must have an energy greater than the ionization energy of the molecule, M, being studied. This process can be written as

$$M + e^- \rightarrow M^{+\cdot} + 2e^- \tag{15-7}$$

The positive ions formed, $M^{+\cdot}$, are referred to as odd-electron molecular ions or radical cations. Because the ionization energy for most molecules is only about 5 eV, the molecular ion produced usually has an excess of energy and fragments. The fragmentation pattern provides a "fingerprint" for each molecule. Libraries of fragmentation patterns are available for the identification of unknown compounds. The primary drawback of this method is that relatively stable and volatile compounds are required. In practice, this limits the molecular weight of samples to less than about 1000 Da.

The technique of fast ion bombardment (FAB) was developed in the 1980s and permits substances with molecular weights of 4000 or greater to be analyzed routinely. The principle of the method is to place a sample that is dissolved into a matrix on the tip of a metal probe. The sample is then bombarded with a stream of fast (high temperature) Ar or Xe atoms, and molecular ions are produced, predominately by the loss or gain of an H atom. (In some cases, a beam of Cs^+ is used.) The key to this method is the matrix, which is a viscous liquid and relatively inert. Typical examples are *m*-nitrobenzyl alcohol and glycerol. The sample is dissolved in a solvent that is miscible with the matrix. The matrix absorbs most of the high-energy atoms and produces a high-temperature, high-density gas within a small volume. Various matrix ions are produced, protonated, and deprotonated, and these react with the sample molecule, M, to produce ions:

$$\begin{aligned} M + \text{matrix–H}^+ &\rightarrow MH^+ + \text{matrix} \\ MH + \text{matrix}^- &\rightarrow M^- + \text{matrix–H} \end{aligned} \tag{15-8}$$

Thus, both negative and positive sample ions are produced. The energy of the ions is considerably less than that produced by electron ionization, but some fragmentation occurs. A schematic representation of FAB is shown in Figure 15-3.

The concept behind MALDI is similar to FAB (Fig. 15-3). The sample is embedded in a matrix, and an external source is used to convert the sample to ions. In the case of MALDI, the sample is embedded into a solid crystalline compound, and a laser, operating at a wavelength at which the matrix strongly absorbs, is used to desorb and ionize the sample. A laser with a high-energy output is needed: both UV and IR lasers have been used. The matrix is essentially a "solvent" for the sample, and the sample forms a microcrystal with the matrix. In practice, the sample is dissolved into a small amount of solvent (water or water plus organic solvents, typically), and the sample is then mixed with a concentrated solution of the matrix. Just a picomole of sample is required and the ratio of sample to matrix is typically 1/1000. A variety of materials has been used for matrices, for example, cinnamic acid, succinic acid, and urea. The mixture is then evaporated and introduced into the mass spectrometer. Short laser pulses, typically 10–20 ns duration, with about 10^6 W/cm^2 power are used. The exact mechanism for producing ions is not well understood, but the general idea is that the matrix is ionized, and the ions produced ionize the sample molecules through a series of proton transfer reactions involving the matrix molecules and the anylate. As with FAB, both positive and negative ions are produced. This method can be used to study

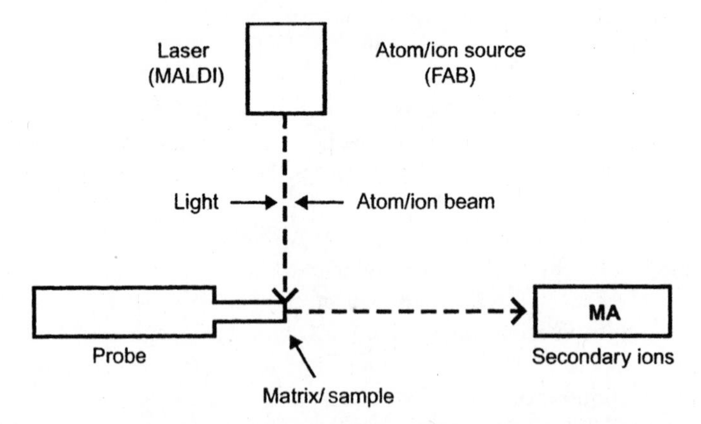

FIGURE 15-3. Schematic representation of FAB (right) and MALDI (left) methods of ionization. In both cases, the sample is embedded in a matrix. In the case of FAB, the ions are created by bombardment with an atom or ion beam, and in the case of MALDI by a pulsed laser. The secondary ions are then injected into the mass analyzer.

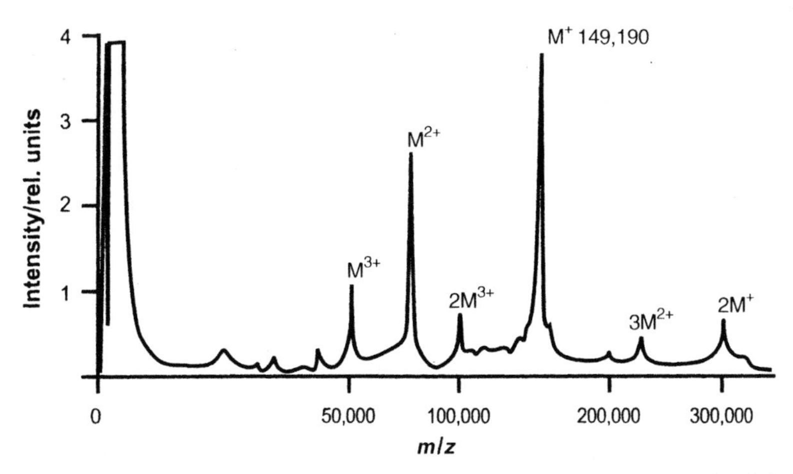

FIGURE 15-4. The MALDI mass spectrum of a monoclonal antibody. Reprinted from F. Hillenkamp and M. Karas, Mass spectrometry of peptides and proteins by matrix assisted ultraviolet laser desorption/ionization, *Methods Enzymol.* **193**, 280 (1990). © 1990, with permission from Eisevier.

macromolecules with molecular weights in excess of 300,000 Da. Moreover, the sensitivity is excellent, in the pico- to femtomole range, and little fragmentation occurs. A typical mass spectrum of a protein obtained with MALDI is shown in Figure 15-4.

Finally, we consider *electrospray ionization* (*ESI*). With this method, a fine spray of highly charged droplets is created by spraying the sample onto the tip of a metal nozzle at approximately 4000 volts. This is done at atmospheric pressure. The solvent is removed through a series of Coulombic explosions. The droplets can also be heated to facilitate evaporation. As a droplet becomes smaller, the electric field density increases, until finally mutual repulsion between the ions causes the ions to leave the droplet. In essence, the electrostatic forces become greater than the surface tension. The ions are then directed into the mass analyzer with an electrostatic lens. An important and unique characteristic of this method is that highly charged ions are produced so that a wide range of m/z values are observed for a single analyte. For example, the mass spectrum of myoglobin obtained with ESI is shown in Figure 15-5. If two adjacent peaks are assumed to differ by only one charge and one proton, the molecular weight can be calculated. This method can be used for molecules of about 100,000 Da or less and has good sensitivity, in the pico- to femtomole range.

15.6 SAMPLE PREPARATION/ANALYSIS

Before discussing applications of mass spectrometry, some general comments should be made. Table 15-2 summarizes the most frequently used methodology for biomolecules. The amount of sample required depends on the specific methods

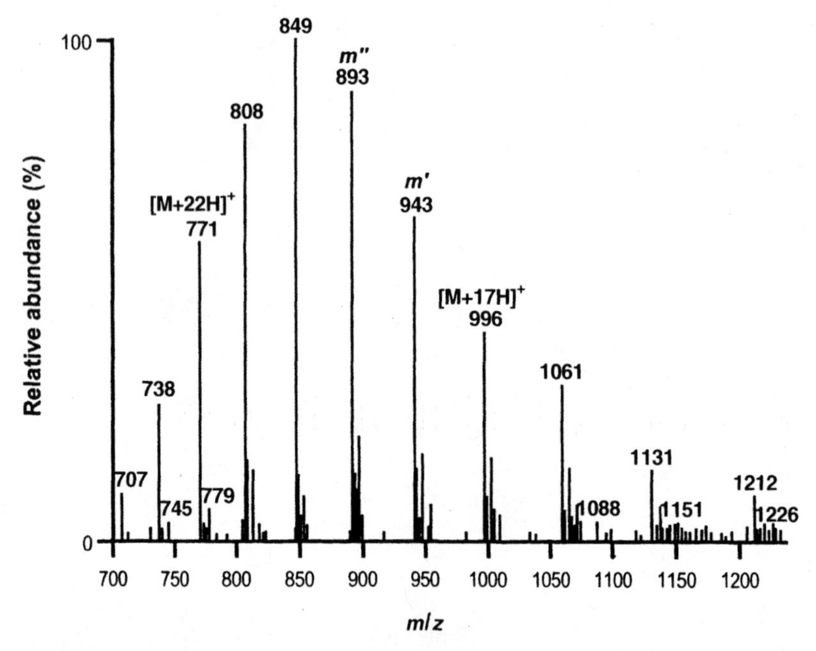

FIGURE 15-5. An electrospray ionization (ESI) mass spectrum of horse myoglobin. Each peak is characteristic of myoglobin with a different charge and number of protons. From C. Dass, *Principles and Practice of Biological Mass Spectrometry*, Wiley, New York, 2001, p. 43. Reprinted with permission of Wiley © 2001.

used, but generally 5–50 µl of 10–100 µM solutions are adequate. Frequently, a mass spectrometer is coupled to purification procedures such as gas chromatography, high-performance liquid chromatography, or capillary electrophoresis. This provides a high throughput of samples and rapid analysis of unknown mixtures. In some cases, an in-depth study of a single substance is carried out. In this case, the purity is important and extraneous background material should be eliminated, notably salts, if possible.

TABLE 15-2. Ionization Techniques for Biomolecules[a]

Compound	Ionization method	Ionization mechanism
Peptides	FAB, MALDI, and ESI	Protonation, deprotonation
Proteins	MALDI and ESI	Protonation
Carbohydrates	FAB, MALDI, and ESI	Protonation, deprotonation, cationization[b]
Oligonucleotides	MALDI and ESI	Protonation, deprotonation, cationization[b]
Small biomolecules	FAB, MALDI, and ESI	Protonation, deprotonation, cationization[b], electron ejection

[a]Adapted from Ref. 2.
[b]Addition of a cation other than H^+.

In order to obtain accurate mass measurements with mass spectrometry, the instrument must be calibrated with standard compounds. This is sometimes done by running a series of calibration curves, but the best procedure is to include an internal mass standard with the sample. This will ensure that fluctuations in instrument response are adequately taken into account. Mass measurements can be made to within 1–500 parts per million. Thus, for example, the accuracy can be within a few tenths of a dalton for a macromolecule of molecular weight of 100,000.

15.7 PROTEINS AND PEPTIDES

With the advent of soft ionization methods, the determination of the molecular weight of proteins with mass spectrometry has become routine and has become the method of choice because of its accuracy and the relatively small amount of material required (femtomoles). The method works best if a pure protein is obtained by conventional methods such as chromatography. With a pure protein, mass spectrometry, usually with MALDI or ESI, can detect differences in a single amino acid quite readily. Thus, for example, mass spectrometry is a confirmatory tool for site-specific mutagenesis. Also if post-translational modifications occur, such as phosphorylation or carbohydrate addition, mass spectrometry can identify and in some cases quantify the modification.

Proteomics is a rapidly developing field that systematically characterizes gene products. Proteomics is concerned with determining the identities of proteins, both known and unknown, and their functions. Tissues or fluids are subjected to protein purification protocols, usually by simple chromatography or gel electrophoresis, and the molecular masses and amino acid sequences of the proteins are then determined by mass spectrometry. This represents the first step in the characterization of gene products. Molecular weight by itself is rarely sufficient to identify a protein so that once proteins have been identified and their molecular weight determined, further studies are carried out. The next step is to take a specific protein, which could be a spot taken from two-dimensional gel electrophoresis of the crude starting material, and subject it to proteolysis with an enzyme such as trypsin. Before the use of mass spectrometry, the resultant peptides had to be separated, often by laborious procedures. However, with mass spectrometry, the trypsin digest can be analyzed directly and the molecular weights of the peptides determined. Since proteins usually have a unique digest, this procedure is often sufficient to determine what the starting protein is. Data banks of trypsin digests for hundreds of proteins and sophisticated software are available to help in the identification process. An example of the mass spectrum of a mixture of peptides is shown in Figure 15-6.

Although molecular weight characterization of peptide fragments after proteolysis is very useful for identifying proteins, the surest identification is to determine the amino acid sequence of the peptides. The amino acid sequence of the protein can then be compared with the vast database of known proteins. In addition to identifying specific proteins, the databases can be searched for homologous proteins, that is, proteins with similar but not identical amino acid sequences.

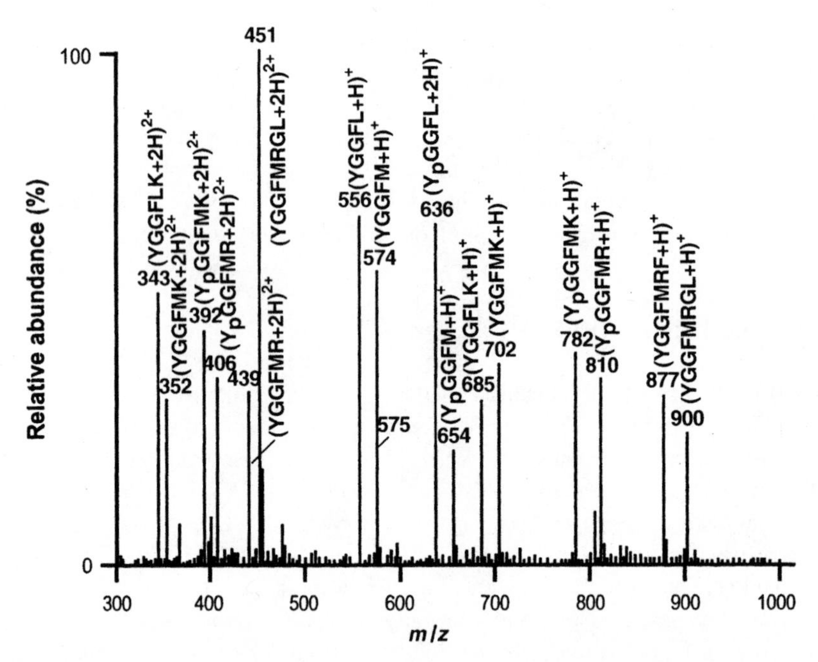

FIGURE 15-6. The positive ion ESI mass spectrum of a mixture of 10 peptides. The masses and amino acid sequences of the major peaks are indicated. Reproduced with permission from C. Dass in B. S. Larsen and C. N. McEwen (eds), *Mass Spectrometry of Biological Materials*, Marcel Dekker, New York, 1998, pp. 247–280.

Determining the sequence of a peptide with mass spectrometry is not as easy as making molecular weight measurements. Basically, the principle is to use fragmentation of the parent peptide as a unique identifier of the sequence. The most frequently used technology is tandem mass spectrometry. With this method, a single peptide can be selected by the initial ion separation, and this peptide can then be subjected to fragmentation and analysis by the second mass analyzer. The amino acid sequence of peptides with molecular weights of up to about 3000 can be readily determined with this technique. If the peptide is large, the amount of useful sequence information that can be generated in the mass spectrometer is limited because of the complexity and large number of fragments produced.

An effective strategy for amino acid sequencing is the use of peptide ladders. With this methodology, a peptide is treated with an exopeptidase (carboxypeptidase or aminopeptidase). These enzymes take off one amino acid at a time from the C- or N-terminus. The products of these digestions are then analyzed with mass spectrometry, and the reduction in molecular weight can be used to deduce the amino acid sequence. An example of ladder sequencing is shown in Figure 15-7, where a peptide from HIV protease was sequenced by degradation from the amino terminus. The letter by each mass spectrum peak indicates the amino acid that was found on the N-terminus as the peptide decreased in length (4).

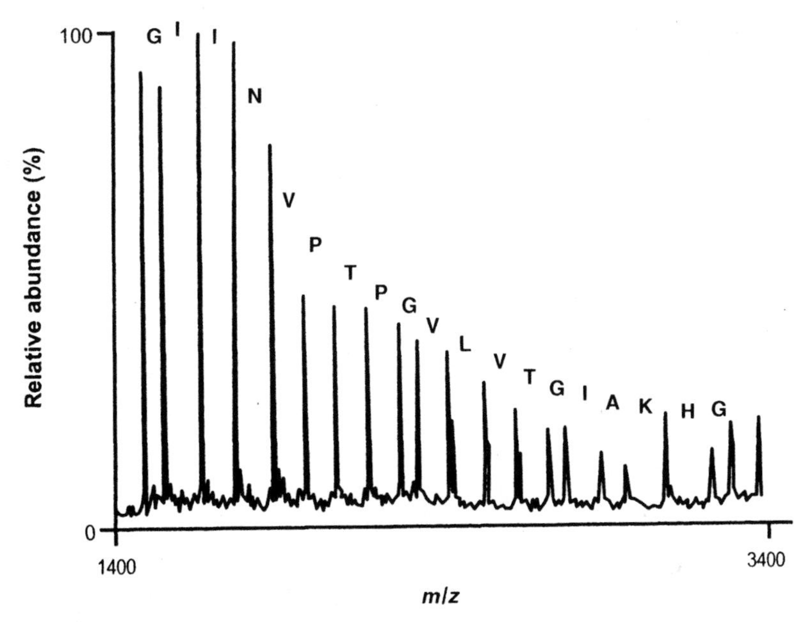

FIGURE 15-7. Mass spectrum of a mixture of peptides from an HIV protease following sequential removal of an amino acid from the N-terminus of the peptide. The letters indicate the amino acid that was removed from N-terminus. Adapted with permission from B. T. Chait, R. Wang, R. C. Beavis, S. B. H. Kent, *Science* **262**, 89 (1993). © 1993 AAAS.

The above sequence of events has not taken into account the fact that proteins often contain disulfide linkages. If this is the case, the protein must first be subjected to a reduction process that converts the disulfides to cysteines, and the cysteines are then typically alkylated so that the disulfides will not reform. Because the molecular weights of the reduced/alkylated and oxidized enzymes are different, mass spectrometry can be used to determine exactly how many disulfide linkages are present.

As the fields of genomics and proteomics expand, mass spectrometry undoubtedly will be a central analytical technique for the characterization of gene products. Both the methodology and database searching techniques are improving rapidly.

15.8 PROTEIN FOLDING

We have previously discussed the transformation of proteins between native and denatured states (Chapters 3, 9 and 12). Mass spectrometry provides a unique tool in such studies. Proteins have many hydrogens that can exchange with water hydrogens. This is a dynamic process and takes place continuously in aqueous solution. The amide hydrogens, that is, those associated with peptide bonds, are of particular interest because their rate of exchange with the solvent is dependent on structure and generally occurs over a time range that is readily accessible experimentally. All of the amide hydrogens are replaced by deuterium at approximately the same rate when a denatured protein is put into D_2O. However, for native proteins the rate depends on the environment of each amide hydrogen. For example, if the amide hydrogen is involved in a stable α-helix, the rate of exchange will be slower than that if it is freely exposed to the solvent. Amide hydrogens directly exposed to the solvent exchange rapidly with the solvent. The remaining amide hydrogens can be divided roughly into two classes. One class exchanges due to the making and breaking of local structures: this is often called breathing motions of the protein. The second, slower, exchanging class of hydrogens is associated with the global structure of the protein. They usually do not exchange unless the native structure is denatured. Within each of these classes, subgroups can be found, especially the class associated with local structures (cf. Refs. 5 and 6 for reviews of this subject).

For each subgroup, the exchange process can be described by the mechanism

$$NH_{cl} \underset{k_{cl}}{\overset{k_{op}}{\rightleftharpoons}} NH_{op} \xrightarrow{k_{int}} ND \tag{15-9}$$

Here, NH_{cl} and NH_{op} are the closed and open forms of the protonated enzyme: closed means hydrogen exchange cannot occur and open means hydrogen exchange can occur. ND is the deuterated amide(s), and the rate constants are for the opening, closing, and exchange of the unprotected amide reactions. If the intermediate state is assumed to be in a steady state, the rate constant for the overall exchange reaction is

$$k_{ex} = \frac{k_{op}k_{int}}{k_{cl} + k_{int}} \tag{15-10}$$

Two limiting cases exist. If the rate constant for closure is much faster than the rate constant for the replacement of hydrogen by deuterium ($k_{cl} \gg k_{int}$), $k_{ex} = k_{op}k_{int}/k_{cl}$. If $k_{int} \gg k_{cl}$, then $k_{ex} = k_{op}$. These are called the EX2 and EX1 limits. In the former case, because k_{ex} can be estimated from studies with model compounds, the equilibrium constant k_{op}/k_{cl} can be calculated and is a direct measure of the thermodynamic stability of the structural element under consideration.

If deuterium is substituted for hydrogen, the molecular weight of the protein increases so that mass spectrometry is an excellent tool for studying hydrogen/deuterium exchange. NMR has also been used extensively because it can monitor specific individual amide protons in a single experiment. However, it cannot easily detect multiple populations of the same conformation. The combination of NMR and mass spectrometry provides a powerful experimental approach to the study of protein folding, but only a few selected examples studied with mass spectrometry are considered here.

In the EX2 limit, the open and closed forms of the protein are in equilibrium so that, as deuterium is substituted for hydrogen, only a single protein species of increasing molecular weight is observed during the time course of exchange. The starting protein species would have all amide hydrogens and the final protein species all deuteriums, with intermediate species having a specific number of hydrogens and deuteriums. This is shown schematically in Figure 15-8. In the EX1 limit,

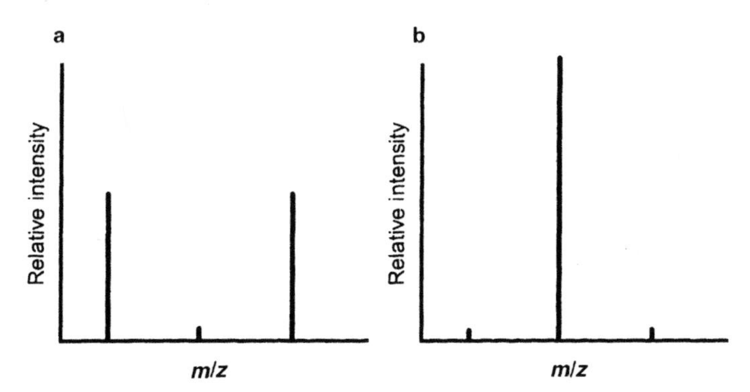

FIGURE 15-8. Schematic representation of mass spectra for the EXI (a) and EX2 (b) exchange limits midway through the hydrogen/deuterium exchange reaction. The mass of the protein or peptide increases as the amide hydrogen is replaced by deuterium. In the EX1 limit, each protein or peptide molecule has either all hydrogens or all deuterons, whereas in the EX2 limit, each protein or peptide molecule contains both hydrogens and deuterons.

the opening reaction is rate determining, with rapid exchange following. Therefore, two species should be present during the time course of the exchange, one completely deuterated and the other completely hydrogenated at the amide position. This situation is also shown schematically in Figure 15-8. The shape of the mass spectrum peaks in the EX1 limit can provide information about correlated exchange reactions and other reaction dynamics, but we will not discuss this more complex situation (6).

An extensive study of the incorporation of deuterium into rabbit muscle aldolase has been carried out (7). The protein was incubated in D_2O for a specified length of time. The exchange reaction was quenched by lowering the pD to 2.5 and the temperature to 0°C. Under these conditions, the rate of the exchange reaction (k_{ex}) becomes very slow. Denaturant was used to increase the initial rate of exchange—it promotes the formation of open forms. After quenching, the protein was proteolyzed with pepsin, and the peptides were subjected to mass spectral analysis by ESI. The time dependence of the exchange reaction was determined, and it was found that two different peptides incorporated deuterium at different rates. Both protonated and deuterated peptides were found, but intermediate states with both protons and deuterons were not found. This indicated that the EX1 limit is operative. Thus, the mass spectra showed that the EX1 mechanism of exchange was occurring and that the hydrogen/deuterium exchange for the two peptides took place at different rates, indicating differences in kinetic stability within the protein.

Mass spectrometry can also be used to study the exchange reaction in the EX2 limit which provides direct information about the thermodynamic stability of the protein. Mass spectrometry coupled with hydrogen exchange provides a means of rapidly scanning a library of proteins for relative stability (8). The stability of overexpressed proteins in crude cell lysates can also be assessed (9). In the latter case, the increase in mass of the overexpressed protein after exposure to D_2O for a fixed time was determined at different guanidine chloride concentrations. The reaction was quenched by addition to a MALDI matrix. The increase in mass for a number of variants of the protein λ repressor followed the typical sigmoidal behavior for a two-state denaturation, and the relative stability of seven mutants was determined. The results correlated well with those from circular dichroism measurements. The same method can be used to estimate binding constants as ligand binding generally stablizes the native protein configuration. In addition, binding of a ligand or macromolecule may selectively alter the exchange rate for specific regions of the target protein, thus providing information about the region of the protein involved in binding. The ability to rapidly screen many samples is a unique advantage of mass spectrometry for such studies.

15.9 OTHER BIOMOLECULES

Although the mass spectrometry literature has been dominated by the study of proteins in recent years, largely because of the genome project and proteomics, many other biomolecules have been studied. Post-translational modification of proteins is a good example. After proteins are synthesized in the cell, they are often modified, for example, by glycosylation and phosphorylation.

The attachment of a carbohydrate to a protein is often important for its function and localization. Carbohydrates also play a significant role in the regulation of physiological processes. Determination of the structure of glycoproteins is challenging, and mass spectrometry can play an important role. If a protein is not homogeneous with respect to glycosylation, multiple molecular weights will be observed. MALDI is especially useful as it generally produces a single charged species and permits the multiple species to be easily sorted. ESI produces multiple ions for each species but has better mass resolution. The molecular mass of the attached carbohydrate can be determined by treating the protein with glycosidases to free the protein of the carbohydrate. If the extent of glycosylation is not great, the nature of the attached group can sometimes be inferred directly. If similar experiments are carried out in conjunction with peptide mapping (proteolysis and separation of peptides), the attachment sites of the carbohydrate can be ascertained. Ultimately, sequencing of the carbohydrate is required for complete characterization, and MS/MS methods are useful in this regard. The coupling of database searching and mass spectrometry is a promising approach for the characterization of carbohydrates. Although the characterization of carbohydrate structures is difficult and not yet well developed, its importance in biology is well recognized (1).

Finally, mention should be made of the application of mass spectrometry to characterizing oligonucleotides and lipids. In the case of lipids, it is a primary tool for structural characterization of the many diverse types of lipids that occur in nature. For nucleic acids, it is especially useful for determining modification of bases. We also have not dwelt on the characterization of metabolites; mass spectrometry is a tool of choice for small molecules because of the small amount of sample required to obtain a complete structure.

REFERENCES

1. C. Dass, *Principles and Practice of Biological Mass Spectrometry*, Wiley, New York, 2001.
2. G. Siuzdak, *Mass Spectrometry for Biotechnology*, Academic Press, San Diego, CA, 1996.
3. G. Siuzdak, *The Expanding Role of Mass Spectrometry in Biotechnology*, MCC Press, San Diego, CA, 2003.
4. B. T. Chait, R. Wang, R. C. Beavis, and S. B. H. Kent, *Science* **262**, 89 (1993).
5. C. K. Woodward, *Curr. Opin. Struct. Biol.* **4**, 112 (1994).
6. D. M. Ferraro, N. D. Lazo, and A. D. Robertson, *Biochemistry* **43**, 587 (2004).
7. Y. Deng, Z. Zhang, and D. L. Smith, *Am. Soc. Mass Spectrom.* **10**, 675 (1999).
8. D. M. Rosenbaum, S. Roy, and M. H. Hecht, *J. Am. Chem. Soc.* **121**, 9509 (1999).
9. S. Ghaemmaghami, M. C. Fitzgerald, and T. G. Oas, *Proc. Natl. Acad. Sci. USA* **97**, 8296 (2000).

PROBLEMS

15.1. The mass spectrum of the enzyme lysozyme was determined using ESI, and several positive ion peaks were observed. Peaks at $m/z = 1432$ and 1592 were found adjacent to each other. Calculate the molecular weight of lysozyme. Assume that the charges are due to protonation of the protein and that z differs by one unit for the two peaks.

15.2. A peptide was subjected to degradation from the N-terminus (Edman degradation). The resultant mixture was subjected to MALDI/MS analysis. The following ladder of m/z was observed: 977.2, 1064.3, 1151.4, 1222.6, 1378.9, and 1465.9. Assume that $z = 1$ in all cases and derive the sequence of the N-terminal region of the peptide. The identical peptide was enzymatically phosphorylated with protein kinase C. For this peptide, the ladder of m/z was as follows: 977.3, 1064.3, 1231.6, 1302.6, 1458.9, and 1545.9. Explain these results.

15.3. Derive an equation for the time of flight of an ion with a mass/charge ratio of m/z that travels a distance L in the mass spectrometer. Calculate the time of flight for a particle with $m/z = 200$ amu that was accelerated by 3000 volts over a distance of 30 cm. (*Hint*: Start with Eq. 15-6 in your derivation. Be careful of the units in making your calculation.)

15.4. Avidin is a glycoprotein found in egg white. It binds an important biomolecule, biotin, very tightly. Mass spectrometry of avidin using ESI displayed the usual multiple-peak spectrum due to multiple charges on a single avidin molecule. Two of the peaks had m/z values of 4002 and 4251. When the avidin was reacted with biotin prior to mass spectrometry, these peaks had m/z values of 4058 and 4312. Determine how many molecules of biotin are bound per molecule of avidin. The molecular weight of biotin is 244.

APPENDICES

APPENDIX 1

Useful Constants and Conversion Factors

Avogadro's number, N_0	6.0221×10^{23} mole^{-1}
Gas constant, R	8.3144×10^{7} erg K^{-1} mole^{-1}
	8.3144 Joule K^{-1} mole^{-1}
	1.9872 cal K^{-1} mole^{-1}
	0.082057 l atmosphere K^{-1} mole^{-1}
Boltzmann's constant, k_B	1.3806×10^{-23} Joule K^{-1} molecule^{-1}
Planck's constant, h	6.6262×10^{-34} Joule/s
Speed of light, c	2.9979×10^{8} m s^{-1}
Standard gravity, g	9.8066 m s^{-2}
Electronic charge, e	1.6022×10^{-19} Coulomb
Electron mass, m_e	9.1094×10^{-31} kg
Proton mass, m_p	1.6276×10^{-27} kg
Faraday constant, F	9.6485×10^{4} Coulomb mol^{-1}

1 calorie = 4.184 Joule
1 Joule = 10^7 erg = 1 Volt-Coulomb
1 electron volt = 1.602×10^{-19} Joule

Physical Chemistry for the Biological Sciences by Gordon G. Hammes

APPENDIX 2

Structures of the Common Amino Acids at Neutral pH

Aliphatic	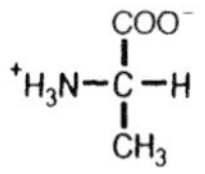	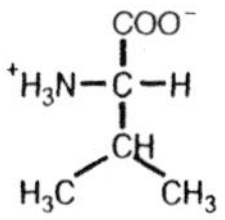	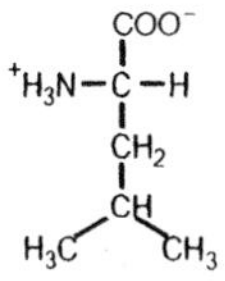	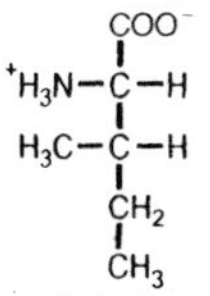
	Alanine (Ala) (A)	**Valine (Val) (V)**	**Leucine (Leu) (L)**	**Isoleucine (Ile) (I)**
Nonpolar	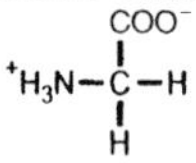	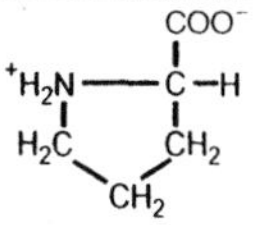	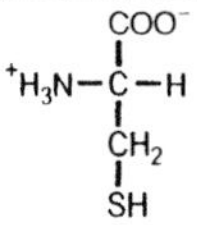	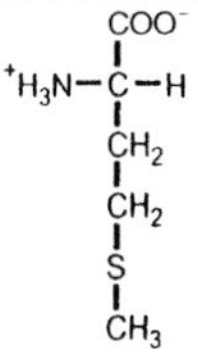
	Glycine (Gly) (G)	**Proline (Pro) (P)**	**Cysteine (Cys) (C)**	**Methionine (Met) (M)**
Aromatic	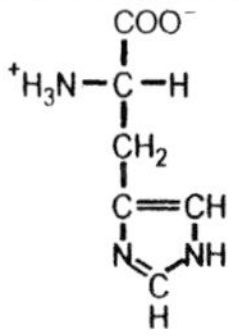	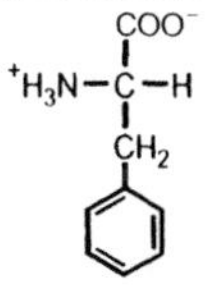	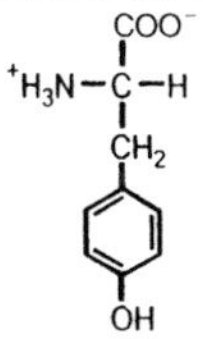	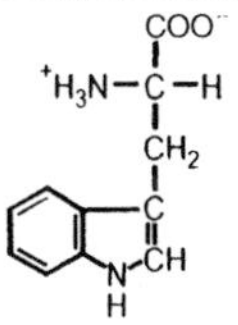
	Histidine (His) (H)	**Phenylalanine (Phe) (F)**	**Tyrosine (Tyr) (Y)**	**Tryptophan (Trp) (W)**
Polar	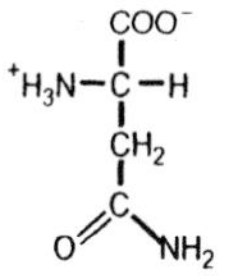	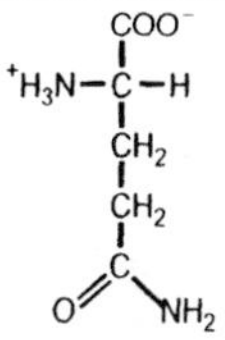	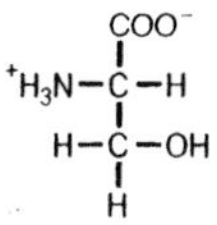	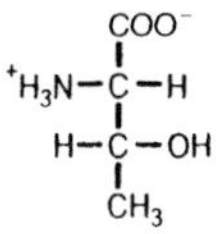
	Asparagine (Asn) (N)	**Glutamine (Gln) (Q)**	**Serine (Ser) (S)**	**Threonine (Thr) (T)**
Charged	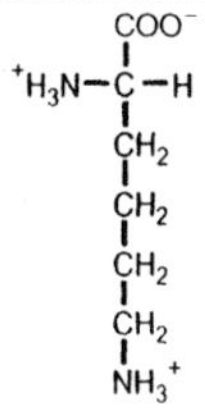	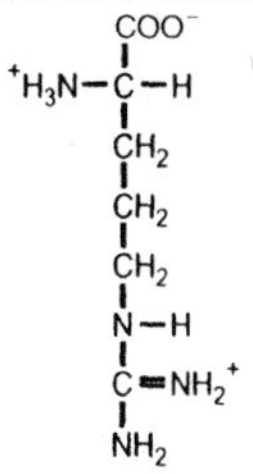	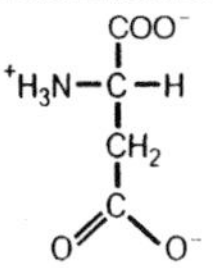	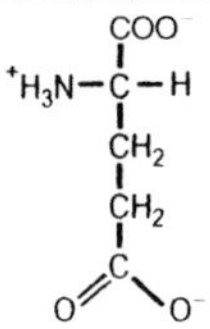
	Lysine (Lys) (K)	**Arginine (Arg) (R)**	**Aspartate (Asp) (D)**	**Glutamate (Glu) (E)**

APPENDIX 3

Common Nucleic Acid Components

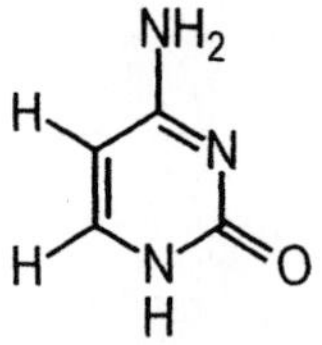

Cytosine
(C)

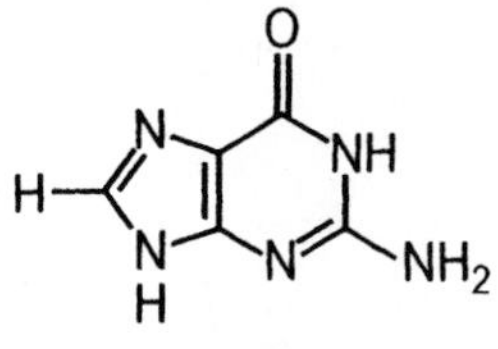

Guanine
(G)

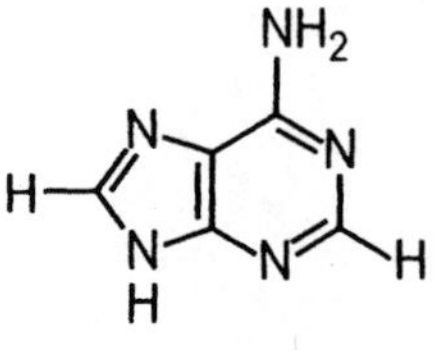

Adenine
(A)

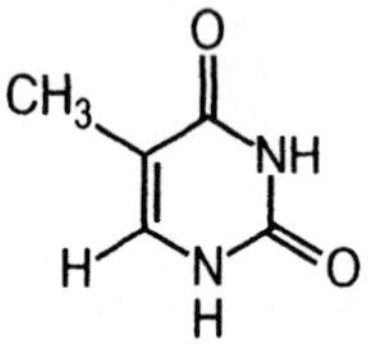

Thymine
(T)

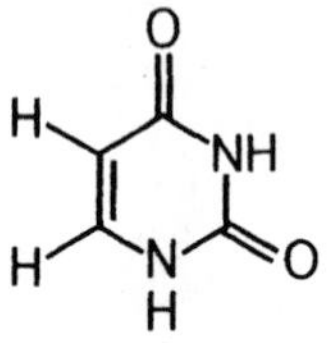

Uracil
(U)

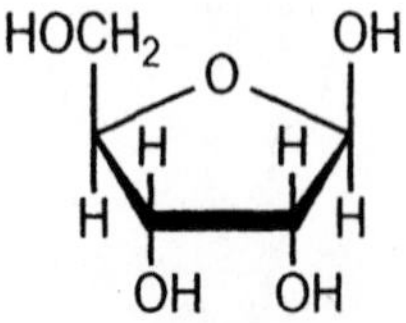

β-D-Ribofuranose

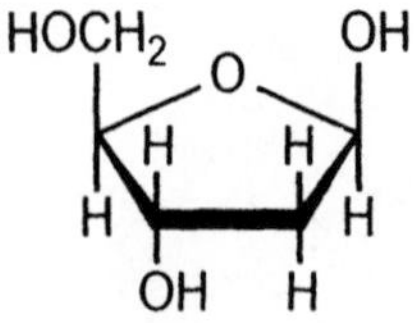

β-D-2-Deoxyribofuranose

Physical Chemistry for the Biological Sciences by Gordon G. Hammes

APPENDIX 4

Standard Free Energies and Enthalpies of Formation at 298 K, 1 atm, pH 7, and 0.25 M Ionic Strength

Substance	ΔG° (kJ/mol)	ΔH° (kJ/mol)
ATP	−2097.89	−2995.59
ADP	−1230.12	−2005.92
AMP	−360.29	−1016.88
Adenosine	529.96	−5.34
P_i	−1059.49	−1299.39
Glucose-6-phosphate	−1318.92	−2279.30
Glucose	−426.71	−1267.11
H_2O	−155.66	−286.65
NAD_{ox}	1059.11	−10.26
NAD_{red}	1120.09	−41.38
$NADP_{ox}$	1011.86	−6.57
$NADP_{red}$	1072.95	−33.28
Acetaldehyde	24.06	−213.97
Acetate	−247.82	−486.83
Alanine	−85.64	−557.67
Ammonia	82.94	−133.74
Ethanol	62.96	−290.76
Pyruvate	−350.78	−597.04
Formate	−311.04	−425.55
Sucrose	−667.85	−2208.90
Total CO_2	−547.10	−692.88
2-Propanol	140.90	−334.11
Acetone	84.89	−224.17
Glycerol	−171.35	−679.84
Lactose	−670.48	−2242.11
Maltose	−677.84	−2247.09
Succinate	−530.62	−908.68

(*Continued*)

Physical Chemistry for the Biological Sciences by Gordon G. Hammes

Substance	$\Delta G°$ (kJ/mol)	$\Delta H°$ (kJ/mol)
Fumarate	−523.58	−776.57
Lactate	−313.70	−688.28
Glycine	−176.08	−525.05
Urea	−39.73	−319.29
Ribulose	−328.28	−1027.12
Fructose	−426.32	−1264.31
Ribose	−331.13	−1038.10
Ribose 5-phosphate	−1219.22	−2042.40
Aspartate	−452.10	−945.46
Glutamate	−372.16	−982.77
Glutamine	−120.36	−809.11
Citrate	−966.23	−1513.66
Isocitrate	−959.58	—
cis-Aconitate	−802.12	—
Malate	−682.83	—
2-Oxoglutarate	−633.59	—
Oxalosuccinate	−979.06	—
Oxaloacetate	−714.99	—
Glycerol 3-phosphate	−1077.14	—
Fructose 6-phosphate	−1315.74	—
Glucose 1-phosphate	−1311.89	—
$CO_2(g)$	−394.36	−393.51
$O_2(g)$	0	0
$O_2(aq)$	16.40	−11.70
$H_2(g)$	81.53	−0.82
$H_2(aq)$	99.13	−5.02

This table is based on the conventions that $\Delta G° = \Delta H° = 0$ for the species H^+, adenosine, NAD^-, and $NADP^{3-}$ at zero ionic strength. These data are from R. A. Alberty, *Arch. Biochem. Biophys.* **353**, 116 (1998).

APPENDIX 5

Standard Free Energy and Enthalpy Changes for Biochemical Reactions at 298 K, 1 atm, pH 7.0, pMg 3.0, and 0.25 M Ionic Strength

Reaction	ΔG° (kJ/mol)	ΔH°(kJ/mol)
$ATP + H_2O \rightleftharpoons ADP + P_i$	−32.48	−30.88
$ADP + H_2O \rightleftharpoons AMP + P_i$	−32.80	−28.86
$AMP + H_2O \rightleftharpoons adenosine + P_i$	−13.55	−1.22
$2\ ADP \rightleftharpoons ATP + AMP$	−0.31	+2.02
$G6P + H_2O \rightleftharpoons Glu + P_i$	−11.61	−0.50
$ATP + Glu \rightleftharpoons ADP + G6P$	−20.87	−30.39

Data from R. A. Alberty, *Arch. Biochem. Biophys.* **353**, 116 (1998).

INDEX

Page references followed by t indicate material in tables.

Physical Chemistry for the Biological Sciences by Gordon G. Hammes